T0202818

Communications in Computer and Information Science 1733

More information about this series at https://link.springer.com/bookseries/7899

Vladimir Jordan · Ilya Tarasov · Ella Shurina ·
Nikolay Filimonov · Vladimir Faerman (Eds.)

High-Performance Computing Systems and Technologies in Scientific Research, Automation of Control and Production

12th International Conference, HPCST 2022
Barnaul, Russia, May 20–21, 2022
Revised Selected Papers

 Springer

Editors
Vladimir Jordan
Altai State University
Barnaul, Russia

Ilya Tarasov 🆔
MIREA - Russian Technological University
Moscow, Russia

Ella Shurina 🆔
Novosibirsk State Technical University
Novosibirsk, Russia

Nikolay Filimonov 🆔
Lomonosov Moscow State University
Moscow, Russia

Vladimir Faerman 🆔
Tomsk State University of Control Systems
and Radio-Electronics
Tomsk, Russia

ISSN 1865-0929 ISSN 1865-0937 (electronic)
Communications in Computer and Information Science
ISBN 978-3-031-23743-0 ISBN 978-3-031-23744-7 (eBook)
https://doi.org/10.1007/978-3-031-23744-7

This Springer imprint is published by the registered company Springer Nature Switzerland AG
The registered company address is: Gewerbestrasse 11, 6330 Cham, Switzerland

Preface

The 12th International Conference on High-Performance Computing Systems and Technologies in Scientific Research, Automation of Control and Production (HPCST 2022) took place at the Altai State University during May 21–22, 2022. Altai State University (AltSU) is in the center of Barnaul city – the capital of the Altai region in the southwestern part of Siberia.

HPCST is a regular scientific meeting that has been held annually since 2011. It attracts specialists in the various fields of modern computer and information science, as well as their applications in the automation of control and production, mathematical modeling, and computer simulation of processes and phenomena in natural sciences by means of parallel computing. Last year, a subsection on information security was also established.

The goal of the conference is to present state-of-art approaches and methods for solving contemporary scientific problems and to exchange the latest research results obtained by scientists from both universities and research institutions. All the reported results are valuable contributions to the field of applied information and computer science. Sessions of the conference are devoted to the relevant scientific topics:

- architecture and design features of high-performance computing systems;
- digital signal processors (DSPs) and their applications;
- IP-cores for field-programmable gate arrays (FPGAs);
- technologies for distributed computing using multiprocessors;
- GRID-technologies and cloud computing and services;
- high-performance and multiscale predictive computer simulation;
- computing in information security services; and
- control automation and mechatronics.

The conference was introduced at the international level back in 2017 since there were more than 120 researchers from Russia, China, Ukraine, Kazakhstan, Kyrgyzstan, Uzbekistan, Tajikistan, Vietnam, and Brazil participating in the conference. The average number of participants in a single year was about 60 for a long time. The last event was attended virtually or personally by more than 140 scholars.

This year the conference was attended virtually or in person by 92 scholars with 73 accepted reports. The reports were chosen by the Program Committee from 82 qualified submissions (nine submissions were not considered to be full papers and were not sent for review). This year, due to complex political situation in Russia, the internationalization trend was interrupted. However, five international reports were presented by authors from Kazakhstan, Kyrgyzstan, and Uzbekistan. There were a dozen international participants in total this year. Geographical distribution of domestic participants remains the same: Siberian regions and Moscow predominates among the participants.

The most significant reported studies were thoroughly reviewed and included in this volume. Only half of the reported papers were invited to these internationally published post-proceedings. All invited papers then were revised, extended to full-paper format, and submitted for a second round of review.

After the review, 23 full papers featuring original studies in the field of computing, mathematical simulation, and control science and information security were accepted for publication. Among them, two papers were accepted as is, 12 papers were accepted after minor revision, and nine papers were accepted after major revision. No guest papers were accepted this year. Out of 11 rejected papers, four were declined after the revision. The acceptance rate for this volume is 28%. The vast majority of the conference reports (43 research papers) that were not included in this volume were published as regular proceedings.

Only 23 full papers featuring original studies in the field of computing, mathematical simulation, and control science are featured in this volume. The papers cover such topics as:

- hardware for high-performance computing and signal processing;
- information technologies and computer simulation of physical phenomena;
- computing technologies in data analysis and decision-making;
- information and computing technologies in automation and control science; and
- computing technologies in information security applications.

To select the best papers among those presented at the conference, the following procedure was applied.

1. Session chairs prepared a shortlist of the most significant original reports, which had a clear potential to be extended to full-paper format.
2. The editorial board comprised of the session chairs and corresponding editor made a list of 35 reports.
3. Authors of the selected manuscripts then were contacted and asked to extend their papers and resubmit them review.

Every paper, with no exceptions, was reviewed by at least three experts. In addition to a routine plagiarism check with iThenticate.com, we applied another check with Antiplagiat.ru. The goal was to detect and decline the papers that were already published in Russian. A single-blind review method was applied using the following criteria:

1. technical content;
2. originality;
3. clarity;
4. significance;
5. presentation style; and
6. ethics.

The organizing committee would like to express their sincere appreciation for the organizational support to the administration of Altai State University and to the staff of the Institute of Digital Technologies, Electronics and Physics of Altai State University.

Only the outstanding effort of the technical staff made the conference possible in the time of travel restrictions.

The editors would like to express their deep gratitude to the Springer editorial and production teams for the opportunity to publish the best papers as post-proceedings and for their great work on this volume.

November 2022

Vladimir Jordan
Ilya Tarasov
Ella Shurina
Nikolay Filimonov
Vladimir Faerman

Organization

General Chair

Vladimir Jordan   Altai State University, Russia

Program Committee Chairs and Section Chairs

Ella Shurina   Novosibirsk State Technical University, Russia
Ilya Tarasov   Russian Technological University, Russia
Nikolay Filimonov   V.A. Trapeznikov Institute of Control Sciences,
                                 RAS, Russia
Vladimir Faerman   Tomsk State University of Control Systems and
                                 Radioelectronics, Russia

Organizing Committee

Vasiliy Belozerskih   Altai State University, Russia
Alexander Kalachev   Altai State University, Russia
Vladimir Pashnev   Altai State University, Russia
Viktor Sedalischev   Altai State University, Russia
Yana Sergeeva   Altai State University, Russia
Igor Shmakov   Altai State University, Russia
Petr Ulanov   Altai State University, Russia

Program Committee

Viktor Abanin   Biysk Technological Institute, Russia
Darya Alontseva   Serikbayev East Kazakhstan Technical University,
                                 Kazakhstan
Valeriy Avramchuk   Tomsk State University of Control Systems and
                                 Radioelectronics, Russia
Sergey Beznosyuk   Altai State University, Russia
Alexander Filimonov   MIREA - Russian Technological University,
                                 Russia
Pavel Gulyaev   Yugra State University, Russia
Ishembek Kadyrov   Kyrgyz National Agrarian University, Kyrgyzstan
Alexander Kalachev   Altai State University, Russia
Vladimir Khmelev   Altai State Technical University, Russia

Shavkat Fazilov	Research Institute for the Development of Digital Technologies and Artificial Intelligence, Uzbekistan
Vladimir Kosarev	Khristianovich Institute of Theoretical and Applied Mechanics, SB RAS, Russia
Nomaz Mirzaev	Tashkent University of Information Technologies, Uzbekistan
Lyudmilla Kveglis	Amanzholov East Kazakhstan State University, Kazakhstan
Roman Mescheryakov	V.A. Trapeznikov Institute of Control Sciences, RAS, Russia
Aleksey Nikitin	Altai State University, Russia
Viktor Polyakov	Altai State University, Russia
Leonid Mikhailov	Al-Farabi Kazakh National University, Kazakhstan
Oleg Prikhodko	Al-Farabi Kazakh National University, Kazakhstan
Sergey Pronin	Altai State Technical University, Russia
Alisher Saliev	Kyrgyz State Technical University, Kyrgyzstan
Viktor Sedalischev	Altai State University, Russia
Vitaliy Titov	Southwest State University, Russia
Pedro Filipe do Prado	Federal University of Espirito Santo, Brazil

External Reviewers

Alexey Saveliev	Tomsk Polytechnic University, Russia
Alexey Tsavnin	Tomsk Polytechnic University, Russia
Anatoliy Gulay	Belarusian National Technical University, Belarus
Andrey Kutyshkin	Yugra State University, Russia
Andrey Malchukov	Tomsk State University of Control Systems and Radioelectronics, Russia
Andrey Russkov	Yandex, Russia
Bibigul Koshoeva	Kyrgyz State Technical University, Kyrgyzstan
Elena Luneva	Tomsk Polytechnic University, Russia
Eugeniy Kostuchenko	Tomsk State University of Control Systems and Radioelectronics, Russia
Evgeniy Mytsko	Tomsk Polytechnic University, Russia
Fedor Garaschenko	Kyiv National University, Ukraine
Gambar Guluev	Institute of Control Systems of National Academy of Sciences of Azerbaijan, Azerbaijan
Kseniya Zavyalova	Tomsk State University, Russia
Leonid Mikhaylov	Al-Farabi Kazakh National University, Kazakhstan
Maksim Pushkaryov	Tomsk Polytechnic University, Russia

Contents

Computing Technologies in Data Analysis and Decision Making

Information and Computing Technologies in Automation and Control Science

Computing Technologies in Information Security Applications

Hardware for High-Performance Computing and Signal Processing

Hierarchical Encoder-Decoder Neural Network with Self-Attention for Single-Channel Speech Denoising

Rauf Nasretdinov⬤, Ilya Ilyashenko⬤, Jacob Filin⬤, and Andrey Lependin$^{(\boxtimes)}$ ⬤

AltSU – Altai State University, Lenin Ave. 61, 656049 Barnaul, Russia

Abstract. In this paper, we present a new approach to effective speech denoising based on deep learning methods. We used encoder-decoder architecture for the proposed neural network. It takes a noisy signal processed by windowed Fourier transform as an input and produces a complex mask which is the ratio of clean and distorted audio signals. When this mask is multiplied element-wise to the spectrum of the in-put signal, the noise component is eliminated. The key component of the approach is usage of hierarchical structure of the neural network model which allowed one to process input signal in different scales. We used self-attention layers to take into account the non-local dependencies in the time-frequency decomposition of the noisy signal. We used spatial reduction attention modification to reduce time and memory complexity. The scale-invariant signal-to-disturbance ratio (SI-SDR) was used as the loss function in the developed method. The DNS Challenge 2020 dataset, which includes samples of clean voice records and a representative set of various noise classes, was used to train the network. To compare performance with the best existing models several standard quality metrics (WB-PESQ, STOI, etc.) was used. The proposed method had shown its effectiveness on all the metrics and can be used to improve the quality of speech audio recording.

Keywords: Signal processing · Speech enhancement · Noise masking · Encoder-decoder architecture · Self-attention

1 Introduction

The quality of modern information services related to the use of speech depends on the level of noise in the speech signal. The signal can be distorted by many different sources. Non-stationary background noises can be superimposed on the speaker's voice, the room can introduce reverberation, the registration channel itself can distort the signal. All these factors together can significantly reduce the intelligibility of recorded, transmitted, or processed speech.

One of the most common types of speech signal distortion is the background noise, which can be considered as an additive to a useful speech signal. Previously developed denoising methods usually assumed that noise is a stationary or quasi-stationary random process [1]. But this hypothesis is not true in real life scenarios. Voice recording by users of modern digital devices is often done under less than ideal conditions. Background

© The Author(s), under exclusive license to Springer Nature Switzerland AG 2022
V. Jordan et al. (Eds.): HPCST 2022, CCIS 1733, pp. 3–14, 2022.
https://doi.org/10.1007/978-3-031-23744-7_1

noise does not change slowly compared to the speaker's speech. Noises often come from non-stationary sources with different characteristics. The relativity of what is noise in general is also essential. Background speech, children scream or somebody's singing may well be considered as distortion if the task is to separate the voice of a particular speaker. In this case, a significant part of noise energy overlaps with human speech in the signal spectrum. Thus, standard methods for speech denoising, usually introduce significant distortions into the restored signal.

Modern speech denoising methods see noise reduction as a supervised machine-learning task [2–7]. Machine learning model approximate end-to-end transformation of mixed noisy audio signals to its clean speech component. In such a problem statement, there are no restrictions on noise statistical characteristics.

In the majority of supervised algorithms for denoising a mask-based strategy is used. Mask-based methods estimate real or complex masks that need to be applied on the signal spectrum to filter noise components. The main advantage of such an approach is fast coverage over a wide range of noise types with various statistical characteristics. During development, masking methods went through the following steps: approaches based on ideal binary masks [8], real ideal ratio masks [9], and spectral amplitudes [10]. The current step in development is complex ratio masks [11], which utilizes the phase component of the signal to significantly improve the quality of the retrieved clean speech.

Deep learning methods of speech denoising often use encoder-decoder architecture [2, 3]. Convolution recurrent networks are the standard usage of this architecture [2]. The convolution layers retract compressed representations of local time-frequency portions or reconstruct low-level signal features from compressed representations, while recursion layers work with long-time dependencies. However, accurate modeling of the complex spectrum of a noisy signal can only be done by accounting for non-local correlations of individual signal components. Utilizing self-attention mechanisms could be a solution in this situation [11]. For encoding correlations in autonomous natural language processing tasks, the self-attention has been proposed. In situations where approaches based on recurrent neural networks could not be trained because of "forgetting" past internal states, we made it possible to handle lengthy text fragments. Due to the much greater volumes of data being processed than those in text processing, the use of self-attention transformations in speech processing tasks has been very limited up until recently. A generalization of this method for image processing was put out in [12–14]. They applied convolution and self-attention transformations to image fragments in a hierarchical neural network design to reduce the representation of the input data. The processing of time-frequency representations of an input speech can be performed easily using this method.

In this paper, a neural network model of the encoder-decoder architecture is proposed to handle the problem of speech denoising. This model employs a hierarchical approach and self-attention transformation to compute complex ratio masks.

2 Proposed Approach

2.1 Encoder-Decoder Neural Architecture

In this work, it was assumed that a noisy audio signal could be represented in the time-frequency domain as follows:

$$Y(f, t) = X(f, t) + N(f, t), \tag{1}$$

where the indices $f = 0,\ldots,$ F-1 correspond to the indices of the frequency bands and $t = 0,\ldots,$ T-1 are the consecutive fragments (frames) of the signal in the time domain. The components $X(f, t)$ are a clean signal, $N(f, t)$ – a noise additive.

The clean audio signal was restored as an element-wise multiplication of the complex mask $M(f, t)$ estimated by the proposed model and the noisy signal:

$$\hat{X} = Y \odot M = (Y_r \odot M_r - Y_i \odot M_i) + j \cdot (Y_r \odot M_i + Y_i \odot M_r). \tag{2}$$

where $\hat{X} = \hat{X}(f, t)$ is the reconstructed audio signal in the time-frequency domain, the indices "r" and "i" denote the real and imaginary parts of the mask and noisy signals respectively, "j" denote the imaginary unit, and the operation $\odot$ is the element-wise multiplication.

The complex mask $M(f, t)$ was calculated by the deep neural network shown in Fig. 1. It should be noted that in this work, the representation of the processed signal at all stages, except for self-attention transformations, was carried out in the form of a three-dimensional tensor[1] with dimensions F $\times$ T $\times$ C. The first and second dimensions of the tensor corresponded to frequency bands and frame numbers. The third dimension was used to encode each frequency-time "cell" of the signal with a vector of length C.

Thus, the input of the proposed neural network model was the representation of the noisy signal $Y(f, t)$ in the form of a three-dimensional tensor of size F $\times$ T $\times$ 2. Each frequency-time "cell" of this tensor consisted of the real $Y_r(f, t)$ and imaginary $Y_i(f, t)$ components of the complex spectrum. The input tensor $Y(f, t)$ was concatenated along the third dimension with the frequency encoding tensor of size F $\times$ T $\times$ K. The result of this operation was fed to the convolution layer, labeled in Fig. 1 as "InputConv".

Thus, the input of the proposed neural network model was the representation of the noisy signal $Y(f, t)$ in the form of a three-dimensional tensor of size F $\times$ T $\times$ 2. Each frequency-time "cell" of this tensor consisted of the real $Y_r(f, t)$ and imaginary $Y_i(f, t)$ components of the complex spectrum. The input tensor $Y(f, t)$ was concatenated along the third dimension with the frequency encoding tensor of size F $\times$ T $\times$ K. The result of this operation was fed to the convolution layer, labeled in Fig. 1 as "InputConv".

The initial computations were followed by three main blocks of the neural network, fully consistent with the idea of the U-Net encoding-decoding architecture [15]: the encoder that compresses the representation of the input data, the compressed representation processing bottleneck block, and the decoder for unfolding of the compressed representation of the signal into an output tensor. After applying the output linear convolution (labeled in Fig. 1 as "OutputConv") on the result of the decoder, the mask tensor

---

[1] Instead of the more accurate term "three-dimensional matrix", the term "tensor" is used in this paper, as is customary in the terminology of deep neural network libraries.

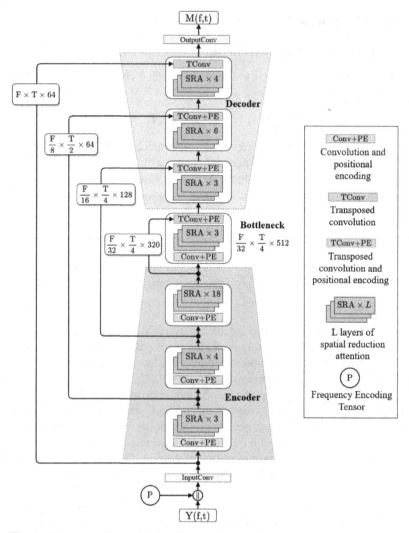

Fig. 1. The proposed encoder-decoder neural network model for speech denoising.

M(f, t) with dimensions F × T × 2 was calculated. This tensor was used to calculate the denoised signal according to (2).

The encoder consisted of three stages of tensor transformations. Each stage consisted of the convolution operation that compresses the input tensor and the positional encoding (labeled in Fig. 1 as "Conv+PE") followed by the sequence of several self-attention transformations (labeled in Fig. 1 as "SRA × L", where L is the number of applied transformations on each stage). Compression meant a decrease in the frequency and time resolution of tensors.

The bottleneck block of the neural network was responsible for processing the compressed representation of the noisy signal and consisted of positional encoding, several self-attention transforms, and transposed convolution followed by positional encoding (labeled in Fig. 1 "TConv+PE"). The size of the output tensor of this block was the same as the input.

The decoder was a symmetrical reflection of the encoder. Each of the three stages of decoding, except the last one, consisted of a sequence of self-attention transformations and positional encoding with transposed convolutions. At the last stage, no positional encoding vector was added. Due to the use of the transposed convolutions, the size of the data tensor in terms of frequency and time dimensions increased from stage to stage.

The sizes of the output tensors of each encoder stage are given in Fig. 1 above the long connections that passed the values of these tensors to the inputs of the corresponding decoding stages. These connections ensured the preservation of significant multi-scale details of the processed signal.

2.2 Frequency Encoding

The complex coefficients of the time-frequency representation of the noisy signal were considered as tensors of size $F \times T \times 2$, where the third dimension corresponds to the real and imaginary parts of the coefficients. the input signal Y was concatenated along the third dimension with a frequency-position encoding tensor P of size $F \times T \times K$, whose elements were calculated as follows [14]:

$$P_{ftk} = \cos(2^k \pi f)/F), \tag{3}$$

where $k = 0,..., K-1$ is the index of the frequency coding vector, f is the index of the corresponding frequency band, and F is the number of frequency bands. The tensor coefficients P did not depend on the frame index.

2.3 Positional Encoding

Each "Conv+PE" layer in the encoder transformed the input data in the following way (Fig. 2). First, the input tensor of size $F_{in} \times T_{in} \times C_{in}$ was divided into $N = F_{in}T_{in}/F_pT_p$ nonoverlapping patches of size $F_p \times T_p \times C_{in}$. These patches were fed into a linear convolution layer to obtain patch embeddings as a tensor of size $N \times C_e$, where C_e is the embedding vector size. This transformation was implemented as convolution with the kernel of size $F_p \times T_p$ with strides F_p and T_p of first and second dimensions. Then the resulting tensor was normalized and summed up with the tensor PE of learnable positional encodings. These encodings were not shared between different layers, for each stage they were trained independently.

As it was mentioned earlier, the structure of the "TConv+PE" layer was like the encoder ones. The linear convolution with the kernel of size $F_p \times T_p$ was replaced by the transposed convolution with the kernel of size $F_p \times T_p$ in "TConv+PE". This allowed expanding the output representation after each decoding transformation, except the last stage of decoding.

The merge of the tensor transmitted over long connection from the encoder and the result of transposed convolution was carried out by the concatenation along the third dimension of the vector representation, while the first and the second dimensions of the tensors coincided in size.

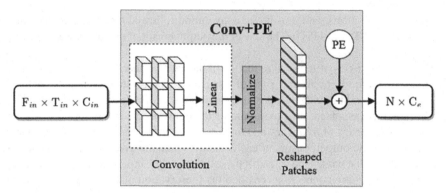

Fig. 2. Convolution and positional encoding layer.

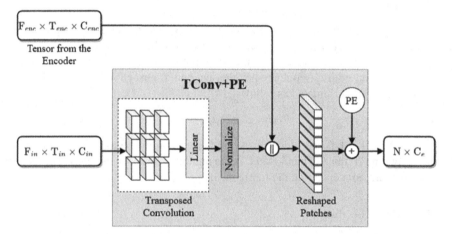

Fig. 3. Transposed convolution and positional encoding layer.

2.4 Self-attention Layer

The self-attention [11] estimated the significance of the encoded patches and nonlinearly reweighted them ("payed attention") with coefficients depending on the input tensor itself ("self-" in self-attention). It should be noted that in this work, we used a modification of the spatial-reduction attention (SRA) version of self-attention [12], which made it possible to reduce the amount of memory used by the attention mask and the number of FLOPs for the compotation. SRA for the input tensor E of size $N \times C_e$ is calculated as follows:

$$SRA(E) = Concatenate(head_1, head_2, \ldots, head_H)W^O, \tag{4}$$

where Concatenate($\cdot$) is the concatenation operation, head, $i = 1, \ldots, H$ are the so-called heads of the self-attention transformation and W^O is the trainable matrix of size $C_e \times C_e$. The idea of such calculation is that each head of self-attention transformation is searching for its own information in the input tensor E. Heads for SRA were calculated as follows:

$$head_i = Attention\left(EW_i^Q, SpatialReduction(E)W_i^K, SpatialReduction(E)W_i^V\right), \quad (5)$$

where the index $i = 1, \ldots, H$ is the number of the head, $W_i^Q, W_i^K, W_i^V, i = 1, \ldots, H$ are the trainable matrices of size $C_e \times d$. These matrices determine the set of linear transformations from C_e-dimensional embeddings tod-dimensional internal representations with $d = C_e/H$ for every head of SRA. Spatial reduction operation in (5) was defined as follows:

$$SpatialReduction(E) = LayerNorm\left(Reshape(E, R)W^S\right), \quad (6)$$

where, E is the input tensor of size $N \times C_e$, R is a reduction ratio, $Reshape(E, R)$ is an operation that reshapes input E to shape $(N/R^2) \times (C_eR^2)$, W^S is the matrix of linear projection of size $C_eR^2 \times C_e$ used to reduce the dimension of the reshaped tensor; $(N/R^2) \times C_e$. $LayerNorm(\cdot)$ is the layer normalization operation [12]. The attention operation was calculated as:

$$Attention(Q, K, V) = Softmax\left(\frac{QK^T}{\sqrt{d}}\right)V, \quad (7)$$

where Q, K and V are so-called query, key, and value tensors respectively, and $Softmax(\cdot)$ is the normalized exponential function [11]. This function takes as input a 2-dimensional tensor $Z = QK^T/\sqrt{d}$ and the result is calculated element-wise as follows:

$$Softmax(Z_{ij}) = \frac{\exp(Z_{ij})}{\sum_j \exp(Z_{ij})}. \quad (8)$$

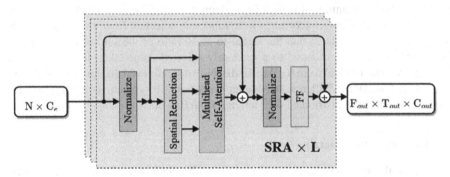

Fig. 4. Spatial-reduction attention layer.

Every SRA transformation was preceded by the normalization layer and followed by another normalization and the feed-forward neural layer with two skip-connections (Fig. 4). After a sequence of L repeats these transformations, the elements of the result were reshaped into a tensor of size $F_{out} \times T_{out} \times C_{out} = (F_{in}/F_p) \times (T_{in}/T_p) \times C_e$ (Fig. 4).

2.5 Loss Function

The mask M was calculated by linear convolution of the decoder output. Then, according to (2), the signal $\hat{X}$ was calculated, and it was used as an argument of the loss function. On training of the network, we used the scale-invariant signal-to-distortion ratio (SI-SDR) as the loss function $\mathcal{L}(\hat{X}, X)$ [16] was:

$$\mathcal{L}(\hat{X}, X) = -\text{SI-SDR}(\hat{X}, X) = -10 \log_{10} \left(\frac{\|\alpha X\|^2}{\|\hat{X} - \alpha X\|^2} \right). \tag{9}$$

SI-SDR is invariant to the loudness level of the signals $\hat{X}$ and X. This property was achieved by rescaling the clean signal with a coefficient $\alpha = \hat{X}^T X / \|X\|^2$, selected so that the vector projections of the reconstructed signal $\hat{X}$ onto the clean signal X and the clean signal X onto the reconstructed one $\hat{X}$ were equal [16].

3 Experiments and Results

3.1 Dataset Description

The training dataset was built on the basis of samples of clean and noisy audio recordings of the DNS Challenge 2020 open dataset [17]. The clean signals in this database, which included 500 h of voice recordings from 2150 speakers, were collected from the Librivox open set. The selection criterion was the quality according to the MOS evaluation metric [18]. Records from the upper quartile of the distribution of this metric were used. Noisy recordings were generated based on the Audioset, Freesound and DEMAND sets. At least 500 different samples of each of 150 noise classes were selected from them, lasting 10 s each. During training, to form a noisy sample, a clean voice recording and a set of randomly selected noise recordings were randomly selected. Next, the noise records were scaled in amplitude so that the signal-to-noise ratio of the sum of the clean and noise components corresponded to a randomly selected value taken from a uniform distribution in the interval from 0 dB to 40 dB. No artificial reverberation was used. The test subset of the DNS Challenge 2020 set consisted of 150 synthetic noisy signals with a signal-to-noise ratio distributed over a range from 0 to 20.

3.2 Implementation Details

In this work, the duration of each sample was 2 s with sampling rate 16 kHz. Each sample was divided into overlapping frames 32 ms long with 50% overlap. After the short-time Fourier transform with Hamming window, the input tensor of size $256 \times 128 \times 2$ was calculated. For the frequency, the coding (3) vector with K = 10 was used. The kernel size of input and output convolutions was 3×5 with stride 1. Table 1 shows the detailed information of internal configuration of the encoder, the bottleneck and the decoder. The kernel size of convolution and transposed convolution in the bottleneck block were the same with size 1×1.

The developed neural network model was implemented in the Python programming language using the PyTorch deep learning library. The training was carried out in 3 weeks on a workstation with two Nvidia GeForce 1080Ti GPU. The Adam optimization method with the learning rate parameter 10^{-3} was used.

Table 1. Hyperparameters of the proposed neural network.

Stage		$F_p \times T_p$	C_e	H	R	L
Encoder	1	8×2	64	1	8	3
	2	2×2	128	2	4	4
	3	2×1	320	5	2	18
Bottleneck		1×1	512	8	1	3
Decoder	1	2×1	320	8	1	3
	2	2×2	128	4	2	6
	3	8×2	64	2	4	4

3.3 Performance Evaluation and Discussion

The choice of metrics for performance evaluation was since it was necessary to avoid using naive "physical" estimates such as the signal-to-noise ratio. The latter did not reflect the quality of the noise reduction method perceived by the listener. This consideration explains why the following quality assessment indicators were used:

1. Perceptual evaluation of speech quality (PESQ) [19]. This value simulates MOS expert assessment on a shifted five-point scale from -0.5 (lowest quality) to 4.5 (highest quality). Two types of PESQ metrics were used: a narrow-band one (NB-PESQ), designed for signals with a sampling frequency of 8 kHz, and a wide-band one (WB-PESQ) for signals with a sampling frequency of 16 kHz.
2. Short-time objective intelligibility measure (STOI) [20], measured as a percentage. The intelligibility was evaluated using the reference and distorted resampled signals with a sampling rate of 10 kHz based on a comparison of their spectral characteristics.
3. The scale-invariant signal-to-distortion ratio SI-SDR [16]. This metric is the version of the signal-distortion rate, which is robust to rescaling of the signal.

Table 2 shows performance metrics of the proposed model compared to the best alternative approaches. The top row, marked as "noisy signal", was the starting point – the values of the quality metrics on the noisy signals, not being processed. The higher the values of the metrics, the better the quality of a model. Three models in this comparison (FullSubNet [4], two-stage LSTM [21] and NSNet [22]) are based on the recurrent neural networks. This allowed them to process the signal in real time fashion with minimal latency. PoCoNet model [3] use own version of encoder-decoder architecture with so-called dense attention blocks and frequency-positional encoding. Conv-TasNet [6] is an end-to-end fully convolution all-time-domain network with minimal latency and small model size. DCCRN-E [23] is a mixed recurrent-convolution architecture with modifications, which allowed it to handle operations on complex values. Table 2 shows that the proposed model outperforms competing approaches in all calculated metrics.

The proposed neural network model was trained to directly optimize the SI-SDR ratio. This explains the significant gap in the values of this metric from alternatives. Also, this fact indirectly indicates the importance of the SI-SDR for the perception of speech by listeners, since other metrics increased synchronously, albeit to a relatively lesser extent.

The following should be noted. When evaluating the quality of our model, the factor of computational efficiency was not considered. Any approach based on convolutions or recurrent layers is significantly "lighter" in terms of the number of parameters. The proposed model requires large computational resources both in training and in use. This limitation will be overcome by more efficient implementation of self-attention transformations, optimization of the neural network structure and fine tuning of network hyperparameters.

Table 2. Performance of the proposed method in comparison with existing approaches.

Model	WB-PESQ	NB-PESQ	STOI	SI-SDR
Noisy signal	1.582	2.454	91.52	9.071
FullSubNet	2.777	3.305	96.11	17.3
Two-Stage LSTM	2.832	3.338	96.05	17.6
NSNet	2.145	2.873	94.47	15.61
PoCoNet	2.748	–	–	–
Conv-TasNet	2.730	–	–	–
DCCRN-E	–	3.266	–	–
Proposed approach	**2.99**	**3.442**	**97.05**	**19.2**

4   Conclusion

In this paper, a new method for noise reduction based on the use of a deep neural network of the encoder-decoder structure with intermediate layers of self-attention was proposed. This approach demonstrated a significant improvement in the quality of work compared to the best modern analogues. The implemented method can be used in a wide range of tasks related to digital speech processing where the perceptual quality of the audio signal is significant.

Acknowledgments. This work was supported by the grant from the Russian Science Foundation, project no. 22–21-00199, https://rscf.ru/en/project/22-21-00199/.

References

1. Loizou, P.C.: Speech Enhancement: Theory and Practice. CRC Press, Boca Raton, FL, USA (2007)
2. Tan, K., Wang, D.A.: Convolutional recurrent neural network for real-time speech enhancement: In: Proc. Interspeech 2018, 2–6 September 2018, Hyderabad, India, pp. 3229–3233 (2018)
3. Umut, I., Giri, R., Phansalkar, N., Valin, J.-M., Helwani, K., Krishnaswamy, A.: PoCoNet: better speech enhancement with frequency-positional embeddings, semi-supervised conversational data, and biased loss. In: Proc. of Interspeech 2020, 25–29 October 2020, Shanghai, pp. 2487–2491 (2020)
4. Hao, X., Su, X., Horaud, R, Li, X.: FullSubNet: a full-band and sub-band fusion model for real-time single-channel speech enhancement. In: IEEE International Conference on Acoustics, Speech, and Signal Processing (ICASSP), 6–11 June 2021, Toronto, Canada, pp. 6633–6637 (2021)
5. Xu, R., Wu, R., Ishiwaka, Y., Vondrick, C., Zheng, C.: Listening to sounds of silence for speech denoising. In: Conference on Neural Information Processing Systems (NeurIPS 2020), 6–12 December 2020, Vancouver, Canada, pp. 9633–9648 (2020)
6. Luo, Y., Mesgarani, N.: Conv-TasNet: surpassing ideal time-frequency magnitude masking for speech separation. IEEE/ACM Trans. Audio Speech Lang. Process. 27(8), 1256–1266 (2019)
7. Hu, G., Wang, D.: Monaural speech segregation based on pitch tracking and amplitude modulation. IEEE Trans. Neural Netw. 15(5), 1135–1150 (2004)
8. Xia, S., Li, H., Zhang, X.: Using optimal ratio mask as training target for supervised speech separation. In: 2017 Asia-Pacific Signal and Information Processing Association Annual Summit and Conference (APSIPA ASC), 12–15 December 2017, Kuala Lumpur, pp. 163–166 (2017)
9. Liu, Y., Zhang, H., Zhang, X., Yang, L.: Supervised speech enhancement with real spectrum approximation. In: IEEE International Conference on Acoustics, Speech, and Signal Processing (ICASSP), 12–17 May 2019, Brighton, UK, pp. 5746–5750 (2019)
10. Williamson, D.S., Wang, Y., Wang, D.: Complex ratio masking for monaural speech separation. IEEE/ACM Trans. Audio Speech Lang. Process. 24(3), 483–492 (2016)
11. Vaswani, A., et al.: Attention is all you need. In: Conference on Neural Information Processing Systems (NIPS 2017), 4–9 December 2017, Long Beach, CA, USA., 11 p. (2017)
12. Dosovitskiy, A., et al.: An image is worth 16×16 words: transformers for image recognition at scale. In: Proc. of International Conference on Learning Representations (ICLR 2021), 3–7 May 2021, virtual only, 21 p. (2021)
13. Wang, W., et al.: Pyramid vision transformer: A versatile backbone for dense prediction without convolutions. In: IEEE/CVF International Conference on Computer Vision (ICCV), 11–17 October 2021, virtual only, pp. 548–558 (2021)
14. Cao, H., et al.: Swin-Unet: Unet-like pure transformer for medical image segmentation (2020). Preprint at https://arxiv.org/abs/2105.05537
15. Ronneberger, O., Fischer, P., Brox, T.: U-Net: convolutional networks for biomedical image segmentation. In: Navab, N., Hornegger, J., Wells, W.M., Frangi, A.F. (eds.) MICCAI 2015. LNCS, vol. 9351, pp. 234–241. Springer, Cham (2015). https://doi.org/10.1007/978-3-319-24574-4_28
16. Roux, J.L., Wisdom, S., Erdogan, H., Hershey, J.R.: SDR – half-baked or well done? In: IEEE International Conference on Acoustics, Speech and Signal Processing, 12–17 May 2019, Brighton, UK, pp. 626–630 (2019)

17. Reddy, C., Beyrami, E., Dubey, H.: Deep noise suppression challenge: Datasets, subjective testing framework, and challenge results. In: Proc. of Interspeech 2020, 25–29 October 2020, Shanghai, China, pp. 2492–2496 (2020)
18. International Telecommunication Union: ITU-T P.808 Subjective Evaluation of Speech Quality with a Crowdsourcing Approach (2018)
19. Rix, A.W., Beerends, J.G., Hollier, M.P. Hekstra, A.P.: Perceptual evaluation of speech quality (PESQ) – a new method for speech quality assessment of telephone networks and codecs. In: IEEE International Conference on Acoustics, Speech, and Signal Processing. Proceedings, 7–11 May 2001, Salt Lake City, UT, USA, pp. 749–752 (2001)
20. Taal, C.H., Hendriks, R.C., Heusdens, R., Jensen, J.: A short-time objective intelligibility measure for time-frequency weighted noisy speech. In: IEEE International Conference on Acoustics, Speech and Signal Processing, 15–19 March 2010, Dallas, TX, USA, pp. 4214–4217 (2010)
21. Nasretdinov, R., Ilyashenko, I., Lependin, A.: Two-stage method of speech denoising by long short-term memory neural network. In: HPCST 2021: High-Performance Computing Systems and Technologies in Scientific Research, Automation of Control and Production, Communications in Computer and Information Science, Springer, 1526, pp. 86–97 (2022)
22. Braun, S., Tashev, I.: Data augmentation and loss normalization for deep noise suppression. In: Karpov, A., Potapova, R. (eds.) SPECOM 2020. LNCS (LNAI), vol. 12335, pp. 79–86. Springer, Cham (2020). https://doi.org/10.1007/978-3-030-60276-5_8
23. Hu, Y., et al.: DCCRN: deep complex convolution recurrent network for phase-aware speech enhancement. In: Proc. of Interspeech 2020, 25–29 October 2020, Shanghai, China, pp. 2472–2476 (2020)

Architecture and CAD Software Solutions for PCB Design

Ilya Tarasov$^{(\boxtimes)}$ ⓘ, Dmitry Mirzoyan, and Peter Sovietov

MIREA – Russian Technological University, Vernadsky Avenue 78, 119454 Moscow, Russia
tarasov_i@mirea.ru

Abstract. The article discusses the CAD architecture for PCB design, designed to create a new generation design system, focused on a large number of components, additional routing layers. The identified disadvantage of modern CAD printed circuit boards is a large load on the CPU when constructing a two-dimensional image in the construction area. Zooming while viewing the PCB results in image rebuilding, where elements such as circles, arcs, and polygon cutouts put additional CPU load and slow down system performance. The article discusses the CAD architecture based on the preparation of data on design elements in a format suitable for transfer to the GPU using the API of the OpenGL library. This approach transfers a significant part of the computational load to the GPU, which increases the overall performance of the CAD graphics subsystem. Additionally, the issues of using scripting languages for the automated creation of regular structures, such as external component pins, buses and vias for connecting signals in a regular way, are considered.

Keywords: CAD · Electronics design · GPU · OpenGL · Scripting language

1 Introduction

Currently, a number of CAD systems for the design of electronic products are available to developers, combining the development of circuit diagrams and printed circuit boards. Since these project components are closely related, they are usually developed as part of a single software tool. An analysis of the current situation shows that in addition to professional design tools (Altium Designer, PCB Expedition, Allegro, etc.), for simpler projects, freely distributed CAD implementations (EasyEDA, Eagle, CircuitMaker, Eagle), including those with open source code, are used (KiCAD) [1].

Modern CAD projects for electronics design usually use separate graphics engines for 2D and 3D graphics [2]. As a rule, the OpenGL library is used to build a three-dimensional image of a printed circuit board with installed components, and the main mode of operation is two-dimensional, implemented mainly in software based on the CPU [3]. Practical CAD implementations use WxWidgets, Qt libraries, and the online version of EasyEDA CAD uses web-based graphics [4]. The practical disadvantage of this solution is the high load on the CPU, which manifests itself in the form of a decrease in performance in the PCB design mode when the scale of the two-dimensional

V. Jordan et al. (Eds.): HPCST 2022, CCIS 1733, pp. 15–21, 2022.
https://doi.org/10.1007/978-3-031-23744-7_2

scene is changed. A feature of the display of a printed circuit board is the need to programmatically cut off invisible objects that may be outside the display area, or be covered by other layers of the printed circuit board [5]. The complication of printed circuit boards occurs both in the direction of increasing the number of components, and in the direction of increasing the number of routing layers. In addition, support for a large number of graphic primitives, including circles and arcs, further increases the load on the CPU. An example image of a multilayer printed circuit board used to verify support for the ODB++ format is shown in Fig. 1.

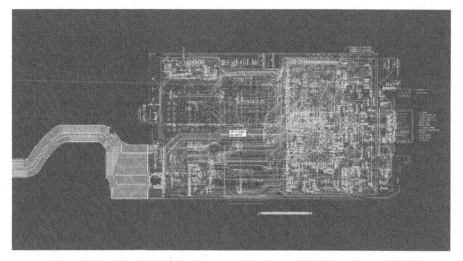

Fig. 1. An example image of a multilayer PCB.

Thus, in order to meet modern requirements in the field of electronic design, it is required to increase the performance of drawing a two-dimensional image of a printed circuit board design. An obvious way to achieve this goal is the active use of the GPU, which is reflected in a number of publications [6, 7].

The levels of formation and presentation of a two-dimensional image can be represented as follows:

1. The level of project structural elements.
   At this level, the design is represented in domain terms, as components, footprints, pads, wires, vias, polygons, and so on.
2. The level of graphic primitives.
   Since objects of the "electronic component pad" type are not supported by graphics libraries and GPUs, these objects must be decomposed into elementary graphic primitives beforehand.
3. The level of the two-dimensional image.
   Rendering graphics primitives, taking into account the display area, the display order of the layers, transparency, and the presence of components in the upper layers,

results in a bitmap that can be generated completely programmatically. In this case, the role of the GPU is reduced to displaying the PCB.

The article discusses ways to increase the performance of building a two-dimensional image of an object, an example of which is a multilayer printed circuit board with a large number of components, including such graphic primitives as circles, arcs, and polygons with cutouts.

In general, the interaction of CPU and GPU in CAD can be illustrated by the following diagram. In the basic version, the preparation of a 2D image is performed entirely on the basis of the CPU, with the subsequent transfer of the contents of the construction area to the GPU. The goal is to prepare graphics primitives on the CPU that are sent to the GPU using high-level graphics library APIs such as OpenGL (Fig. 2).

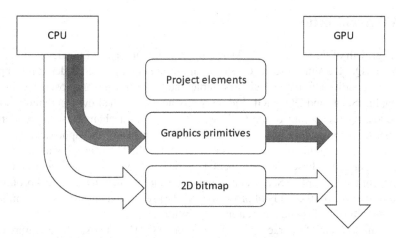

Fig. 2. Interaction of CPU and GPU in the proposed CAD architecture.

To determine practical ways to use OpenGL, you should consider the capabilities of this library and the API options as part of CAD.

2 Features of the OpenGL Library

The easiest way to pass data about objects to the OpenGL library is to use the glBegin.. glEnd sequences, between which drawing commands for individual graphical primitives are passed. This approach saves a large amount of data transfer, since the possible primitives are elements such as GL_POINTS, GL_TRIANGLES, GL_QUADS, and so on. [eight]. Since the circles and arcs of PCB elements require decomposition into such basic components, the total amount of data transferred is still excessive.

The next step is to use OpenGL display lists. Using this mechanism, you can significantly increase the display speed of components that do not change in the scene, but require display in different coordinates [9].

Further, it is possible to use arrays of vertices previously loaded into GPU memory. Such array can be transferred to the GPU as a large data packet, which effectively uses the capabilities of modern graphics accelerators and utilizes the capabilities of high-speed PCI Express interfaces. The need to rebuild the array of vertices when changing the project does not create significant problems, since changing the vertices is a consequence of the actions of the CAD operator, therefore, it occurs at a moderate speed.

The general approach proposed for the CAD architecture is the active use of GPU memory to store pre-prepared data about the vertices of graphic primitives. To do this, the structural elements of the project (pads, conductors, polygons, holes, silk-screen symbols) must be converted into graphic primitives compatible with OpenGL data formats and a vertex array must be formed for its subsequent display on the basis of the GPU.

3   CAD Architecture

Since the growth of the volume of tasks is almost an integral part of the development of any field, the Qt platform using the C++ programming language is considered for implementation. In addition to being highly portable, the Qt platform was chosen because of the QGraphicsScene and QOpenGLWidget components, which allow you to move some of the rendering processing to the GPU without significant changes to the rendering mechanisms. Unfortunately, the results of development using the approaches implemented in these components gave an unsatisfactory result. Therefore, it was decided to consider visualization based on the OpenGL library without additional intermediaries in the form of the QGraphicsScene component, especially since the technical prototype implemented as part of the study of promising rendering mechanisms showed promising results both in terms of speed and scaling capabilities.

The main problem of interaction between modern CAD systems, which require efficient rendering of a large number of objects, and modern hardware-accelerated rendering libraries, is the difference in tasks and interfaces on both sides. CAD systems operate mainly with primitives of a flat geometric nature (lines, circles, arcs, Bezier curves, etc.), while modern libraries for working with a graphics processor are mainly focused on gaming tasks and operate with triangles in three-dimensional space.

To combine these interfaces, an additional layer was included in the CAD system, which converts the primitives of one side (lines, arcs) into the primitives of the other side (sets of triangles). Potentially, this approach allows you to use high-level program logic with any rendering engine – it is possible to implement rendering not only using OpenGL, but also DirectX, Vulkan, or program mechanisms – depending on the requirements for performance and portability of the target system.

It is proposed to use a visualization mechanism based on this approach, which made it possible to increase the volume of tasks that are comfortable for the user to be solved up to approximately 10 million elements. The main problem in the implementation of the rendering mechanism is the almost complete absence of approaches, techniques and examples of solving such problems in the world literature. Most of the articles on working with hardware-accelerated graphics are devoted to efficient rendering of three-dimensional objects, special graphic effects, ray tracing and clipping mechanisms, and so on. The available articles on working with 2D graphics consider it mainly as an addition to 3D graphics (usually for visualizing the user interface) and the approaches described in them are unacceptable for solving 2D problems of the volume under consideration.

The main mechanism for solving problems in hardware-accelerated 3D graphics is the preparation of the necessary data set in the form of a vertex structure and the implementation of an algorithm for processing this data on a GPU using its capabilities through a shader language. Since modern GPUs are general purpose computing devices and are suitable for solving a wide range of tasks, working with two-dimensional graphics is not a problem for them from an architectural point of view. Therefore, having formed suitable data structures, it is possible to shift the task of visualizing a large number of two-dimensional objects to the graphics processor, having compiled a suitable implementation of the algorithms necessary for this in the shader language.

At the same time, it should be taken into account that 2D graphics, unlike 3D graphics, often depend on pixel processing mechanisms, which are an important addition to working with rasterized vector primitives. Fortunately, in the field of game graphics there are examples of the implementation of a similar technology – the so-called Post-Processing, which processes the results of the initial rendering of the scene to implement various visual effects (for example, blurring, depth of field, some lighting models). This technique allows you to implement the necessary pixel processing in solving problems of visualization of CAD objects.

To implement the necessary pixel effects (transparency and translucency of layers, masking), a layer-by-layer rendering mechanism was implemented, followed by mixing and post-processing.

At the same time, depending on the properties of the elements, they are divided into a set of sublayers. Each sublayer consists of a set of vector primitives with specified visualization parameters (colors, transparency, fill, etc.). Each sublayer is rendered separately into a separate buffer. After processing all sublayers, the results are reduced to a layer image, while taking into account the properties of individual sublayers (for example, the mask sublayer image is subtracted from the image of the underlying sublayer). The finished set of layers is brought together, taking into account the parameters of the layers (transparency), forming the finished image.

This mechanism is similar in its properties to the standard software drawing algorithms used in modern CAD for PCB design. However, the developed implementation is executed completely (with the exception of the formation of a set of initial data) on the GPU, which allows processing tasks of up to 10 million elements, the possible use of additional optimization and promising GPUs will increase the amount of tasks to be solved up to the amount of memory available to the GPU, today comprising about 25–30 million elements.

4 Design Process Using Scripting Languages

Managing a large number of components in a regular layout or behavior reduces the effectiveness of the GUI. A number of typical scenarios can be listed when a CAD operator must perform a sequence of the same actions on a large number of identical design elements.

4. Creating a footprint for a high pin count chip package where certain pins are displaced, removed, or modified. Such operations are typical for BGA packages.
5. Connecting power to the pins of the microcircuit with a serial implementation of the connection pattern – for example, in the form of a conductor connecting the pin of the BGA package with a via located diagonally relative to the center of the pin.
6. Connecting resistors to match the impedance to the group of lines corresponding to the data bus.

It is more convenient to implement these and similar operations on the basis of scripts launched as part of CAD. Similar solutions are used in practice – for example, the Tcl language is built into some microelectronic CAD systems [10]. In addition, there are alternative options for building scripting languages adapted to the subject area of developing electrical circuits [11].

5 Conclusions

The PCB design CAD architecture discussed in the article offers the transfer of the computational load from the CPU to the GPU to speed up the display of a two-dimensional representation of a printed circuit board. For this, an intermediate level of project presentation is used, which is formed on the basis of the decomposition of the elements of the printed circuit board project and the creation of a list of graphic primitives. Loading these elements into GPU memory is done in batch mode, which speeds up the subsequent generation of a two-dimensional representation that is performed almost entirely on the basis of the GPU. This approach allows to significantly increase the performance of the graphics subsystem of PCB CAD when displaying a large number of complex elements located in different layers.

Acknowledgements. The work is carried out within the framework of the state task of the Ministry of Science and Higher Education of the Russian Federation (subject No. FSFZ-2022-0004 Architectures of specialized computing systems, methods, algorithms, and tools for designing digital computing devices).

References

1. Kanagachidambaresan, G.: Introduction to KiCad design for breakout and circuit designs. In book: Role of Single Board Computers (SBCs) in rapid IoT Prototyping (2021). https://doi.org/10.1007/978-3-030-72957-8_8
2. Islam, M.N., Alam, M.S., Haque, M.A.S.: Development of Eagle Multi-layer Printed Circuit Board with CAD and CAM. In: "International Conference on Electronics and Informatics-2021" organized by the Bangladesh Electronics and Informatics Society (BEIS); Atomic Energy Centre, Dhaka (AECD) & Zoom Online Platform, Paper ID: 20170, pp. 99, Dhaka, Bangladesh, Poster Presentation, 27–28 Nov 2021
3. Jurado, D., Jurado, J.M., Ortega, L., Feito, F.R.: 3D environment understanding in real-time using input CAD models for AR applications. In: CEIG – Spanish Computer Graphics Conference, pp. 81–84 (2019)

4. Jasani, K., Mehta, S., Mehta, J.: Altium: a fast schematic designer. Indian J. Appl. Res. **5**, 60–63 (2015)
5. Sharma, P.: Software overview. In book: PCB Design for Absolute Beginners (2022). https://doi.org/10.1007/978-1-4842-8040-9_3
6. Park, S.: A real-time rendering algorithm of large-scale point clouds or polygon meshes using GLSL. Trans. Soc. CAD/CAM Eng. **19**, 294–304 (2014). https://doi.org/10.7315/CADCAM.2014.294
7. Mendoza, J., López, Z., Sotelo, J.: Methods for rendering polylines with geospatial coordinates in OpenGL ES 2.0. Cuban J. Inform. Sci. **15**, 74–91 (2021)
8. Kilgard, M.: NVIDIA OpenGL Extension Specifications NVIDIA Corporation (2004)
9. Wright Jr., R.S., Lipchak, B., Haemel, N.: OpenGL Superbible: Comprehensive Tutorial and Reference (2010)
10. Lata Tripathi, S., Kumar, A., Pathak, J.: Tcl-Tk for EDA Tool. In: Programming and GUI Fundamentals: TCL-TK for Electronic Design Automation (EDA), pp. 185–210, IEEE (2023). https://doi.org/10.1002/9781119837442.ch10
11. Li, H., He, Y., Xiao, Q., Tian, J., Bao, F.S.: BHDL: a lucid, expressive, and embedded programming language and system for PCB designs. In: 2021 58th ACM/IEEE Design Automation Conference (DAC), pp. 355-360 (2021)

Frequency-Domain Generalized Phase Transform Method in Pipeline Leaks Locating

Vladimir Faerman$^{(\boxtimes)}$ ⃝, Kirill Voevodin ⃝, and Valeriy Avramchuk ⃝

Tomsk State University of Control Systems and Radioelectronics, 40 Lenina Avenue,
Tomsk 634050, Russia
fva@fb.tusur.ru

Abstract. This article considers the problem of instrumental-assisted determination of the location of leaks in water pipes using acoustic correlation leak detectors (correlators). The aim of the work is to adapt the algorithm of time delay estimation (TDE) in the frequency domain to use it as part of the correlator's software. The significance of the problem of effective and fast locating of the leak is significant and obvious. That is why there is a large variability of instrumental solutions designed to solve it. Correlators, which are characterized by high accuracy and portability, occupy an important niche among the hardware leak-finding tools. We analyze the factors that affect the accuracy of correlators, in particular their software. The article reviews various TDE techniques used in correlators, including time domain, frequency domain, time-frequency domain, and those based on adaptive regression models. Despite the lack of clear advantages in terms of accuracy, the TDE in the frequency domain provides an additional information channel to the leak detector operator. We propose and describe an adapted algorithm for generalized TDE in the frequency domain with the least squares regression model. The prototype of the program is implemented in Mathcad and is used to analyze signals received with a correlator in the field. The presented results testify to the applicability of the method both in combination with the traditional correction technique and independently.

Keywords: Time-delay estimation · Leak noise correlator · Generalized phase spectrum · Acoustic testing · Water pipes

1 Introduction

Leaks in utility pipelines are a significant problem for all urbanized communities in the world. According to the estimates of various researchers [1], the costs associated with the loss of processed liquid from heat and water supply systems amount to "multiple millions of dollars worldwide". According to a special report [2] released with the support of the United States Agency for International Development, the total drinking water losses in water supply systems around the world exceed 32 billion cubic meters annually. Estimated volume of water loss accounts for about 35% of the total volume of supply.

In several countries, the situation diverges from the global one in a negative way. In particular, in Russia, water losses in regional water distribution networks are up to 45% high. [3] This is due to the enormous scale of the pipeline infrastructure and its deterioration. According to official figures in [4], the total length of water pipelines in single-pipe terms is more than 600 thousand km. At the same time, the resource of the pipes has been worked out by more than 60%. [5].

Reduction in water losses is hardly possible without the replacement of pipes. The only viable option is fast location of leaks, with their consequent repair. Since the most of major pipelines are located underground, rapid detection of the liquid outflow sites is possible only with the use of modern instrumental appliances. As for now, various leak detection systems are known and employed by sanitary services. Articles [6–8] study the diversity of leak detection systems and propose their classifications.

Leak noise correlators are an important class of instruments designed for leak detection. [9, 10] The advantages of such devices are portability [9–11], reliability [9, 12, 13] and relatively high accuracy [9, 12, 14]. They are normally used "on demand" to locate leaks on a separate linear section of the pipe and provide a high level of accuracy comparable to expensive stationary systems. The correlators usually comprised by [15] a pair of vibroacoustic sensors, analog-to-digital converters (ADC), telecommunication equipment for point-to-point connection, and a signal processor. An example of the device is shown in Fig. 1.

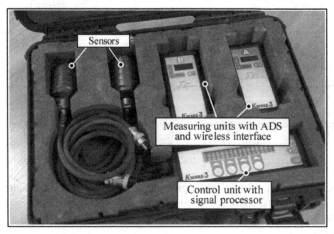

Fig. 1. Leak noise correlator KASKAD-3 developed by Moscow-based company RentTechnologies Ltd. [16].

Correlator principle of operation is shown in Fig. 2. It is based on the method of passive location via a vibroacoustic channel. [10] Acoustic emission is a result of the interaction between the pressurized fluid flow, the walls of the pipe and the environment outside of it. [17] The leak can be considered as the source S of the vibroacoustic signal. The receivers A, B of the signal are vibroacoustic sensors supplied with ADC. The received and sampled signals are then transmitted wirelessly to the signal processor, which computationally estimates the delay time. The signal propagates omnidirectionally along the pipe, therefore

$$d_{A,B} = \frac{d \pm T_D \cdot V}{2}$$

where d – the distance between receivers A and B;

$d_{A,B}$ – the distance from the leak to the corresponding receiver A or B;

V – signal propagation velocity;

T_D – time delay between the copies of the signal received by A and B.

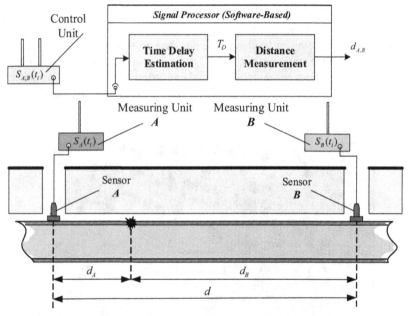

Fig. 2. Scheme for determining leaks in underground water supply pipes by correlation technique. Here $s_A(t)$, $s_B(t)$ – acoustical signals received by sensors A, B and sampled by corresponding measuring units.

The efficiency of the practical application of correlators is determined by a variety of hardware and software factors: the type of pipe [10, 11], characteristics of the sensors and the way they are fitted [12, 18], the error in estimation of velocity [19], the TDE algorithm. It should be noted that TDE is normally software-based, and hence its algorithmic implementation is of great research interest in the field of applied mathematics and digital signal processing.

A multitude of works are devoted to both velocity estimation [19–21], and TDE [22–25]. The accuracy of the estimation of the signal propagation velocity has a decisive value on the accuracy of determining the leakage position [26]. However, the velocity is reproducible for pipes of a certain type, and its estimation with sufficient accuracy for practice is possible with the use of premade lookup tables. In this paper, we focus on TDE and propose an algorithm of frequency domain TDE. The proposed technique is based on the estimation of phase shift. The applicability of the algorithm is demonstrated for signals received in the field by the CASCADE-3 correlator shown in Fig. 1.

2   Materials and Methods

2.1   Time Delay Estimation

Algorithms of TDE are standard tool of statistical signal processing and are used in such applications as distance estimation by sonar or radar, indoor positioning systems, faults detection in non-destructive testing [27]. Figure 2 shows a passive TDE scenario, which is described mathematically as follows

$$
\begin{aligned}
s_A(t) &= \alpha \cdot s_0(t - d_A/V) + n_A(t), \\
s_B(t) &= \beta \cdot s_0(t - d_B/V) + n_B(t),
\end{aligned}
\tag{1}
$$

where $s_A(t)$, $s_B(t)$ – signals received by A and B;

$s_0(t)$ – emitted signal in the closest proximity of the leak S;

$n_A(t)$, $n_B(t)$ – random additive noises in channels A and B;

α, β – attenuation coefficients in channels A and B.

The difference from the active TDE scenario is that the reference signal $s_0(t)$ cannot be rectified in any way. The classical approach to solving problem (1) is the well-known correlation technique developed in [28]. The use of correlation techniques to determine the position of pipeline leaks has been well studied, for example, in [25] and especially in [24]. The method described in [28] and its variants are known as time-domain methods. Methods that mathematically equivalent to correlation [29], but based on the analysis of the frequency responses are called frequency-domain methods. The applicability of both time-domain methods and frequency-domain in relation to determining the position of pipeline leaks is demonstrated and discussed in [30].

Time-frequency TDE methods are the subject of studies as well. Several studies propose the wavelet transform as an alternative to the Fourier transform in estimating the correlation between the channels in problems of determining the position of leaks [31, 32], Few other papers propose using the wavelet transform as a discrete filter [14] applied prior to the correlation function is computed. Other less conventional approaches can be based on the use of adaptive regression methods [22, 33, 34] and neural network [35, 36].

Despite the variety of methods, none of them are universally applicable for all cases. From case to case, spectrum of $s_0(t)$ can be different, the distortions introduced by the pipeline filter are nontrivial, the origin and characteristics of the noises $n_A(t)$, $n_B(t)$ are a priori unpredictable as well. In a significant part of practical cases, determining the position of the peak on the correlogram is not trivial due to the specifics of the working conditions in the field. In such circumstances [37], the result obtained independently by an alternative method is useful for the leak detector operator interpreting the results of the correlogram. Frequency-domain methods could be employed as an addition to conventional correlation technique. Therefore, frequency-domain methods are of practical interest, but not so well studied as time-domain methods. This work is devoted to the adaptation of the frequency domain TDE method to search for leaks in pipelines.

2.2   Phase Transform TDE

The TDE technique based on the analysis of the cross phase spectrum of signals was proposed in Piersol in 1981 [38]. It was further developed in [29], where the theoretical equivalency between TDE in the frequency and time domains was mathematically proven. The study of the frequency domain TDE in application to pipeline leaks detection was carried out in [30]. Here it was empirically and analytically shown that the accuracy of the estimation is practically identical to those of correlation methods. In [37], the frequency domain TDE was used to provide an operator with an additional information channel. That idea was further developed in [39], where the TDE in the frequency domain was used to refine the preliminary result obtained by the correlation method. Finally, in [23], an adaptive algorithm of frequency domain TDE was proposed. In particular, the adaptive least mean square filter was employed to determine the slope coefficient of the phase characteristic, that resulted in significant increase in accuracy.

Let us describe the basic frequency domain TDE algorithm initially proposed in [38]. First, we calculate the cross-correlation function $R_{AB}(\tau)$ for the signals $s_A(t)$, $s_B(t)$

$$R_{AB}(\tau) = E[s_A(t) \cdot s_B(t + \tau)], \tag{2}$$

where $E[\ ]$ – mathematical expectation.

Expression (2) can be transformed using (1)

$$R_{AB}(\tau) = \alpha\beta \cdot E\left[s_0(t) \cdot s_0\left(t - \left(\frac{d_A}{V} - \frac{d_B}{V}\right) + \tau\right)\right]. \tag{3}$$

All other terms in (3) are random, hence their expected value will be zero. Therefore (3) can be simplified

$$R_{AB}(\tau) = \alpha\beta \cdot R_{00}\left(\tau - \left(\frac{d_A}{V} - \frac{d_B}{V}\right)\right). \tag{4}$$

If we apply the direct Fourier transform to left and right sides of (4), we will get

$$X_{AB}(f) = \alpha\beta \cdot X_{00}(f) \cdot e^{-2\pi f\left(d_A/V - d_B/V\right)}, \tag{5}$$

where $X_{AB}(f)$ – cross-spectrum of signals $s_A(t)$, $s_B(t)$;

$X_{00}(f$ – auto-spectra of $s_0(t)$.

As we know that auto-spectra $X_{00}(f)$ must be real-valued, thus corresponding phase-spectra $\varphi_{00}(f)$ is identically equal to zero.

$$\varphi_{00}(f) = \arg(X_{00}(f)) \equiv 0.$$

It means that

$$\varphi_{AB}(f) = \arg(X_{AB}(f)) = 2\pi f\left(d_A/V - d_B/V\right). \tag{6}$$

Therefore, the lag value can be determined as the constant slope coefficient of the cross phase spectrum $\varphi_{AB}(f)$. Regression models, in particular the least mean squares method, can be used to determine the slope.

$$\overline{T}_{AB} = 2\pi\frac{\int f \cdot \varphi(f)df}{\int f^2 \cdot df}, \tag{7}$$

where $\overline{T}_{AB}$ – the estimation of $T_{AB} = (d_A - d_B)/V$;

Since contaminating noises are present in the received signals according to (1), some parts of $\varphi_{AB}(f)$ may be significantly distorted. This will negatively affect the accuracy of T_{AB} estimation. The solution to this problem was proposed in [29] and is known as the generalized method of frequency domain TDE. The main feature is that instead of (7) the following formula is applied

$$\overline{T}_{AB} = 2\pi\frac{\int f \cdot W(f) \cdot \varphi(f)df}{\int W(f) \cdot f^2 \cdot df}, \tag{8}$$

where $W(f)$ – frequency weight function, $0 \leq W(f) \leq 1, \forall f$.

The weight function $W(f)$ is chosen in such a way that its values are close to 1 at those frequency intervals at which the influence of noise is minimal. On the contrary, if the effect of noise is significant, then the values of $W(f)$ should be close to 0.

2.3 Practical Frequency-Domain TDE Algorithm

The practical generalized frequency domain TDE algorithm used in this work is described in detail in [40]. Core designations and relations are described below

$$\varphi_{AB}(f_k) = U\big(\arg[X_{AB}(f_k)]\big), \ k = 0, 1, ..., N/2, \tag{9}$$

where $\varphi_{AB}(f_k)$ – unwrapped cross phase spectrum of sampled signals $s_A(t_i)$, $s_B(t_i)$;
  $X_{AB}(f_k)$ – estimation cross-spectrum of sampled signals $s_A(t)$, $s_B(t)$;
  U – phase unwrapping operator;
  N – length of time series in a single fast Fourier transform (FFT) window.
  The estimation $X_{AB}(f_k)$ is obtained by averaging multiple instances of instantaneous cross spectrums

$$X_{AB}(f_k) = E\big[F_D^*(s_A(t_i) \times w\ (t_i)) \times F_D(s_B(t_i) \times w\ (t_i))\big], \ i = 0, 1, ..., N - 1,$$

where F_D – FFT operator;
  F_D^* – FFT with element-wise conjugation operator;
  $w\ (t_i)$ – sampled window function;
  $\times$ – element wise product.
  To extract $\overline{T}_{AB}$ from $\varphi_{AB}\ (f_k)$ the least squares method is used as follows

$$\overline{T}_{AB} = \frac{\Delta \cdot N}{2\pi} \cdot \frac{\Lambda \cdot K - A \cdot \Theta}{K \cdot B - A^2}, \tag{10}$$

$$K = \sum_k W(f_k), \ A = \sum_k k, \ B = \sum_k k^2, \ \Theta = \sum_k \varphi_{AB}(f_k) \cdot W(f_k),$$

$$\Lambda = \sum_k k \cdot \varphi_{AB}(f_k) \cdot W(f_k).$$

A set of the following weight functions is used.

$$W_{PHAT}(f_k) = 1, \ W_{BCC}(f_k) = \frac{|X_{AB}(f_k)|}{\max(|X_{AB}(f_k)|)}, \ W_{SCOT}(f_k) = \gamma_{AB}(f_k),$$

$$W_{ML}(f_k) = \frac{\gamma_{AB}^2(f_k)}{1 - \gamma_{AB}^2(f_k)}, \ W_{COH}(f_k) = \gamma_{AB}^2(f_k),$$

where $\gamma_{AB}\ (f_k)$ – coherence functions [41].
  The described computational scheme is shown in Fig. 3. The advantage of this algorithm is the possibility to use an arbitrary (possibly not consecutive) set of frequency bins in (10). This will allow taking into account in the calculations only those frequency intervals at which the leakage signal is expected to dominate over the in-pipe noise.

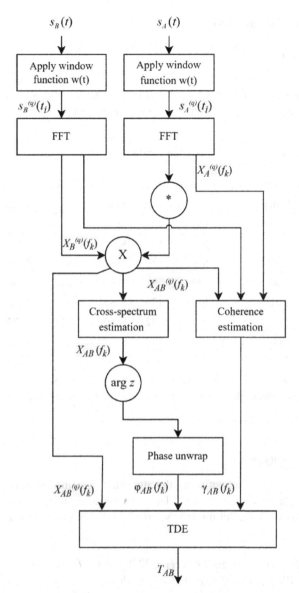

Fig. 3. Computational scheme for frequency-domain TDE. Simple element-wise operations (algebraic or arithmetic) are depicted as circles. Complex operations are depicted as rectangles. Everywhere on the scheme $i = 0, 1, \ldots, N - 1, k = 0, 1, \ldots, N/2$.

3 Results and Discussion

3.1 Experimental Setting

During experimental study, we used recordings of signals obtained during the study of water pipes using the KASKAD-3 leak detector. The characteristics of the signals and the context of their acquiring are summarized in Table 1 [14].

Table 1. Parameters of recordings and corresponding practical cases.

Case	Recording parameters		Pipe parameters			Reference parameters		
	$1/\Delta$, Hz	Duration, min	D, mm	L, m	Material	d_A, m	V, m/s	T_{AB}, ms
1	22050	1.02	500	139	Steel	46	1121	41.93
2	21362	3.00	700	138	Steel	46	1047	43.93
3	21362	3.00	100	75	Steel	51	1268	−21.29
4	21362	3.00	100	92	Steel	86	1268	−63.09
5	22050	1.01	300	126	Steel	33	1199	50.04
6	21333	2.00	90	32	Steel	14	1273	3.14
7	21333	2.00	100	137	Steel	40	1268	44.95
8	21333	2.00	200	49	Steel	23	1225	2.45

[*] Sound velocity V obtained with the formula in [26]. We took the pipe wall thickness and steel type those are most common for metal pipes of corresponding diameter used in Russia.

3.2 Software Prototype

Signal processing was carried out using the algorithm described in Sect. 2. We implemented it in the Mathcad environment. The source signals were in wav format. The processing parameters were set directly via the code in Mathcad and used for all the experiments:

- two alternative sets of frequency bins $f_k \in [200, 1200]$, $f_k \in [400, 700]$ Hz;
- number $N = 8192$ of ticks in a single FFT window;
- number of non-overlapping windows $Q = 30$ in a single fragment;
- high-resolution time window $w(t)$.

In the course of processing, the signals were divided into overlapping fragments, which were then studied independently. When calculating the window functions, we use an estimate of the coherence function obtained using the complete recording. Subsequently, statistical processing of the results was carried out, as proposed in [39]. To refine the results, we were segregating up to 20 median values of TDE, and then we average them to obtain the final estimation. We used Mathcad for statistical processing of the results, as well as to draw graphs.

3.3   Estimation of Leak Position

TDEs for various frequency weight functions $W(f_k)$ are in Table 2.

Table 2. TDE obtained with different frequency weight functions for each of the reported practical cases.

Case	Reference time delay	Frequency range	Time delay $\overline{T}_{AB}$, ms for various weight functions				
	T_{AB}, ms	$[f_{min}, f_{max}]$, Hz	W_{BCC}	W_{PHAT}	W_{SCOT}	W_{ML}	W_{COH}
1	41.93	[200, 1200]	38.90	30.07	34.28	40.61	39.92
2	43.93	[400, 700]	40.50	42.40	42.34	43.79	43.58
3	−21.29	[400, 700]	−18.15	−19.65	−20.37	−21.56	−21.46
4	−63.09	[200, 1200]	−47.54	−40.59	−49.08	−53.48	−51.57
5	50.04	[400, 700]	49.04	46.42	48.85	50.28	50.05
6	3.14	[200, 1200]	3.53	3.55	3.36	2.98	3.06
7	44.95	[200, 1200]	39.88	37.35	38.60	39.27	39.08
8	2.45	[400, 700]	2.35	2.65	2.60	4.06	3.59

Cross phase spectrums for all the cases are shown in Fig. 4. The corresponding cross-correlograms are given in Fig. 5. The presented graphs are obtained without signal fragmentation with $N = 8192$.

3.4   Discussion

The frequency domain TDE algorithm is more complex implementation-wise compared to conventional time domain TDE algorithms. Frequency domain TDE requires the employment of a certain regression model to estimate the slope of unwrapped cross phase spectrum. Implementation of phase unwrapping as well as selection of the regression model could affect the accuracy and computational complexity of the method. Another important feature is that the frequency domain TDE requires more complex preprocessing of the signals. However, if the preprocessing is right and the frequency interval is adequate, frequency domain TDE is less susceptible to the error than time domain TDE with machine correlogram interpretation. At the same time, TDE obtained via frequency domain methods is generally less accurate.

Experiments have shown that the use of extended time windows ($N > 8192$), does not contribute to an increase in accuracy, as it normally works with the time domain methods [37]. Actual TDE values could vary greatly from one fragment of the signal to another. However, statistical processing allows us to make a relatively accurate final estimation. The method proposed in Sect. 3.2 is designed to use only such fragments where the effect of contaminating noises is limited. For the reference, Fig. 6 shows the value of estimation over time for various fragments of the case 6 recording. Figure 7 shows forms of the various weight functions for the same case.

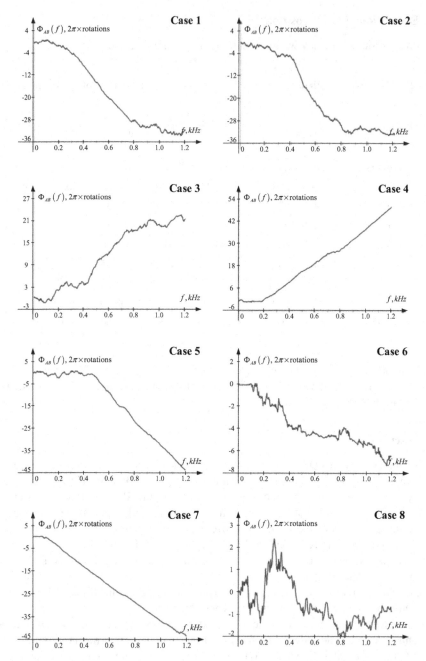

Fig. 4. Sample phase cross spectrums Φ_{ab} (f_k) for all the cases. Estimations are obtained with the full recordings. Cases 2, 3, 5 and 8 require using reduced frequency range for estimation.

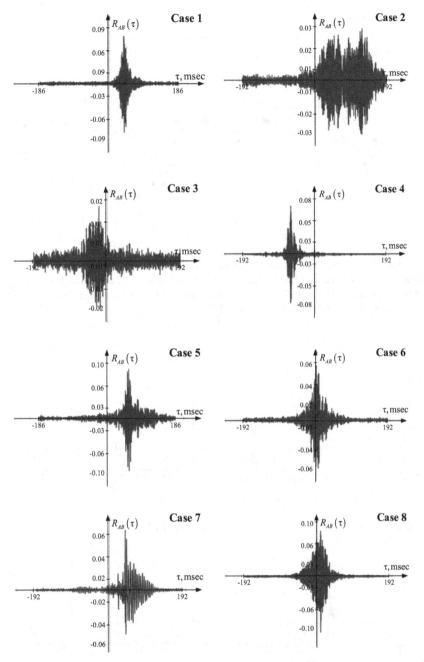

Fig. 5. Cross-correlation functions $R_{ab}(\tau_j)$ for all presented cases. Cross-correlation estimations are obtained with the full recordings.

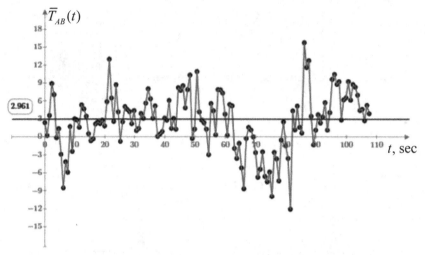

Fig. 6. Time-delay estimation obtained with the various fragments of the case 6 recording. Each estimation is determined from a fragment of 30 with time windows 8192 ticks in each. Overlapping rate of the adjusted fragments is 95%. A horizontal marker indicates the median value. Therefore, an increment on a time shift scale is about 1.152 s.

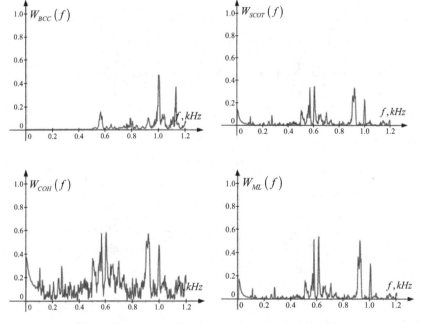

Fig. 7. Weighting functions $W(f_k)$ for the case 6.

The accuracy of determining the leak position for some of the presented cases is not adequate. In case 4 we have an estimated error of more than 10 ms that could imply

15 m error in estimated leak position. At the same time in Fig. 5 case 5, there is the distinguishable accurate peak in correlogram that corresponds to correct the delay of − 64 ms. As we can see in Fig. 4 case 4, the cross phase spectrum is composed of two linear parts with different slope. First slope in localized in (200, 650) Hz frequency interval and corresponds to −54.2 ms delay. The second one is localized in (800, 1200) Hz frequency interval and corresponds to delay of −63.9 ms. There is no clear way to determine which of the slopes is associated with the major peak on a correlation function. It should be noted, that between those two components, there is a highly contaminated with noises fragment in (650, 800) Hz frequency interval.

We have an opposite situation with the case 2. Correlogram in Fig. 5 case 2 does not have distinguishable peak. At the same phase spectrum in Fig. 4 case 2 has an almost linear fragment in (400, 700) Hz frequency interval that corresponds to correct time delay. Even though phase spectrums in Fig. 4 case 6 and case 8 seem to be highly contaminated, we still have rather adequate delay estimates. It demonstrates that weight functions are efficient tool to mitigate noise impact.

Minor errors in the obtained TDE values could be the result of insufficient data on cases we have. Unfortunately, we do not have complete information on the pipe type and field measurements that were produced to supplement the record. So, provided distances between sensors could possibly be measured with the mistake. Alternatively, studied pipes could have less thick walls than conventional pipes of this diameter.

4   Conclusion

In this paper, we proposed a practical technique of frequency domain TDE for application to detection of pipeline leaks. We had studied the proposed method on a set of cases and proved its applicability. Although, in general, frequency domain TDE is inferior both accuracy-wise and implementation-wise to time domain TDE, this technique still has the potential for practical application.

Despite the theoretical equality of the frequency domain and time domain TDE methods, the practice of their implementation is different: fragmentation of the recording, interpretation, processing of interim results. For this reason, the estimation obtained with the use of frequency domain TDE can be considered by the operator of the correlator as an additional channel of information. Therefore, the implementation of frequency domain TDE along with conventional correlation technique could improve the quality of decision-making by the operator.

References

1. Mashford, J., De Silva, D., Marney, D., Burn, S.: An approach to leak detection in pipe networks using analysis of monitored pressure values by support vector machine. In: 2009 Third International Conference on Network and System Security, pp. 534–539 (2009). https://doi.org/10.1109/NSS.2009.38
2. Farley, M., Wyeth, G., Ghazali, Z.B., Istandar, A., Singh, S.: The Manager's Non-Revenue Water Handbook: A Guide to Understanding Water Losses. United States Agency for International Development (USAID), Bangkok (2008)

3. Poryadin, A.V.: Water Supply and Sanitation in Russia [Vodosnabzhenie i vodootvedenie v Rossii]. Sanitarnay Tekhnika. **3–2**, 31–34 (2013)
4. Masakova, I., Vlasenko, N.: ред: Housing in Russia: Statistical Collection 2019. Moscow
5. Shlychkov, D.M.: Problems of the technical condition of present pipeline systems [Problemy tekhnicheskogo sostoyaniya dejstvuyushchih truboprovodnyh sistem]. Inpvatsii i Investitsii. **4**, 207–210 (2020)
6. Puust, R., Kapelan, Z., Savic, D.A., Koppel, T.: A review of methods for leakage management in pipe networks. Urban Water J. **7**(1), 25–45 (2010). https://doi.org/10.1080/157306210036 10878
7. Mohd Ismail, M.I., et al.: A review of vibration detection methods using accelerometer sensors for water pipeline leakage. IEEE Access. **7**, 51965–51981 (2019). https://doi.org/10.1109/ACCESS.2019.2896302
8. Wang, X., Lambert, M.F., Simpson, A.R.: Leak detection in pipeline systems and networks: a review. B: Conference on Hydraulics in Civil Engineering. pp. 1–10. The Institution of Engineers, Australia, Hobart (2001)
9. Glentis, G.O., Angelopoulos, K.: Leakage detection using leak noise correlation techniques – Overview and implementation aspects. B: PCI'19: 23rd Pan-Hellenic Conference on Informatics, pp. 50–57. ACM, New York (2019)
10. Fuchs, H.V., Riehle, R.: Ten years of experience with leak detection by acoustic signal analysis. Appl. Acoust. **33**, 1–19 (1991). https://doi.org/10.1016/0003-682X(91)90062-J
11. Bracken, M., Hunaidi, O.: Practical aspects of acoustical leak location on plastic and large diameter pipe. B: Leakage 2005 Conference Proceedings. pp. 448–452. NRCC, Halifax (2005)
12. Lapshin, B.M., Ovchinnikov, A.L., Chekalin, A.S.: Development and application of acoustic emission leak detectors [Razrabotka i primenenie akustiko-emissionnyh techeiskatelej]. V Mire NK **44**, 18–22 (2009)
13. Ozevin, D., Harding, J.: Novel leak localization in pressurized pipeline networks using acoustic emission and geometric connectivity. Int. J. Press. Vessels Pip. **92**, 63–69 (2012). https://doi.org/10.1016/j.ijpvp.2012.01.001
14. Faerman, V., Sharkova, S., Avramchuk, V., Shkunenko, V.: Towards applicability of wavelet-based cross-correlation in locating leaks in steel water supply pipes. J. Phys.: Conf. Ser. **2176**(1), 012067 (2022). https://doi.org/10.1088/1742-6596/2176/1/012067
15. Iwanaga, M.K., Brennan, M.J., Almeida, F.C.L., Scussel, O., Cezar, S.O.: A laboratory-based leak noise simulator for buried water pipes. Appl. Acoust. **185**, 108346 (2022). https://doi.org/10.1016/j.apacoust.2021.108346
16. Acoustical tomograph KASKAD-3 [Akusticheskij tomograf «KASKAD-3»] (2020). https://watersound.ru/products/akusticheskiy-tomograf-kaskad-3-s-funktsiey-korrelyatsionnogo-techeiskatelya.php
17. Papastefanou, A.S., Joseph, P.F., Brennan, M.J.: Experimental investigation into the characteristics of in-pipe leak noise in plastic water filled pipes. Acta Acust. Acust. **98**, 847–856 (2012). https://doi.org/10.3813/AAA.918568
18. Almeida, F., Brennan, M., Joseph, P., Whitfield, S., Dray, S., Paschoalini, A.: On the acoustic filtering of the pipe and sensor in a buried plastic water pipe and its effect on leak detection: an experimental investigation. Sensors (Switzerland). **14**, 5595–5610 (2014). https://doi.org/10.3390/s140305595
19. Scussel, O., Seçgin, A., Brennan, M.J., Muggleton, J.M., Almeida, F.C.L.: A stochastic model for the velocity of leak noise propagation in plastic water pipes. J. Sound Vib. **501**, 116057 (2021). https://doi.org/10.1016/j.jsv.2021.116057
20. Glentis, G.O., Angelopoulos, K.: Using generalized cross-correlation estimators for leak signal velocity estimation and spectral region of operation selection. In: Conference Record – IEEE Instrumentation and Measurement Technology Conference (2022). https://doi.org/10.1109/I2MTC48687.2022.9806476

21. Glentis, G.O., Angelopoulos, K.: Sound velocity measurement in acoustic leak noise correlation systems. In: Conference Record – IEEE Instrumentation and Measurement Technology Conference (2021). https://doi.org/10.1109/I2MTC50364.2021.9459921
22. Faerman, V.A., Avramchuk, V.S.: Comparative study of basic time domain time-delay estimators for locating leaks in pipelines. Int. J. Netw. Distrib. Comput. **8**, 49–57 (2020). https://doi.org/10.2991/IJNDC.K.200129.001
23. Ma, Y., Gao, Y., Cui, X., Brennan, M., Almeida, F., Yang, J.: Adaptive phase transform method for pipeline leakage detection. Sensors **19**(2), 310 (2019). https://doi.org/10.3390/s19020310
24. Kousiopoulos, G.-P., Papastavrou, G.-N., Kampelopoulos, D., Karagiorgos, N., Nikolaidis, S.: Comparison of time delay estimation methods used for fast pipeline leak localization in high-noise environment. Technologies **8**, 27 (2020). https://doi.org/10.3390/technologies8020027
25. Gao, Y., Brennan, M.J., Joseph, P.F.: A comparison of time delay estimators for the detection of leak noise signals in plastic water distribution pipes. J. Sound Vib. **292**, 552–570 (2006). https://doi.org/10.1016/j.jsv.2005.08.014
26. Becker, D.: Leak detection in water distribution networks by correlation: Sound velocity as a possible source of error. Guttersloh (2015)
27. Chen, T.: Highlights of statistical signal and array processing. IEEE Signal Process. Mag. **15**, 21–64 (1998). https://doi.org/10.1109/79.708539
28. Knapp, C.H., Carter, G.C.: The generalized correlation method for estiation of time delay. IEEE Trans. Acoust. Speech Signal Process. **24**, 320–327 (1976)
29. Zhen, Z., Zi-qiang, H.: The generalized phase spectrum method for time delay estimation. In: ICASSP'84. IEEE International Conference on Acoustics, Speech, and Signal Processing, pp. 459–462. IEEE, San Diego, CA (1984)
30. Brennan, M.J., Gao, Y., Joseph, P.F.: On the relationship between time and frequency domain methods in time delay estimation for leak detection in water distribution pipes. J. Sound Vib. **304**, 213–223 (2007). https://doi.org/10.1016/j.jsv.2007.02.023
31. Ting, L.L., Tey, J.Y., Tan, A.C., King, Y.J., Faidz, A.R.: Improvement of acoustic water leak detection based on dual tree complex wavelet transform-correlation method. IOP Conf. Ser.: Earth and Environ. Sci. **268**(1), 012025 (2019). https://doi.org/10.1088/1755-1315/268/1/012025
32. Ahadi, M., Bakhtiar, M.S.: Leak detection in water-filled plastic pipes through the application of tuned wavelet transforms to Acoustic Emission signals. Appl. Acoust. **71**, 634–639 (2010). https://doi.org/10.1016/j.apacoust.2010.02.006
33. Wen, Y., Li, P., Yang, J., Zhou, Z.: Adaptive leak detection and location in underground buried pipelines. Int. J. Inform. Acquisition **01**, 269–277 (2004). https://doi.org/10.1142/s0219878904000240
34. Guo, C., Wen, Y., Li, P., Wen, J.: Adaptive noise cancellation based on EMD in water-supply pipeline leak detection. Measurement **79**, 188–197 (2016). https://doi.org/10.1016/j.measurement.2015.09.048
35. Na, S., Liu, J., Li, Q., Liu, Y., Tie, Y.: An adaptive method for water pipeline leak localization. In: 4th Workshop on Advanced Research and Technology in Industry Applications (WARTIA 2018), vol. 173, pp. 146–153 (2018). https://doi.org/10.2991/wartia-18.2018.24
36. Bykerk, L., Miro, J.V.: Vibro-acoustic distributed sensing for large-scale data-driven leak detection on urban distribution mains. Sensors **22**, 6897 (2022). https://doi.org/10.3390/s22186897
37. Ovchinnikov, A.L., Lapshin, B.M., Chekalin, A.S., Evsikov, A.S.: Experience of using the TAK-2005 leak detector for pipeline monitoring [Opyt primeneniya techeiskatelya TAK-2005 v gorodskom truboprovodnom hozyajstve], pp. 196–202 (2008)
38. Piersol, A.G.: Time delay estimation using phase data. IEEE Trans. Acoust. Speech Signal Process. **29**, 471–477 (1981). https://doi.org/10.1109/TASSP.1981.1163555

39. Firsov, A.A., Terentiev, D.A.: An algorithm for improving the accuracy of location in correlation leak detection based on the analysis of the phase function of the mutual spectrum. Test. Diagn. **8**, 23–27 (2014)

40. Faerman, V., Avramchuk, V., Voevodin, K., Sidorov, I., Kostyuchenko, E.: Study of generalized phase spectrum time delay estimation method for source positioning in small room acoustic environment. Sensors **22**(3), 965 (2022). https://doi.org/10.3390/s22030965

41. Carter, G.C.: Coherence and time delay estimation. Proc. IEEE **75**, 236–255 (1987). https://doi.org/10.1109/PROC.1987.13723

Information Technologies and Computer Simulation of Physical Phenomena

Inverse Problem Regularization of Parametric Identification of the Particle Temperature Distribution of Gas-Thermal Flow with Optimization of Its Solution

Vladimir Jordan[1,2(✉)] [ID] and Denis Kobelev[3]

[1] Altai State University, Lenin Avenue 61, 656049 Barnaul, Russia
jordan@phys.asu.ru

[2] Khristianovich Institute of Theoretical and Applied Mechanics, SB RAS, Institutskaya Street 4/1, 630090 Novosibirsk, Russia

[3] Joint-Stock Company "Radiy TN", Krupskaya Street 99a, 656031 Barnaul, Russia

Abstract. The inverse problem of parametric identification of the model function of the temperature distribution density of heterogeneous flow particles, which form functional powder coatings on the surface of technical products in the process of their thermal spraying, is considered. The regularization of this ill-posed inverse problem consists in choosing a physically adequate parameterized model function of the density of the particle temperature distribution using the registered spectrum of the particle thermal radiation. Structural diagrams of the complex for recording the spectrum of particle thermal radiation and the complex for its calibration are given. As the objective function of parametric optimization, the criterion function of the least squares method (LSM) is used. This function is defined as the sum of the squared deviations of the experimental readings of the thermal radiation spectrum of particles from the calculated values of its model spectrum corresponding to these readings. The model spectrum of the thermal radiation of particles is determined by the integral Fredholm operator of the first kind with the kernel of the operator based on the Planck function and the model parametrized function of the density of the particle temperature distribution. The Planck function depends on the temperature and wavelength of the thermal radiation of particles. Computational experiments were carried out to analyze the accuracy and numerical stability of the parameters being optimized when solving the inverse problem. The influence of the error aperture in the registered spectrum of the thermal radiation of particles on the estimate of the "shift" of the optimized parameters of the model density function of the particle temperature distribution is studied. Numerical modeling and experimental verification have shown that the accuracy and numerical stability of the regularized solution to measurement errors of the recorded spectrum of particle thermal radiation can be improved.

Keywords: Regularization · Inverse problem · Parametric identification · Particle temperature distribution · Fredholm integral operator of the 1st kind · Thermal radiation spectrum · Gas-thermal flow · Coating

© The Author(s), under exclusive license to Springer Nature Switzerland AG 2022
V. Jordan et al. (Eds.): HPCST 2022, CCIS 1733, pp. 41–61, 2022.
https://doi.org/10.1007/978-3-031-23744-7_4

1 Introduction

In various industries, including modern mechanical engineering, gas-thermal technologies for spraying protective coatings (plasma, detonation-gas (DGN), supersonic gas-flame HVOF and other types of spraying) are used [1]. Their use makes it possible to obtain sufficiently dense, thermal barrier, wear-resistant and corrosion-resistant coatings. In the field of thermal spraying of metal and composite powder coatings on products, the relevance of the technological problem is associated with the improvement of functional characteristics (bond strength with the base, porosity structure, crystal structure, etc.), providing a high resource of their operation. Powder coatings are formed by applying a finely dispersed powder of metal particles or their oxides to the surface of the product, sprayed by the dispenser of the technological spraying equipment during powder loading, for example, into the gas-thermal jet of the plasma torch (or DGN jet) flowing out of the output "nozzle" of the spraying equipment.

The functional characteristics and properties of coatings during their formation are largely determined by the temperature and velocity parameters of the particles of the dispersed phase, which are distributed dynamic parameters of the heterophase flow [1]. In any cross section of the flow (and in general in the entire flow), one can speak of time-varying distributions of particles according to their certain parameters (sizes, velocities, temperatures, etc.). Therefore, measuring only one average temperature value near the sprayed coating is not enough, since the properties of the coating are affected not only by particles with average temperatures, but also by all other particles of the flow with different temperatures. Consequently, the distribution law and the values of its parameters have a significant impact on the functional characteristics of the sprayed coatings.

In the problems of diagnosing gas-thermal particle flows during the spraying of functional coatings, an urgent problem is the development of computerized instrumental complexes [2–7] designed to measure and control the temperature and velocity parameters of sprayed particles. Measurement and control of the temperature and velocity parameters of the sprayed particles provides optimal control of the technological regime of coating spraying. The correct solution of the problem of diagnosing gas-thermal particle flows is associated with the use of methods of "non-destructive testing" of spraying processes, for example, methods of remote sensing of the thermal radiation of a particle flow. To measure the temperature of a particle heterogeneous flow, spectral pyrometry methods are effective [2–7]. The linear and matrix CCD receivers [8] used in modern spectrophotometers [9] as a detector make it possible to record the emission spectra of objects of various nature with high resolution and transfer measurement data via a USB port to a computer using a control program. The program allows the primary calibration of the device, taking into account the number of pixels in the detector and the linear dispersion of the diffraction grating [9].

The method for measuring the distributed temperature parameter of solid phase particles in a heterogeneous gas-thermal flow, considered earlier [10, 11] and in this paper, is based on the regularization solution of inverse problem of the parametric identification of particle temperature distribution. The regularization of this ill-posed inverse problem consists in choosing a physically adequate parameterized model function of the density of the particle temperature distribution and using the registered spectrum of the particle thermal radiation. As the objective function of parametric optimization, the criterion function of the least squares method (LSM) is used. This function is defined as the sum of the squared deviations of the experimental readings of the thermal radiation spectrum of particles from the calculated values of its model spectrum corresponding to these readings. The model spectrum of the thermal radiation of particles is determined by the integral Fredholm operator of the first kind with the kernel of the operator based on the Planck function and the model parametrized function of the density of the particle temperature distribution. The Planck function depends on the temperature and wavelength of the thermal radiation of particles. The following sections of the paper provide information on the hardware for recording the thermal radiation spectrum of particles, methodological and algorithmic aspects of solving the inverse problem of parametric identification of the temperature distribution of particles, and analysis of the results of the numerical solution of the inverse problem.

2   Optoelectronic System for Recording the Spectrum of Thermal Radiation of Gas-Thermal Flow Particles and Digital Equipment for Its Calibration

The registration of the thermal radiation spectrum of particles of powder material transported by a gas-thermal (in particular, gas-plasma) flow to the surface of a technical product (on a spraying base or on a substrate) is carried out using the spectral diagnostics complex shown in Fig. 1.

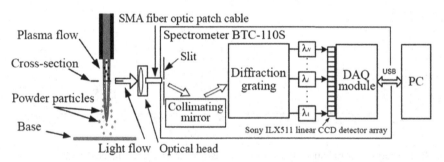

Fig. 1. Structural diagram of the complex for spectral diagnostics of particle temperature in a gas-plasma flow: $\lambda_1, ..., \lambda_i, ..., \lambda_N$ – wavelengths of the particle radiation spectrum projected onto the CCD detector.

As shown in Fig. 1, the luminous flux of radiation corresponding to the spatial cross section of the gas-plasma jet injected with powder particles (in our case, particles of zirconium dioxide ZrO_2) is supplied to the input of the optical head and through the SMA fiber optic patch cable is transferred to the BTC-110S spectrometer. In the spectrometer, using a collimating mirror, the light flux is redirected to a diffraction grating, from which the radiation decomposed into a spectrum is projected onto a Sony ILX511 linear CCD detector array. The signal taken from the CCD receiver is digitized in the DAQ module using a 16-bit Digitizer into an array of 65535 samples and from the output of the serial interface via 9-pin Serial Communication Cable and a USB-to-Serial

Adapter is transmitted to a PC for further processing (save and load in the csv file format) [9].

Below are the main features of the BTC-110S spectrometer and its DAQ module [9]:

- 16 bit Digitizer (65535 samples);
- RMS read noise of 50 counts (typical);
- Sony ILX511 linear CCD detector array;
- 350 ms readout time (20 Hz analog output possible);
- external 5 V power supply;
- overall dimensions: $5.75 \times 3.75 \times 1.75$ in.

The software Spectrum Studio [9] includes a line identification utility. It is possible to add molecular lines or other spectral features to the database using the simple step-by-step instructions in the Spectrum-Studio's help file.

Figure 2 shows the recorded emission spectrum of a gas-plasma spray jet injected with zirconium dioxide (ZrO_2) powder particles, and which is determined by the sum of the line spectrum of the transporting gas-plasma jet (Fig. 3) and the thermal radiation spectrum of heated powder particles (Fig. 4).

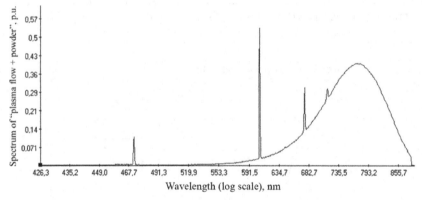

Fig. 2. The registered radiation spectrum of a gas-plasma spray jet injected with particles of zirconium dioxide powder.

As can be seen from Fig. 2, a significant "broadening" of the lines of the line spectrum, based on the contour of the continuous spectrum of thermal radiation, is noticeable. The effect of broadening of the lines of the line spectrum makes it difficult to separate the spectrum of particle thermal radiation into an independent array of its readings. One way to solve this problem is as follows. Additionally, the emission spectrum of the gas-plasma jet, which does not contain sprayed particles, is recorded (Fig. 3).

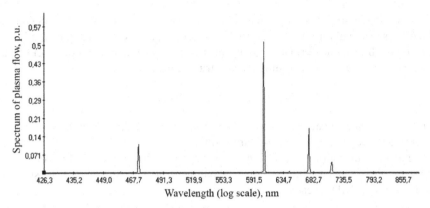

Fig. 3. The registered lines spectrum of the transporting gas-plasma jet (without powder particles).

Then, from the emission spectrum of the plasma jet injected with particles (Fig. 2), the emission spectrum of the plasma jet without particles (Fig. 3) is subtracted. As a result, a signal of the thermal radiation spectrum of powder particles is obtained, shown in Fig. 4.

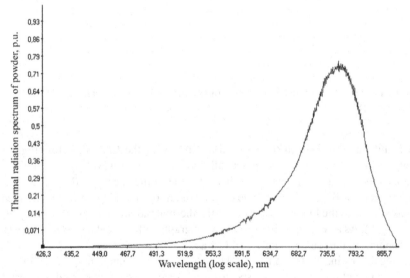

Fig. 4. Thermal radiation spectrum of powder, p.u. (log scale).

Reducing the aperture of residual distortions in the signal of the thermal radiation spectrum of particles (see Fig. 4) can be achieved by mathematical signal processing (for example, filtering methods). This task is discussed in the following sections of the paper.

The registered spectrum of particle thermal radiation (Fig. 4), which we denote as $\hat{B}(\lambda)$, differs in signal form from the "true" spectrum of particle thermal radiation $W(\lambda)$ due to its nonlinear transformation by means of the "apparatus" function $\alpha(\lambda)$ of the optoelectronic channel for recording the spectrum signal, i.e.

$$\hat{B}(\lambda) = k_{scale} \cdot \alpha(\lambda) \cdot W(\lambda), \tag{1}$$

where k_{scale} – is the signal scaling factor, which can be set (changed) by hardware each time during the registration of the spectrum signal.

The main contribution of distortions to the apparatus function $\alpha(\lambda)$ of the optoelectronic recording channel is made by the inhomogeneous characteristic of the spectral sensitivity of the CCD receiver (Fig. 5) [8].

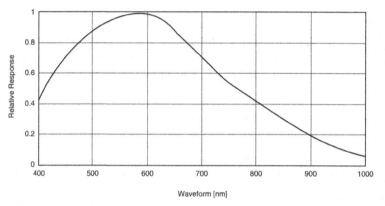

Fig. 5. CCD spectral sensitivity typical pattern (excluding lens specifications and light source specifications) [8].

A significant decrease in the spectral sensitivity of the CCD detector (Fig. 5) in the range from 600 to 1000 nm (practically down to 0 at a wavelength of 1100 nm) determines the change in growth to a sharp decline in the spectrum $\hat{B}(\lambda)$ (see Fig. 4). The true spectrum $W(\lambda)$ of the particle thermal radiation should increase monotonically and look similar to the Planck spectrum for the thermal radiation of an "absolutely black body" (Fig. 6). As an example, in Fig. 6 shows a graph of the calculated Planck spectrum corresponding to the temperature $T = 1118$ K.

Having determined at the calibration stage the apparatus function $\alpha_{cal}(\lambda)$ of the optoelectronic channel for recording the spectrum, the reconstructed (corrected) spectrum of the particle thermal radiation $\hat{W}(\lambda)$ is determined by the expression

$$\hat{W}(\lambda) = \frac{\hat{B}(\lambda)}{k_{scale} \cdot \alpha_{cal}(\lambda)}, \tag{2}$$

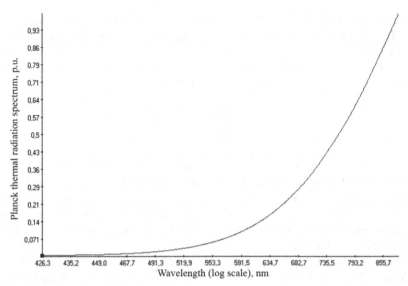

Fig. 6. The Planck spectrum for the thermal radiation of an "absolutely black body" corresponding to the temperature $T = 1118$ K.

The function $\alpha_{cal}(\lambda)$ is evaluated at the stage of calibration of measurements of spectral readings using the complex, the block diagram of which is shown in Fig. 7.

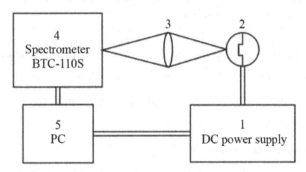

Fig. 7. Block diagram of the complex for calibration BTC-110S spectrometer.

The TRU 1100–2350 incandescent lamp with a tungsten plate is used as a "black body" standard (2 – in Fig. 7), through which a constant current is passed, regulated by a precision highly stable power supply (1 – in Fig. 7) to achieve maximum accuracy when calibrating the spectrometer (4 – in Fig. 7). The instability of highly stable power supply does not exceed 0.05% [12]. Using the calibration characteristic of the lamp (dependence of the temperature of the heated tungsten plate on the current flowing through it), the temperature T_{ref} of the radiation standard (tungsten plate) is determined. Thermal radiation from the reference lamp is focused by lens (3 – in Fig. 7) onto the input of the BTC-110S spectrometer (more precisely, onto the optical head with a fiber optic cable, as shown in Fig. 1). Then, from the output of spectrometer, the spectrum signal $B_{ref}(\lambda, T_{ref})$ recorded for the reference source (radiation standard), similar in shape to the spectrum (Fig. 4), is transmitted to the PC in the form of an array of digital readings $\{B_{ref}(\lambda_i, T_{ref}); \ i = 1, 2, ..., N\}$. Then the array of readings of the apparatus function $\{\alpha_{cal}(\lambda_i); i = 1, 2, ..., N\}$ is calculated by the formula

$$\alpha_{cal}(\lambda_i) = \frac{B_{ref}(\lambda_i, T_{ref})}{k_{scale} \cdot \phi(\lambda_i, T_{ref})}, \tag{3}$$

where $i = 1, 2, ..., N$; $\varphi(1, T) = C_1 \cdot 1^{-5}/\left(e^{\frac{C_2}{1T}} - 1\right)$ is Planck's function for the spectral density of the radiation of the "absolutely black body" (radiation standard).

Note: the values of the apparatus function $\alpha_{cal}(\lambda)$ depend only on the characteristics of the elements and devices that are sequentially included in the optoelectronic channel for recording the spectrum signal, and do not depend on the characteristics and properties of the thermal radiation source.

Considering the note, at the calibration stage, the registration of spectra for different temperatures T_{ref} of the radiation standard is repeated, followed by averaging the values of readings of the apparatus function $\{\alpha_{cal}(\lambda_i); i = 1, 2, ..., N\}$.

3   Methodical Aspects for Solving the Inverse Problem of Determining the Temperature Distribution of Particles

3.1   Integral Fredholm Model for the Thermal Radiation Spectrum of Particles and the Problem of Its Inversion

The papers [10, 11] substantiate the model of the true spectrum of particle thermal radiation $W(\lambda)$ using the integral Fredholm operator of the first kind, namely:

$$W(\lambda) = S_\Sigma \cdot \varepsilon_1(1) \cdot \int_{T_{min}}^{T_{max}} \phi(1, T) \cdot \varepsilon_2(T) \cdot P(T) dT. \tag{4}$$

Then, taking into account formula (1)

$$\hat{B}(\lambda) = k_{scale} \cdot \alpha(\lambda) \cdot W(\lambda) = S_\Sigma \cdot \beta(\lambda) \cdot \int_{T_{min}}^{T_{max}} \phi(\lambda, T) \cdot \varepsilon_2(T) \cdot P(T) dT, \tag{5}$$

$$\beta(\lambda) = k_{scale} \cdot \alpha(\lambda) \cdot \varepsilon_1(1), \tag{6}$$

where $\hat{B}(\lambda)$ is the signal of the thermal radiation spectrum of the particles of the gas-thermal spray jet recorded by a linear CCD detector; S_Σ is the total surface of particles from which thermal radiation is focused at the spectrometer input; $\varphi(\lambda, T) = C_1 \cdot \lambda^{-5}/\left(e^{\frac{C_2}{\lambda T}} - 1\right)$ is Planck's function for the spectral density of radiation of a "absolutely black body" (radiation standard) ($C_1 = 2\pi hc^2$, $C_2 = hc/k = 14.38786 \cdot 10^6$ nm $\cdot$ K);

$P(T)$ is the probability density function of the temperature distribution of particles; $\alpha(\lambda)$ is apparatus function of the optoelectronic channel for the spectrum registration, determined at the calibration stage (see Sect. 2 of this paper).

The relative emissivity of the particles of the powder material is taken into account in expressions (4)–(6) in the model form $\varepsilon(\lambda, T) = \varepsilon_1(\lambda) \cdot \varepsilon_2(T)$, where $\varepsilon_1(\lambda)$ is the temperature-averaged spectral component of the function $\varepsilon(\lambda, T)$, and the second component $\varepsilon_2(T)$ is called the "integral emissivity factor of black" of the gray body (as a rule, it increases with increasing T). For CCD receivers that record the spectrum in the optical range (in a relatively narrow wavelength range), the function $\varepsilon_1(\lambda)$ can be taken as a constant ε_1. In gas-thermal spraying flows of particles, the temperature difference $\Delta T = (T_{max} - T_{min})$ relative to the values of these temperatures is not large, therefore, most often (in the first approximation), the integral emissivity factor $\varepsilon_2(T)$ is also assumed to be a constant ε_2.

In [11], a functional form of the inverse integral operator was obtained, which determines the analytically exact solution of the inverse problem.

The main formula calculations (expressions) are shown below.

We denote the following functions: $\eta(\lambda) = C_1 \cdot S_\Sigma \cdot k_{scale} \cdot \lambda^{-5} \cdot \alpha(\lambda) \cdot \varepsilon_1(\lambda)$, $g(T) = \varepsilon_2(T) \cdot P(T)$. Then instead of (5) we can write:

$$\hat{B}(\lambda) = \eta(\lambda) \cdot \int_{T_{min}}^{T_{max}} J(\lambda, T) \cdot g(T) dT, \tag{7}$$

$$J(\lambda, T) = \frac{1}{e^{\frac{C_2}{\lambda T}} - 1}. \tag{8}$$

Variable substitution looks like this: $\omega = 1/\lambda$; $t = C_2/T$. Then $\lambda = 1/\omega$, $T = C_2/t$; $dT = -C_2 dt/t^2$; $t_{max} = C_2/T_{min}$; $t_{min} = C_2/T_{max}$. Instead of (7) we get

$$\hat{B}(1/\omega) = \gamma(\omega) \cdot \int_{t_{min}}^{t_{max}} J(\omega, T) \cdot f(t) dt, \tag{9}$$

where

$$f(t) = \varepsilon_2(C_2/t) \cdot P(C_2/t)/t^2, \tag{10}$$

$$\gamma(\omega) = C_1 \cdot C_2 \cdot S_\Sigma \cdot k_{scale} \cdot \omega^5 \cdot \alpha(1/\omega) \cdot \varepsilon_1(1/\omega), \tag{11}$$

or

$$\gamma(1/\lambda) = C_3 \cdot k_{scale} \cdot \lambda^{-5} \cdot \alpha(\lambda) \cdot \varepsilon_1(\lambda), \tag{12}$$

$$C_3 = C_1 \cdot C_2 \cdot S_\Sigma, \tag{13}$$

$$J(\omega, t) = \frac{1}{e^{\omega \cdot t} - 1} = \frac{e^{-\omega \cdot t}}{1 - e^{-\omega \cdot t}} = e^{-\omega \cdot t} \cdot \frac{1}{1 - e^{-\omega \cdot t}}. \tag{14}$$

Considering in (14) the conditions: $t \in [t_{min}, t_{max}]$, $t_{min} \neq 0$, $e^{-\omega \cdot t} < 1 e^{-\omega \cdot t} < 1$, as well as the Maclaurin series expansion of the function

$$(1-x)^{-1} = \sum_{n=0}^{\infty} x^n,$$

the core $J(\omega, t)$ takes the form

$$J(\omega, t) = e^{-\omega \cdot t} \cdot \frac{1}{1 - e^{-\omega \cdot t}} = \sum_{n=1}^{\infty} e^{-n \cdot \omega \cdot t}. \tag{15}$$

Considering the condition $e^{-\omega \cdot t} \cong 1 - \omega t$ at $\omega \to 0$, a singularity $J(\omega, t) \cong (1 - \omega t)/(\omega t)$ arises in the core $J(\omega, t)$, but, due to the presence of a factor ω^5 in the function $\gamma(\omega)$, in the integral expression (9), the singularity is eliminated. Dividing both parts of Eq. (9) by $\gamma(\omega)$, taking into account (11) and (12), we obtain [11]:

$$\Phi(\omega) = \frac{\hat{B}(1/\omega)}{\gamma(\omega)} = \int_{t_{min}}^{t_{max}} J(\omega, T) \cdot f(t) dt, \tag{16}$$

Next, we transform $\Phi(\omega)$ to the form [13]

$$\Phi(\omega) = \int_{t_{min}}^{t_{max}} J(\omega, T) \cdot f(t) dt = \sum_{n=1}^{\infty} \int_{t_{min}}^{t_{max}} e^{-n \cdot \omega \cdot t} \cdot f(t) dt = \sum_{n=1}^{\infty} F(n \omega), \tag{17}$$

$$F(n \omega) = \int_{t_{min}}^{t_{max}} e^{-n \cdot \omega \cdot t} \cdot f(t) dt = \int_{0}^{\infty} e^{-n \cdot \omega \cdot t} \cdot f(t) dt. \tag{18}$$

In (18), the condition $f(t) = 0$ outside the range $[t_{min}, t_{max}]$ is taken into account. For $n = 1$ formula (18) takes the form

$$F(\omega) = \int_{0}^{\infty} e^{-\omega \cdot t} f(t) dt. \tag{19}$$

Expression (17), namely $\Phi(\omega) = \sum_{n=1}^{\infty} F(n\omega)$, based on the well-known inversion formula [14], in which the Möbius function $\mu(m)$ is present, allows, taking into account $\omega > 0$, to write down the result of the inversion

$$F(\omega) = \sum_{m=1}^{\infty} \mu(m) \cdot \Phi(m\omega). \tag{20}$$

Using expressions (19) and (20) and performing analytical continuation to the complex plane (transition from a real variable ω to a complex variable s), we can write

$$F(s) = \int_{0}^{\infty} e^{-s \cdot t} f(t) dt = \sum_{m=1}^{\infty} \mu(m) \cdot \Phi(ms), \tag{21}$$

which corresponds to the Laplace transform. Taking into account the inverse Laplace transform [15, 16]

$$f(t) = \frac{1}{2\pi i} \int\limits_{\sigma-i\infty}^{\sigma+i\infty} e^{s \cdot t} F(s) ds,$$

we obtain the function

$$f(t) = \frac{1}{2\pi i} \int\limits_{\sigma-i\infty}^{\sigma+i\infty} e^{p \cdot t} \left(\sum_{m=1}^{\infty} \mu(m) \cdot \Phi(mp) \right) dp = \frac{1}{2\pi i} \int\limits_{\sigma-i\infty}^{\sigma+i\infty} \left(\sum_{m=1}^{\infty} \frac{\mu(m)}{m} \cdot e^{\frac{mp \cdot t}{m}} \Phi(mp) \right) d(mp)$$

$$= \frac{1}{2\pi i} \int\limits_{\sigma'-i\infty}^{\sigma'+i\infty} \left(\sum_{m=1}^{\infty} \frac{\mu(m)}{m} \cdot e^{\frac{s \cdot t}{m}} \right) \Phi(s) ds = \frac{1}{2\pi i} \int\limits_{\sigma'-i\infty}^{\sigma'+i\infty} J_{-1}(s, t) \Phi(s) ds. \tag{22}$$

Thus, the kernel of the inverse integral operator (22) is expressed as a generating function [11, 13]

$$J_{-1}(s, t) = \sum_{m=1}^{\infty} \frac{\mu(m)}{m} \cdot e^{\frac{s \cdot t}{m}}. \tag{23}$$

3.2 Regularization of the Solution of the Inverse Problem Based on a Physically Adequate Model Function of the Temperature Distribution Density of Particles

In the computational aspect, the solution of the inverse problem based on the inverse integral operator (22) with the kernel (23) is quite difficult, since according to (9) and (16) the model spectrum $\Phi(\omega)$ is associated with the experimental "noisy" spectrum $\hat{B}(\lambda) = \hat{B}(1/\omega)$.

In the process of plasma spraying of a coating from a powder of zirconium dioxide particles ZrO_2 (stabilized Y_2O_3), the integral spectrum $\hat{B}(\lambda)$ of thermal radiation of particles ZrO_2 was recorded using a BTC-110S spectrometer in the wavelength range (280–900 nm). As mentioned above, in the first approximation, the averaged function $\varepsilon_1(\lambda) = \varepsilon_1(1/\omega)$ can be considered a constant ε_1. Then, taking into account the constants, formulas (11) and (12) can be written in the form (we introduce the constant $C_4 = C_3 \cdot \varepsilon_1$)

$$\gamma(\omega) = C_4 \cdot k_{scale} \cdot \omega^5 \cdot \alpha(1/\omega) = \gamma(1/\lambda) = C_4 \cdot \lambda^{-5} \cdot k_{scale} \cdot \alpha(\lambda). \tag{24}$$

Similarly, we will replace the second component $\varepsilon_2(T)$ with a constant ε_2. According to eq. (10) and the condition $t = C_2/T$, the function $f(t) = C_5 \cdot T^2 \cdot P(T)$ and

$$P(T) = \frac{C_6}{T^2} \cdot f(C_2/T) = \frac{t^2}{\varepsilon_2} f(t), \tag{25}$$

where are the constants: $C_5 = \varepsilon_2/C_2^2$ and $C_6 = 1/C_5 = C_2^2/\varepsilon_2$. Then, taking into account the conditions $\omega = 1/\lambda$, formulas (1) and (24), expression (16) is transformed to the form

$$\Phi(\omega) = \Phi(1/\lambda) = \frac{\hat{B}(\lambda)}{\gamma(1/\lambda)} = \frac{\hat{B}(\lambda)}{k_{scale} \cdot \alpha(\lambda)} \cdot \frac{\lambda^5}{C_4} = \frac{\lambda^5}{C_4} \cdot W(\lambda), \tag{26}$$

where $W(\lambda)$ is the true spectrum of thermal radiation of particles. We introduce the notation: $\hat{k}_{scale} = k_{scale} \cdot C_4$ and $G(\lambda) = \hat{k}_{scale} \cdot \Phi(1/\lambda)$, then, taking into account (17) and (26), we write

$$G(\lambda) = \hat{k}_{scale} \cdot \Phi(1/\lambda) = \hat{k}_{scale} \cdot \sum_{n=1}^{\infty} F(n/\lambda) = k_{scale} \cdot \lambda^5 \cdot W(\lambda) \qquad (27)$$

$$F(n/\lambda) = \int_{t_{min}}^{t_{max}} e^{-(n/\lambda)\cdot t} \cdot f(t) dt. \qquad (28)$$

In many cases, when solving ill-posed problems, the Tikhonov regularization approach [17–22] is used, which is based on minimizing the "smoothing" functional with the presence of regularization parameters. However, under the condition of a high degree of noise in the experimentally recorded signals, the values of the regularization parameters in the process of solving the optimization problem are not always sufficiently small, and the desired solution may contain negative components of small value.

In this paper, the regularization of the desired function is achieved by choosing a physically adequate parameterized model function of the density of the temperature distribution of particles using the registered spectrum of the thermal radiation of particles.

To conduct computational experiments and search for a regularized solution, a model function $f(t)$ was chosen in the form of a "parabola" (its branches are directed "down") and it is non-negative in the segment $[\tau_1; \tau_2]$. In addition, it is symmetrical with respect to the midpoint of the segment $[\tau_1; \tau_2]$. Therefore [13]

$$f(t) = \begin{cases} f_0(t) \equiv 0, & t \in (-\infty; \tau_1), \\ f_1(t) = -A \cdot (t - \tau_1) \cdot (t - \tau_2), & t \in [\tau_1; \tau_2], \\ f_2(t) \equiv 0, & t \in (\tau_2; +\infty), \end{cases} \qquad (29)$$

where $\tau_1 = t_{min} = C_2/T_{max}$, $\tau_2 = t_{max} = C_2/T_{min}$. The constant A is determined from the normalization condition

$$\int_{T_{min}}^{T_{max}} P(T) dT = 1$$

for the probability density $P(T)$. Taking into account expression (25) and condition $dT = -C_2 dt/t^2$, we obtain

$$\int_{T_{min}}^{T_{max}} P(T) dT = \frac{C_2}{\varepsilon_2} \int_{\tau_1}^{\tau_2} f(t) dt = A \cdot \frac{C_2}{6\varepsilon_2} \cdot (\tau_2 - \tau_1)^3 = 1.$$

Therefore, the parameter A is

$$A = A(\tau_1, \tau_2) = \frac{6\varepsilon_2}{C_2} \cdot \frac{1}{(\tau_2 - \tau_1)^3}. \qquad (30)$$

The density function $P(T)$ of the temperature distribution of particles according to expression (25), namely $P(T) = C_6 \cdot f(C_2/T)/T^2 = t^2 \cdot f(t)/\varepsilon_2$, turns out to be asymmetric with respect to its maximum corresponding to the "most probable" temperature value T_0. The hyperbolic dependence on the variable T in (25) ensures the asymmetry of the graph of the distribution density function $P(T)$ in the form of "stretching" to the right of the most probable value T_0. This property of asymmetry is quite adequate to physical reality. For particle powders used in the spraying process, the particle size distribution is close to the normal distribution law, which is symmetrical about the average particle size (the number of small particles is approximately equal to the number of large particles). However, smaller particles are heated to higher temperatures than larger ones. Therefore, in the temperature distribution of particles, the most probable value T_0 is shifted from the middle of the range $[T_{min}; T_{max}]$ to T_{min}.

Substituting the model function $f(t)$ into (28) and integrating "by parts" (change of variable: $z = -n \cdot (t - \tau_1)/\lambda; z_0 = -n \cdot (\tau_2 - \tau_1)/\lambda \, t - \tau_2 = (t - \tau_1) - (\tau_2 - \tau_1)$, we obtain

$$F(n/\lambda) = -A \int_{\tau_1}^{\tau_2} \left[(t - \tau_1)^2 - (\tau_2 - \tau_1)(t - \tau_1) \right] \cdot e^{\frac{-n \cdot t}{\lambda}} \, dt$$

$$= A \cdot \left(\frac{\lambda}{n} \right)^3 e^{\frac{-n \cdot \tau_1}{\lambda}} \left\{ \int_0^{z_0} z^2 e^z dz - z_0 \int_0^{z_0} z \cdot e^z dz \right\} = A_1 \left(\frac{\lambda}{n} \right)^3 e^{\frac{-n \cdot \tau_1}{\lambda}} \left\{ e^{z_0}(2 - z_0) - (2 + z_0) \right\} \tag{31}$$

$$= A \cdot \left(\frac{\lambda}{n} \right)^2 e^{\frac{-n \cdot \tau_1}{\lambda}} \left\{ \left(\frac{\lambda}{n} \right) e^{z_0}(2 - z_0) - \left(\frac{\lambda}{n} \right)(2 + z_0) \right\}$$

$$= \frac{A \cdot \lambda^2}{n^2} \left\{ (\tau_2 - \tau_1) \cdot \left(e^{\frac{-n \cdot \tau_1}{\lambda}} + e^{\frac{-n \cdot \tau_2}{\lambda}} \right) - \frac{2 \cdot \lambda}{n} \cdot \left(e^{\frac{-n \cdot \tau_1}{\lambda}} - e^{\frac{-n \cdot \tau_2}{\lambda}} \right) \right\}.$$

Summing up the components $F(n/\lambda)$ calculated by formula (31), we obtain, according to (27), an array of spectral readings in the following form ($i = 1, 2,...,$ N)

$$G(\lambda_i, \tau_1, \tau_2) = \hat{k}_{scale} \cdot \sum_{n=1}^{\infty} F(n/\lambda_i) =$$

$$\frac{\tilde{A} \cdot \lambda_i^2}{n^2} \left\{ (\tau_2 - \tau_1) \cdot \left(e^{\frac{-n \cdot \tau_1}{\lambda_i}} + e^{\frac{-n \cdot \tau_2}{\lambda_i}} \right) - \frac{2 \cdot \lambda_i}{n} \cdot \left(e^{\frac{-n \cdot \tau_1}{\lambda_i}} - e^{\frac{-n \cdot \tau_2}{\lambda_i}} \right) \right\}, \tag{32}$$

$$\tilde{A} = \hat{k}_{scale} \cdot A. \tag{33}$$

Further, the spectrum of thermal radiation $\hat{B}(\lambda)$ (see Fig. 4) recorded in the technological experiment is corrected by dividing by the apparatus function $\alpha_{cal}(\lambda)$. Taking into account formula (1) $\hat{B}(\lambda_i)/\alpha_{cal}(\lambda_i) = k_{scale} \cdot \hat{W}(\lambda_i)$ for $i = 1, 2,..., N$. Then you can calculate the array

$$\{ \hat{G}_1(\lambda_i) = \lambda_i^5 \cdot \hat{B}(\lambda_i)/\alpha_{cal}(\lambda_i); \; i = 1, 2, ..., N \}, \tag{34}$$

for whose elements, taking into account (1), the equalities

$$\{ \hat{G}_1(\lambda_i) = \lambda_i^5 \cdot \hat{B}(\lambda_i)/\alpha_{cal}(\lambda_i) = k_{scale} \cdot \lambda_i^5 \cdot \hat{W}(\lambda_i); \; i = 1, 2, ..., N \}.$$

Thus, arrays $\{G(\lambda_i, \tau_1, \tau_2);\ i = 1, 2, ..., N\}$ and $\left\{\hat{G}(\lambda_i);\ i = 1, 2, ..., N\right\}$ corresponding to each other calculated by formulas (32) and (34), respectively, are used to identify the optimal parameters τ_1 and τ_2 in the model function $f(t)$ in accordance with formula (29).

4 Algorithm for Optimizing the Parameters of the Model Density Function of the Particle Temperature Distribution

As the objective function of parametric optimization, the criterion function $S^2(\tau_1, \tau_2)$ of the least squares method (LSM) is used. This function is defined as the sum of the squared deviations of the experimental spectral readings (written to the array $\left\{\hat{G}(\lambda_i);\ i = 1, 2, ..., N\right\}$) from the calculated values (written to the array $\{G(\lambda_i, \tau_1, \tau_2);\ i = 1, 2, ..., N\}$) corresponding to these readings, obtained on the basis of its model spectrum $W(\lambda)$. Let us define the objective function $S^2(\tau_1, \tau_2)$ in the form

$$S^2(\tau_1, \tau_2) = \sum_{i=1}^{N} \left[G(\lambda_i, \tau_1, \tau_2) - \hat{G}(\lambda_i)\right]^2. \tag{35}$$

For an iterative search for the optimal solution, the well-known method of the "golden section" is used, which in this paper is generalized to the two-dimensional case – a two-dimensional area (plane) is used. The golden section parameter ϕ is equal to $\phi = \left(\sqrt{5} + 1\right)/2 \approx 1.618$ or $1/\phi = \left(\sqrt{5} - 1\right)/2 \approx 0.618$. Below is the algorithm.

1. For the parameter τ_1, the initial boundaries are set: $\tau_{10}^{(0)}$ and $\tau_{11}^{(0)}$, and for the parameter τ_2 – boundaries: $\tau_{20}^{(0)}$ and $\tau_{21}^{(0)}$.

2. At each k-th step (starting from $k = 1$), the values are calculated

$$x_1 = \tau_{11}^{(k-1)} - \left(\tau_{11}^{(k-1)} - \tau_{10}^{(k-1)}\right)/\phi,\ x_2 = \tau_{10}^{(k-1)} + \left(\tau_{11}^{(k-1)} - \tau_{10}^{(k-1)}\right)/\phi,$$

$$y_1 = \tau_{21}^{(k-1)} - \left(\tau_{21}^{(k-1)} - \tau_{20}^{(k-1)}\right)/\phi,\ y_2 = \tau_{20}^{(k-1)} + \left(\tau_{21}^{(k-1)} - \tau_{20}^{(k-1)}\right)/\phi.$$

3. According to formula (35), the following values are calculated: $z_{11} = S^2(x_1, y_1)$, $z_{12} = S^2(x_1, y_2)$, $z_{21} = S^2(x_2, y_1)$, $z_{22} = S^2(x_2, y_2)$. Among them z_{max} is determined.

4. New boundaries are determined from the conditions:

a. If $z_{max} = z_{11}$, then $\tau_{10}^{(k)} = x_1$, $\tau_{20}^{(k)} = y_1$, $\tau_{11}^{(k)} = \tau_{11}^{(k-1)}$, $\tau_{21}^{(k)} = \tau_{21}^{(k-1)}$, otherwise b.

b. If $z_{max} = z_{12}$, then $\tau_{10}^{(k)} = x_1$, $\tau_{21}^{(k)} = y_2$, $\tau_{11}^{(k)} = \tau_{11}^{(k-1)}$, $\tau_{20}^{(k)} = \tau_{20}^{(k-1)}$, otherwise c.

c. If $z_{max} = z_{21}$, then $\tau_{11}^{(k)} = x_2$, $\tau_{20}^{(k)} = y_1$, $\tau_{10}^{(k)} = \tau_{10}^{(k-1)}$, $\tau_{21}^{(k)} = \tau_{21}^{(k-1)}$, otherwise d.

d. If $z_{max} = z_{22}$, then $\tau_{11}^{(k)} = x_2$, $\tau_{21}^{(k)} = y_2$, $\tau_{10}^{(k)} = \tau_{10}^{(k-1)}$, $\tau_{20}^{(k)} = \tau_{20}^{(k-1)}$.

5. If $\left| \tau_{11}^{(k)} - \tau_{10}^{(k)} \right| < eps$ and $\left| \tau_{21}^{(k)} - \tau_{20}^{(k)} \right| < eps$, then we calculate the final values of the parameters using the formulas: $\tau_1 = \left(\tau_{10}^{(k)} + \tau_{11}^{(k)} \right)/2$, $\tau_2 = \left(\tau_{20}^{(k)} + \tau_{21}^{(k)} \right)/2$, otherwise $k = k + 1$ and go to step 2.

The final values of the parameters τ_1 and τ_2 correspond to the minimum of the objective function $S^2(\tau_1, \tau_2)$.

5 Results of Numerical Optimization of the Model Density Function Parameters of the Particle Temperature Distribution

As already noted at the end of Sect. 3 of this paper, using the thermal radiation spectrum $\hat{B}(\lambda)$ of the particles ZrO_2 recorded in the technological experiment (see Fig. 4), its correction was carried out taking into account the apparatus function $\alpha_{cal}(\lambda)$ and the array $\left\{ \hat{G}_1(\lambda_i) = \lambda_i^5 \cdot \hat{B}(\lambda_i)/\alpha_{cal}(\lambda_i); \; i = 1, 2, ..., N \right\}$ was calculated using formula (34). His graph shown in Fig. 8.

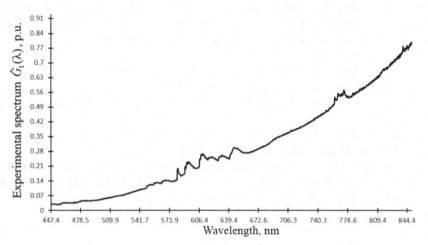

Fig. 8. Corrected signal $\hat{G}_1(\lambda)$.

Two sections of distortion are noticeable in the signal $\hat{G}_1(\lambda)$ with a fluctuation aperture of about 0.06 (in the form of several elevations – noise with a plus sign). Using signal filtering methods (median filter and moving average filter), the fluctuation aperture was reduced by a factor of 3 (the fluctuation is 0.02). The filtering result of signal $\hat{G}_1(\lambda)$ is shown in Fig. 9 and is designated as $\hat{G}_2(\lambda)$.

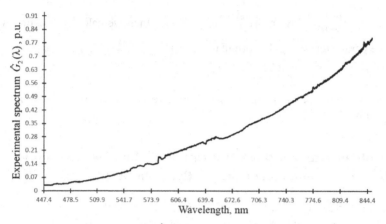

Fig. 9. The spectrum signal $\hat{G}_2(\lambda)$ is the filtering result of signal $\hat{G}_1(\lambda)$.

To calculate the array $\{G(\lambda_i, \tau_1, \tau_2); \; i = 1, 2, ..., N\}$ by formula (32), it is required to set a constant $\tilde{A}$ so that then it would be possible to start the iterative process of optimizing the parameters τ_1 and τ_2 to execute it (the optimization algorithm is described in Sect. 4). The objective function $S^2(\tau_1, \tau_2)$ of the optimization problem in the region of its minimum has several local minima. Tables 1 and 2 selectively reflect only some of the characteristic data.

The initial ranges for optimizing each of the two parameters τ_1 and τ_2 based on the signal $\hat{G}_1(\lambda)$ were the same, namely, [2500; 5500]. Then, at each step, the interval was shifted by 10, i.e. the second interval is [2510; 5510] and so on. The last interval is [2640; 5640].

The same is true for the signal $\hat{G}_2(\lambda)$. The initial intervals for the parameters τ_1 and τ_2 are [2600; 5600], the last ones are [2740; 5740].

Local minima for each variable value of the parameter $\tilde{A}$ corresponded each time to the same initial interval: a) the signal $\hat{G}_1(\lambda)$ corresponded to the range [2580; 5580]; b) for a signal $\hat{G}_2(\lambda)$ – [2660; 5660]. The optimization results are shown below in Tables 1 and 2.

Analyzing Table 1, we determine the optimal values of the parameters τ_1, τ_2 and τ_0: $\tau_1^{opt} = 3965.7$ nm, $\tau_2^{opt} = 4194.1$ nm и $\tau_0^{opt} = 4079.9$ nm. The optimal temperature values for the distribution density function $P(T)$ are:

$$T_{min} = C_2/\tau_2^{opt} = 3430.5 \text{ K}; \; T_{max} = C_2/\tau_1^{opt} = 3628.1 \text{ K}; \; \hat{T}_0 = C_2/\tau_0^{opt} = 3526.5 \text{ K}.$$

$\hat{T}_0$ is the temperature value corresponding to the center τ_0^{opt} of the symmetric function $f(t)$, where $C_2 = hc/k = 14.38786 \cdot 10^6$ nm K. The width of the temperature range, equal to $\Delta T = T_{max} - T_{min} = 190.44$ K, characterizes the domain of definition of the distribution density function $P(T)$ in the first case (based on the signal $\hat{G}_1(\lambda)$).

Table 1. Optimization results of the parameters τ_1 and τ_2 based on the signal $\hat{G}_1(\lambda)$.

$\tilde{A}$	τ_1, nm	τ_2, nm	$\tau_0 = (\tau_1 + \tau_2)/2$, nm	$\sigma = (\tau_2 - \tau_1)/2$, nm	$S^2(\tau_1, \tau_2)$
5	3927.1	4232.7	4079.9	152.8	0.078620
6	3936.1	4223.7	4079.9	143.8	0.078616
7	3943.3	4216.5	4079.9	136.6	0.078612
8	3949.2	4210.6	4079.9	130.7	0.078609
9	3954.2	4205.6	4079.9	125.7	0.078606
10	3958.5	4201.3	4079.9	121.4	0.078605
11	3962.3	4197.5	4079.9	117.6	0.078604
12	3965.7	4194.1	4079.9	114.2	0.078603
13	3968.7	4191.1	4079.9	111.2	0.078604
14	3961.9	4177.9	4069.9	108.0	0.078605
15	3973.9	4185.9	4079.9	106.0	0.078606
16	3976.1	4183.7	4079.9	103.8	0.078607
17	3978.2	4181.6	4079.9	101.7	0.078614
18	3980.1	4179.7	4079.9	99.8	0.078612
19	3981.9	4177.9	4079.9	98.0	0.078613
20	3983.6	4176.2	4079.9	96.3	0.078616

Table 2. Optimization results of the parameters τ_1 and τ_2 based on the signal $\hat{G}_2(\lambda)$.

$\tilde{A}$	τ_1, nm	τ_2, nm	$\tau_0 = (\tau_1 + \tau_2)/2$, nm	$\sigma = (\tau_2 - \tau_1)/2$, Nm	$S^2(\tau_1, \tau_2)$
5	4001.6	4316.6	4159.1	157.5	0.023257
6	4011.0	4307.2	4159.1	148.1	0.023249
7	4018.4	4299.8	4159.1	140.7	0.023247
8	4024.5	4293.7	4159.1	134.6	0.023245
9	4029.7	4288.5	4159.1	129.4	0.023245
10	4034.2	4284.0	4159.1	124.9	0.023244
11	4038.1	4280.1	4159.1	121.0	0.023244
12	4041.6	4276.6	4159.1	117.5	0.023244
13	4044.7	4273.5	4159.1	115.4	0.023241
14	4047.5	4270.7	4159.1	111.6	0.023242
15	4050.1	4268.1	4159.1	109.0	0.023242
16	4052.4	4265.8	4159.1	106.7	0.023246
17	4054.6	4263.6	4159.1	104.5	0.023248
18	4056.6	4261.6	4159.1	102.5	0.023249
19	4058.4	4259.8	4159.1	100.7	0.023249
20	4050.6	4247.8	4149.2	98.6	0.023256

Analyzing Table 2, we determine the optimal values of the parameters τ_1, τ_2 and τ_0: $\tau_1^{opt} = 4044.7$ nm; $\tau_2^{opt} = 4273.5$ nm and $\tau_0^{opt} = 4159.1$ nm. The optimal temperature values for the distribution density function $P(T)$ are

$$T_{min} = C_2/\tau_2^{opt} = 3366.76 \text{ K}; \, T_{max} = C_2/\tau_1^{opt} = 3557.2 \text{ K}; \, \hat{T}_0 = C_2/\tau_0^{opt} = 3459.37 \text{ K}.$$

$\hat{T}_0$ is the temperature value corresponding to the center τ_0^{opt} of the symmetric function $f(t)$. The width of the temperature range, equal to $\Delta T = T_{max} - T_{min} = 190.44$ K, characterizes the domain of definition of the distribution density function $P(T)$ in the second case (based on the signal $\hat{G}_2(\lambda)$).

Comparing the parameters of two temperature distributions obtained on the basis of signals $\hat{G}_1(\lambda)$ and $\hat{G}_2(\lambda)$, we can say that the estimates of the width of their temperature ranges are almost the same, and the shift of one distribution relative to the other is about 67–70 K. The particle temperature distribution density function $P(T)$, calculated on the basis of optimal parameters corresponding to a signal $\hat{G}_2(\lambda)$ with a smaller fluctuation aperture, is more preferable (Fig. 10). According to expression (25), the function $P(T)$ is asymmetric with respect to the symmetric function $f(t)$. Calculations and search by the iterative algorithm for the maximum of the function $P(T)$, which corresponds to the most probable temperature T_0, showed the following. The value of the most probable temperature (see Fig. 10) turned out to be equal $T_0 = 3456.75$ K (it is less by 2.62 K than $\hat{T}_0 = C_2/\tau_0^{opt} = 3459.37$ K).

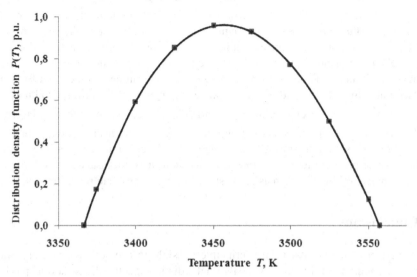

Fig. 10. Density function of the temperature distribution of particles ZrO_2 in a gas-plasma spray jet.

The value T_0 is located to the left of the middle (equal to 3461.98 K) of the temperature range $[T_{min}; T_{max}] = [3366.76; 3557.2]$ by 5.23 K (the asymmetry of the temperature distribution of particles). The width of the temperature range, equal to $\Delta T = T_{max} - T_{min} = 190.44$ K, with respect to the most probable temperature $T_0 = 3456.75$ K is about 5.51%.

Concluding the analysis of the results of optimization of parameters τ_1 and τ_2, it is necessary to note the following:

1. Local minima are characterized by fairly close values, and in relation to them the global minimum differs slightly from them (this is true for each of the two Tables). Therefore, taking into account the set of local minima, the accuracy of optimization of the parameters τ_1 and τ_2 (the uncertainty of their values) is about 0.3–0.5% – in a practical sense, this value is quite acceptable. However, for more "accurate" models of the function $f(t)$, the global minimum may differ from the local minimums more significantly, and the range width $\left[\tau_1^{opt}; \tau_2^{opt}\right]$ for such models may be larger than for the parabolic model.

2. The proposed parabolic model for the function $f(t)$ is physically adequate, but the values of the derivative df/dt at the end points τ_1^{opt} and τ_2^{opt} are not small. It can be assumed that the range $\left[\tau_1^{opt}; \tau_2^{opt}\right]$ corresponding to the parabolic model is the "dominant" in the "true" probability density function $f_{true}(t)$. That is, the area under the graph of the function $f_{true}(t)$, based on the segment $\left[\tau_1^{opt}; \tau_2^{opt}\right]$ of the parabolic

model, can correspond, for example, to about 90–95% of the entire area under the graph of the function $f_{true}(t)$. In other words, the range $\left[\tau_1^{true}; \tau_2^{true}\right]$ corresponding to the function $f_{true}(t)$ in this case is wider than the range $\left[\tau_1^{opt}; \tau_2^{opt}\right]$ corresponding to the parabolic model. Therefore, most likely, a more accurate model can be a model with zero values at the ends of the segment $[\tau_1; \tau_2]$ and similar to the "bell-shaped" model, or similar to the corresponding fragment of the $\sin(t)$ function. For such models, the derivative df/dt, when approaching the ends of the segment $[\tau_1; \tau_2]$, is quite small and gradually decreases to zero. Then it can be assumed that for more "accurate" models the width of their range $[\tau_1; \tau_2]$ in relation to the width of the range $\left[\tau_1^{opt}; \tau_2^{opt}\right]$ corresponding to the parabolic model can be noticeably larger.

3. The use of spectrum detectors not only in the optical and near-IR ranges, but also with an extension to the mid-IR range (and beyond), also makes it possible to solve the inverse problem more accurately. In addition, it is necessary to develop more efficient smoothing functionals at the stage of inverse problem regularization.

6   Conclusions

Thus, we can conclude that it is in principle possible to use software and hardware measuring complexes for recording the thermal radiation spectra of gas-thermal particle flows in the spraying process of functional coatings on technical products. The use of mathematical methods for solving inverse problems of identifying the functions of the particle temperature distribution makes it possible to control the dispersion spread of the particle temperature in the gas-thermal spraying flow. An effective solution to the problem of measuring and controlling fluctuations in the particle temperature of the spraying flow opens up the prospect for solving the problem of optimizing the operating modes of technological equipment and obtaining coatings with efficient performance characteristics.

References

1. Galimzyanov, F.G., Galimzyanov, R.F.: Theory of Internal Turbulent Motion: Monograph. Ufa, Publisher "Expert", 352 p. (1999). (in Russian)
2. Boronenko, M.P., et al.: Estimation of the velocity and temperature of the dispersed phase in the jets of plasma-arc spraying. Basic Res. **11**(10), 2135–2140 (2014). (in Russian)
3. Jordan, V.I., Soloviev, A.A.: Optoelectronic methods for testing systems for measuring the temperature and velocity parameters of particles in plasma spraying of powder coatings. News of the Altai State University. Ser. Phys. **1/2**(65), 168–171 (2010). (in Russian)
4. Dolmatov, A.V., Gulyaev, I.P., Imamov, R.R.: Spectral pyrometer for temperature control in thermosynthesis processes. Bull. Yugra State Univ. **2**(33), 32–42 (2014). (in Russian)
5. Magunov, A.N.: Spectral pyrometry (overview). Devices Tech. Exp. **4**, 5–28 (2009). (in Russian)
6. Mauer, G., Vassen, R., Stöver, D.: Study on detection of melting temperatures and sources of errors using two-color pyrometry for in-flight measurements of plasma sprayed particles. Int. J. Thermophys. **29**(2), 764–786 (2008)

7. Lee, J.: Estimation of emission properties for silica particles using thermal radiation spectroscopy. Appl. Opt. **50**(22), 4262–4267 (2011). https://doi.org/10.1364/AO.50.004262

8. Sony ICX445ALA-E: http://www.altavision.com.br/Arquivos/Sensor/ICX445_datasheet.pdf

9. Spectrometer Model BTC-110S: http://www.science-surplus.com/products/spectrometers

10. Jordan, V.I., Soloviev, A.A.: The reduction of temperature particles distribution of heterogeneous flows by the inverse of their integrated thermal spectrum. St. Petersburg Polytech. Univ. J. Phys. Math. **2**(98), 85–95 (2010). (in Russian). https://physmath.spbstu.ru/en/article/2010.7.13/

11. Jordan, V.I.: Inverse integral transformation for restoring the temperature distribution of particles of a heterogeneous flow from their integral thermal spectrum. Izvestiya vuzov. Physics **56**(8/3), 293–299 (2013). (in Russian)

12. Kobelev, D.I., Jordan, V.I.: Application of a precision programmable DC power supply for spectrometer calibration. J. Phys.: Conf. Ser. **1843**, 012022 (2021). https://doi.org/10.1088/1742-6596/1843/1/012022

13. Jordan, V.I.: Parametric optimization of the particle temperature distribution of the gas-thermal flux by the means of thermal radiation spectrum of particles. News Altai State Univ. **1**(111), 11–17 (2020). https://doi.org/10.14258/izvasu(2020)1-01. (in Russian)

14. Abramovitz, M., Stegan, I.: Help Guide for special functions with formulas, graphs and tables, 832 p. Transl. from English. Ditkin, V.A, Karmazina. L.N. Nauka, Moscow (1979). (in Russian)

15. Shostak, R.Ya.: Operational calculus. Short course. Ed. second, add. Textbook for higher educational institutions, 280 p. Vysshaya shkola, Moscow (1972). (in Russian)

16. Krasnov, M.L., Kiselev, A.I., Makarenko, G.I.: Integral Equations, 192 p. Nauka, Moscow (1968). (in Russian)

17. Tikhonov, A.N., Arsenin, V.Ya.: Methods for Solving Ill-posed Problems, 288 p. Nauka, Moscow (1986). (in Russian)

18. Tikhonov, A.N., Goncharsky, A.V., Stepanov, V.V., Yagola, A.G.: Numerical Methods for Solving Ill-posed Problems, 229 p. Nauka, Moscow (1990). (in Russian)

19. Golub, G.H., Hansen, P.C., O'Leary, D.P.: Tikhonov regularization and total least squares. SIAM J. Matrix Anal. Appl. **21**(1), 185–194 (1999)

20. Voskoboinikov, Yu.E., Litasov, V.A.: A stable image reconstruction algorithm for inexact point-spread function. Avtometrija [Optoelectron., Instrum. Data Process.] **42**(6), 3–15 (2006). (in Russian)

21. Erokhin, V.I., Volkov, V.V.: Recovering images, registered by device with inexact point-spread function, using Tikhonov's regularized least squares method. Int. J. Artif. Intell. **13**(1), 123–134 (2015). https://www.researchgate.net/publication/282059964_Recovering_images_registered_by_device_with_inexact_point-spread_function_using_Tikhonov's_regularized_least_squares_method

22. Sizikov, V.S.: Inverse applied problems and Matlab, 256 p. Saint Petersburg, Lan' Publ. (2011). (in Russian). https://www.researchgate.net/publication/279296999_Inverse_Applied_Problems_and_MatLab. ISBN 978-5-8114-1238-9

Application of the Heterogeneous Multiscale Finite Element Method for Modelling the Compressibility of Porous Media

Anastasia Yu. Kutishcheva[1,2]($\boxtimes$) [iD] and Sergey I. Markov[1,2] [iD]

[1] Trofimuk Institute of Petroleum Geology and Geophysics, SB RAS, Koptug Avenue 3, 630090 Novosibirsk, Russia
KutischevaAY@ipgg.sbras.ru
[2] Novosibirsk State Technical University, Karl Marx Avenue 20, 630073 Novosibirsk, Russia

Abstract. One of the tasks of geology and geophysics is to establish the relationship between the mechanical external load and the change in the density of complexly organized media, which can include porous media of various types. For this purpose, the theory of porous media introduces the concept of compressibility coefficient, which allows us simplifying mathematical models describing, for example, the flows in such media. One of the options for obtaining the compressibility coefficient is the direct modeling of the deformation of heterogeneous media under various external loads. Within the framework of this work, the heterogeneous multiscale finite element method on polyhedron supports is used for the numerical simulation because porous media, as a rule, are non-periodic and essentially multiscale. This method makes it possible to develop parallel algorithms on the basis of singling out special subdomains (macroelements) from the general domain, in each of which the solution is constructed independently. To ensure the continuity of the stress-strain state in the whole modeling domain, special projectors are formed taking into account the macro- and microstructure of the sample. This approach makes it possible to obtain the results of the required accuracy with a minimum expenditure of computational resources.

Keywords: Compressibility · Heterogeneous media · Heterogeneous multiscale finite element method · Porosity · Elastic deformation

1 Introduction

The study of geological media is closely connected with mathematical modeling of various processes [1, 2]. In this case, it is important to take into account the features of geometry and physical properties with the greatest accuracy, which is not always possible due to limited computational resources. Therefore, in geophysical applications different approaches to model reduction are often used, for example, these are upscaling [3, 4] and homogenization [5]. One of the upscaling variants is a transition from simulation of a complex flow in porous media to a model of seepage through an area where porosity is determined by some scalar or tensor coefficient (Darcy model).

© The Author(s), under exclusive license to Springer Nature Switzerland AG 2022
V. Jordan et al. (Eds.): HPCST 2022, CCIS 1733, pp. 62–71, 2022.
https://doi.org/10.1007/978-3-031-23744-7_5

The porosity can be obtained in a laboratory, but a dynamic model is more interesting, i.e. obtaining the dependence of porosity on the deformation of the sample, for example, during its compression. In the theory of porous media there are well-known relations describing this dependence [6–8]. Another way to estimate changes in porosity is direct numerical modeling of elastic deformation of an object. This approach makes it possible to consider different loading modes, which is preferable for simple analytical evaluations, and to draw conclusions about the applicability of theoretical homogenization models in conditions close to the real ones.

As a rule, various mesh approaches are used for numerical simulation of deformation of porous media. In some cases, classical one-level methods such as the finite element method [9] or the finite difference method [10] may be applied. However, to working with media approximating the real ones, it is necessary to develop and adapt various multilevel or multiscale approaches. For example, in [11] the application of stepwise homogenization for modeling two-dimensional media with double porosity is considered. Also, the initial medium can be approximated by a set of some ideal structures with known features in advance [12, 13]. A different approach to offer modern multiscale methods is based on taking into account macroscopic or effective properties of the medium and microscopic features (such as porous formations) at different levels of the formed functional spaces. This significantly reduces the computational costs while maintaining the required accuracy of the solution [14–16]. Besides, this class of methods assumes initial algorithmic parallelism, since the most resource-intensive calculations, namely, the construction of functional subspaces at the microlevel, are performed independently in each of subdomains of the simulated sample [15].

2 Problem Description

We consider the stationary model of elastic deformation of a porous solid without regard to the effect of gravity, and at constant temperature:

$$\nabla \cdot \sigma = 0, \tag{1}$$

where $\sigma = \sigma_{ij}$, $i, j = \{x, y, z\}$ is the symmetric stress tensor of the second rank [Pa].

The absence of initial stresses and strains is assumed, so the defining relation is the generalized Hooke's law:

$$\sigma = \mathbf{D} : \varepsilon, \tag{2}$$

where $\mathbf{D} = d_{ijkp}$, $i, j, k, p = \{x, y, z\}$ is the elasticity tensor of the fourth rank for an anisotropic medium [Pa], which has the following symmetry properties: $d_{ijkp} = d_{jikp} = d_{ijpk} = d_{kpij}$; $\varepsilon = \varepsilon_{kp}$; $k, p = \{x, y, z\}$ is the symmetric strain tensor of the second rank, such that $\varepsilon = \nabla_s u = \frac{1}{2}(\nabla u + \nabla u^T)$; $u = (u_x, u_y, u_z)^T$ is the displacement vector [m].

Taking into account the equilibrium Eq. (1) and defining the relation (2), let us formulate the boundary problem of stationary elastic deformation of a porous solid $\Omega = \bigcup_{k=1}^{M} \Omega_k$ under external loading f_{ext} (Pa) and in the absence of internal forces and initial stresses and strains at constant temperature:

$$-\nabla \cdot (\mathbf{D}(\mathbf{x}) : \nabla_s \mathbf{u}) = 0 \text{ on } \Omega, \tag{3}$$

$$\mathbf{u} = \mathbf{u}_D \text{ on } \Gamma_D,$$

$$\mathbf{n} \cdot \sigma = \mathbf{f}_{ext} \text{ on } \Gamma_N,$$

$$\mathbf{u}^+ = \mathbf{u}^- \text{ on } \Gamma_{in},$$

$$\mathbf{n}^+ \cdot \sigma^+ = \mathbf{n}^- \cdot \sigma^- \text{ on } \Gamma_{in},$$

where Γ_D and Γ_N are the external boundaries of the solid Ω ($\partial\Omega = \Gamma_D \cup \Gamma_N$, $\Gamma_D \cap \Gamma_N = \emptyset$) on which the initial displacement and loading conditions are set, respectively, Γ_{in} is the internal (contact) boundaries of the solid Ω on which the continuity conditions are set.

To solve the problem (3), the computational scheme of the heterogeneous multiscale finite element method in space $H(\boldsymbol{grad})$ is used.

Let us introduce the following functional spaces in the entire computational domain $\Omega \in R^3$:

$$L^2(\Omega) = \left\{ u \Big| \int_\Omega |u|^2 d\Omega < +\infty \right\}, \tag{4}$$

$$\mathbf{H}(\mathbf{grad}, \Omega) = \left\{ u \Big| u \in \left[L^2(\Omega)\right]^3, \nabla u \in \left[L^2(\Omega)\right]^3 \right\}, \tag{5}$$

with the corresponding scalar products:

$$(u, v)_{L^2(\Omega)} = \int_\Omega u \cdot v d\Omega, \forall u, v \in L^2(\Omega), \tag{6}$$

$$(u, v)_{\mathbf{H}(\mathbf{grad}, \Omega)} = (u, v)_{L^2(\Omega)} + (\nabla u, \nabla v)_{L^2(\Omega)}, \forall u, v \in \mathbf{H}(\mathbf{grad}, \Omega). \tag{7}$$

Let us introduce the space $\mathbf{H}_0(\mathbf{grad}, \Omega)$ as the space of the following form:

$$\mathbf{H}_0(\mathbf{grad}, \Omega) = \{\mathbf{u} \in \mathbf{H}(\mathbf{grad}, \Omega) : \mathbf{u}|_{\partial\Omega} = 0\}. \tag{8}$$

The variational formulation at the macrolevel is defined as: find $\boldsymbol{u} \in \boldsymbol{H}_0(\boldsymbol{grad}, \Omega) + \mathbf{u}_D$ such that $\forall v \in \boldsymbol{H}_0(\boldsymbol{grad}, \Omega)$:

$$\int_\Omega \nabla\mathbf{u} : \mathbf{D}(\mathbf{x}) : \nabla\mathbf{v} \, d\Omega = \int_{\Gamma_N} \mathbf{f}_{ext} \cdot \mathbf{v} \, d\Omega. \tag{9}$$

To form a discrete variational formulation, let us define an irregular polyhedral macroelement mesh of the computational domain $\Pi_H(\Omega)$:

$$\Omega = \bigcup_{P_i \in \Pi_H(\Omega)} P_i.$$

The mesh $\Pi_H(\Omega)$: is constructed so that the sizes of macroelements were much larger than the characteristic sizes of inclusions and the number of intersections of the boundaries of macroelements and inclusions was minimal. By the definition of the heterogeneous multiscale method [15]: the finite element P_i is a *quad* consisting of a geometric carrier, a numerical integration formula, and conjugate subspaces of degrees of freedom and corresponding shape functions $\varphi_j^{P_i}, j = \overline{1, n(P_i)}$. The functions of the form satisfy the following requirements: they are differentiable almost everywhere, finite, and satisfy the unit partition condition [15, 17]. Then it is possible to introduce a discrete analogue of the function spaces (5) and (8):

$$\Phi(\Pi_H(\Omega)) = span\left\{\varphi_j^{P_i}, \forall j = \overline{1, n(P_i)}, \forall P_i \in \Pi_H(\Omega)\right\} \subset \mathbf{H}(\mathbf{grad}, \Omega), \quad (10)$$

$$\Phi_0(\Pi_H(\Omega)) = \left\{\mathbf{u}^H \in \Phi(\Pi_H(\Omega)) : \mathbf{u}^H\big|_{\partial\Omega} = 0\right\} \subset \mathbf{H}_0(\mathbf{grad}, \Omega). \quad (11)$$

The discrete variational formulation at the macrolevel looks like: find $\boldsymbol{u}^H \in \Phi_0(\Pi_H(\Omega))$ such that $\forall \, v^H \in \Phi_0(\Pi_H(\Omega))$:

$$\int_{\Omega} \nabla \mathbf{u}^H : \mathbf{D}(\mathbf{x}) : \nabla \mathbf{v}^H \, d\Omega = \int_{\Gamma_N} \mathbf{f}_{ext} \cdot \mathbf{v}^H \, d\Omega. \quad (12)$$

The heterogeneous multiscale method uses numerical integration formulas for calculating the matrix elements of the system of linear algebraic equations (SLAE) to ensure the continuity of the solution. In this case all the numerical integration points should lie inside the macroelements and be consistent on the neighboring macroelements with respect to the common edges. Thus, the elements of SLAE matrix and the right part will have the form:

$$\mathbf{Aq} = \mathbf{F}, \quad (13)$$

$$A_{n,m} = \sum_{P_i \in \Pi_H(\Omega)} \sum_l \frac{\omega_l}{|I(\mathbf{x}_l)|} \int_{I(\mathbf{x}_l)} \nabla \varphi_n^{P_i} : \mathbf{D} : \nabla \varphi_m^{P_i} d\Omega, \quad (14)$$

$$F_n = \sum_{P_i \in \Pi_H(\Omega)\Gamma_N(P_i)} \int \mathbf{f}_{ext} \cdot \varphi_n^{P_i} d\Omega, \quad (15)$$

where $\mathbf{q}$ is vector of weights of the decomposition of the desired solution $\mathbf{u}^H$ by the shape functions $\varphi_j^{P_i}, j = \overline{1, n(P_i)}, \forall P_i \in \Pi_H(\Omega)$, ω_l and $\mathbf{x}_l$ are weights and numerical integration points; $I(\mathbf{x}_l) \subset P_i$ is the small neighborhood of the integration point; n and m are the global indices of the shape functions associated with the mesh nodes $\Pi_H(\Omega)$.

To calculate the elements of SLAE (13)–(15) it is necessary to know the shape functions $\varphi_j^{P_i}, j = \overline{1, n(P_i)}, \forall P_i \in \Pi_H(\Omega)$. Since macroelements may contain inclusions or pores, for the correct construction of the corresponding shape functions $\varphi_j^{P_i}, j = \overline{1, n(P_i)}$, additional subproblems are formulated on each of the macroelements $P_i \in \Pi_H(\Omega)$ (microlevel)

$$\nabla \cdot \left(\mathbf{D}(\mathbf{x})\nabla_s \varphi_j^{P_i}\right) = 0 \text{ on } P_i, \quad (16)$$

$$\left.\varphi_j^{P_i}\right|_{\partial P_i} = !, \tag{17}$$

where η is some function providing finite differentiability, and unit partitioning conditions for the resulting shape functions. For macroelements-tetrahedrons or polyhedrons with homogeneous boundaries (without boundary crossings by inclusions) the function η is equal to the first order Lagrangian basis functions. In the general case it is required to solve a series of nested subproblems on edges and faces, which are constructed in a similar way.

Let us define in each macroelement $P_i \in \Pi_H(\Omega)$ a tetrahedral irregular consistent mesh $T_h(P_i)$, taking into account all internal boundaries Γ_{in} (3). Then the discrete variational formulation for the problem (16)–(17) looks like this: find $\varphi_j^{P_i} \in T_h(P_i) \subset \mathbf{H}_0(\mathbf{grad}, \Omega)$ such that $\forall \tilde{v} \in T_h(P_i) \subset \mathbf{H}_0(\mathbf{grad}, \Omega)$:

$$\int_{P_i} \nabla \varphi_j^{P_i} : \mathbf{D} : \nabla \tilde{\mathbf{v}} \, dP_i = 0. \tag{18}$$

Quasi-vector basis functions of the first order based on barycentric coordinates were chosen as basis functions at the microlevel.

3   Change in Sample Porosity Under External Mechanical Loading

3.1   Case of Loading with Different Initial Porosity

To determine the compressibility coefficient C_φ [Pa^{-1}] of unstructured porous media let's consider the problem of elastic deformation (3) of samples of radius 2.1 mm and height 1.8 mm (Fig. 1) with the bottom face fixed, the force $\mathbf{F}$ [Pa] acting on the top face, directed perpendicularly downwards. For certainty, we will assume that the upper base can move only along the Oz axis. The location of pores is chosen randomly, as well as their diameter (from 0.4 mm to 0.6 mm). Thus, the porosity of the samples before deformation varies from 3.9% to 6.9% (Fig. 1).

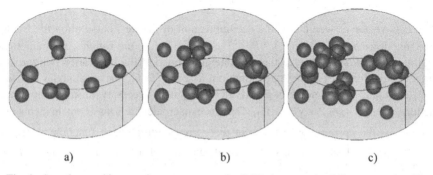

a)   b)   c)

Fig. 1. Samples used in experiments: a – porosity 3.9%, b – porosity 5%, c – porosity 6%.

The Young's modulus of the sample matrix is considered to be 30 MPa, Poisson's ratio is 0.3, which corresponds to some types of sandstone. There is air in the pores. Thus, from the solution of a series of problems with increasing absolute force F, $i = \overline{1, N}$, it is possible to calculate the compressibilities for each of the samples:

$$C_\varphi^i \approx \frac{1}{\varphi_i}\left(\frac{\varphi_i - \varphi_0}{|F_i| - |F_0|}\right), \quad i = \overline{1, N}, \tag{19}$$

$$C_\varphi^I = \frac{1}{N}\sum_{i=1}^{N} C_\varphi^i, \tag{20}$$

$$C_\varphi^{II} = \underset{i=1..N}{\mathrm{med}}(C_\varphi^i), \tag{21}$$

where φ_i, $i = \overline{0, N}$ are porosities when subjected to appropriate forces F_i on the top face of the sample where $|F_0| = 0$, C_φ^I and C_φ^{II} compressibility are calculated in different ways.

For numerical modeling, the heterogeneous multiscale finite-element method described above was used on a macroelement mesh with 80 polyhedrons and a microelement mesh with approximately 200 to 600 thousand tetrahedrons (Fig. 2).

To calculate porosity for three samples (porosity 3.9%, 5%, and 6.9%) under six variants of external loading (10^2 to 10^7 Pa), it took 389 min on a single computational thread. When the number of computational threads was increased to 16, the total solution time was 36 min (Fig. 3).

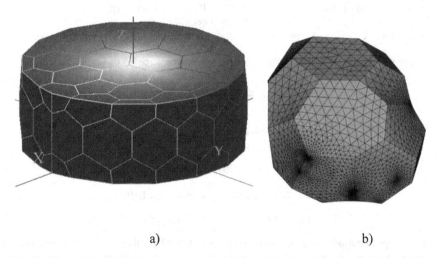

a)   b)

Fig. 2. Macroelement mesh (a) and microelement mesh in one of the inner polyhedra (b).

If the compressibility coefficients are known, then according to [6] the dependence of porosity on external force can be given by the formula:

$$\varphi = \varphi_0 e^{C_\varphi(|F|-|F_0|)}. \tag{22}$$

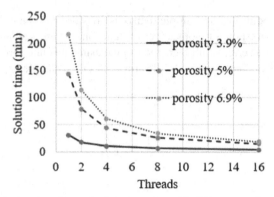

Fig. 3. Problem solution time for each of the samples with different number of threads involved in the calculations.

Figure 4 shows porosity dependence plots obtained by direct numerical simulation and analytically using compressibilities (20) and (21) by formula (22).

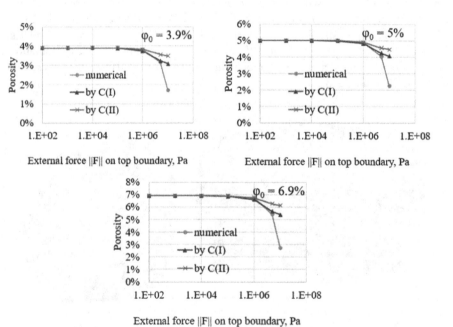

Fig. 4. Change of porosity with increasing loading on the top face of the sample: lines with round markers are porosity obtained from numerical simulations; lines with triangular and x-shaped markers are porosity obtained by formula (22) using and C_φ^I and C_φ^{II}.

The results of comparing the porosity diagrams show a fairly good agreement (relative difference less than 5% for C_φ^I and less than 10% for C_φ^{II}) for small deformations (less than 1% change in porosity). In case of larger deformations, the analytical estimates diverge from the numerical simulations.

3.2   Case of Loading at Different Initial Pore Size

Besides the value of initial porosity the compressibility can be influenced by characteristic pore size, so let's consider two samples (cylinders with 2.1 mm radius and 1.8 mm height) with porosity of 6.9%. The pore diameter in the first sample (Fig. 5.a) varies randomly from 0.2 mm to 0.4 mm, in the second sample (Fig. 5.b) the pore diameter ranges from 0.4 mm to 0.6 mm. The air-filled pores are randomly located in a sandstone matrix (Young's modulus 30 MPa, Pusson's ratio 0.3). Just as in the previous computational experiment to determine the compressibility C_φ [Pa^{-1}] of the described samples, we consider the problem of elastic deformation (3) with boundary conditions: the bottom face is rigidly fixed, a force $\mathbf{F}$ [Пa], acting on the top face, is directed perpendicularly downwards. For certainty, we will assume that the top face can move only along the Oz axis.

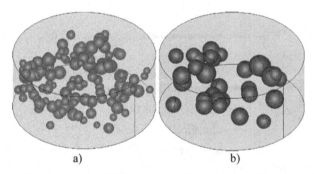

a)   b)

Fig. 5. Samples with the porosity of 6.9%: a – pores diameter from 0.2 mm to 0.4 mm; b – pores diameter from 0.4 mm to 0.6 mm.

Figure 6 shows that both samples involved in the experiment deform in a similar way even under significant loads. This can also be seen in the graph comparing the change in porosity under increasing external loading (Fig. 7). However, the final porosity of the sample with initially larger pores is lower than that of the sample with initially smaller pores, given equal initial porosity and equal loading, which is consistent with the theory of porous materials strength.

As in the previous experiment we varied the initial porosity while maintaining the characteristic pore size. The results of comparing the porosity plots (Fig. 7) were obtained numerically and analytically with the compressibility factor (20) through the ratio (22). These show a fairly good agreement (relative difference less than 4% for the sample with large pores and less than 9% for the sample with small pores for C_φ^I) for small deformations (less than 1% change in porosity). It is important to note the tendency for the error between the numerical and analytical estimates to increase along with increasing external load, which is observed in both experiments.

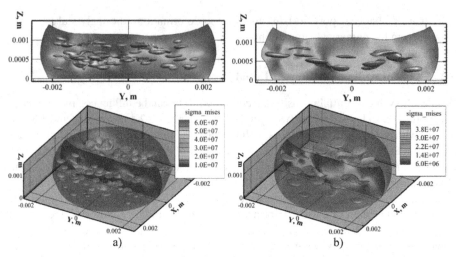

Fig. 6. Mises stress distribution in deformed samples under an external load of 10 MPa: a) is a pores diameter from 0.2 mm to 0.4 mm; b) is a pores diameter from 0.4 mm to 0.6 mm.

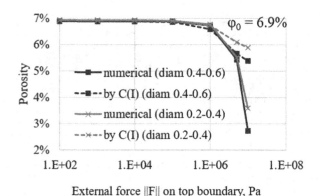

Fig. 7. Change of porosity with increasing pressure on the top face of the sample: red lines are for a sample with 0.2–0.4 mm pores (Fig. 5.a), green lines are for a sample with 0.4–0.6 mm pores (Fig. 5.b).

4 Conclusions

In this work, parallel algorithms of the heterogeneous multiscale finite element method were implemented to solve the problem of elastic deformation of non-periodic porous media under different external loadings. This allowed us to perform computational experiments to obtain the compressibility coefficient. The considered model of porosity change under different external loads can be used for small deformations. The applicability of the model for large deformations requires additional research.

Acknowledgments. This work was carried out with the financial support of Project FWZZ-2022-0030 (porous medium model), Grant of the President of the Russian Federation MK-3230.2022.1.5 (elastic deformation problem).

References

1. Cattin, R., Martelet, G., Henry, P., Avouac, J., Diament, M., Shakya, T.: Gravity anomalies, crustal structure and thermo-mechanical support of the Himalaya of Central Nepal. Geophys. J. Int. **147**(12), 381–392 (2001)
2. Kuznetsov, V.: Geophysical field disturbances and quantum mechanics. E3S Web of Conferences 20 "Solar-Terrestrial Relations and Physics of Earthquake Precursors" (2017)
3. Milad, B., Ghosh, S., Slatt, R., Marfurt, K., Fahes, M.: Practical aspects of upscaling geocellular geological models for reservoir fluid flow simulations: a case study in integrating geology, geophysics, and petroleum engineering multiscale data from the Hunton group. Energies **13**(7), 1604 (2020)
4. Khassanov, D.I., Lonshakov, M.A.: Investigation of the scale effect and the concept of a representative volume element of rocks in relation to porosity. Georesources **22**(4), 55–59 (2020)
5. Wirgin, A.: Dynamic homogenization of a complex geophysical medium by inversion of its near-field seismic response. Wave Motion **81**, 46–69 (2018)
6. Heinemann, Z. and Mittermeir, D.G.: Fluid flow in porous media, PHDG, Austria, 206, (2013).
7. Zhu, S., Du, Z., Li, C., You, Z., Peng, X., Deng, P.: An analytical model for pore volume compressibility of reservoir rock. Fuel **232**, 543–549 (2018)
8. Ashena, R., Behrenbruch, P., Ghalambor, A.: Log-based rock compressibility estimation for Asmari carbonate formation. J. Petrol. Explor. Prod. Technol. **10**(7), 2771–2783 (2020). https://doi.org/10.1007/s13202-020-00934-0
9. Grishchenko, A., Semenov, A., Melnikov, B.: Modeling the processes of deformation and destruction of the rock sample during its extraction from great depths. J. Min. Inst. **248**, 243–252 (2021)
10. Zhao, Y., Zhang, K., Wang, C., Bi, J.: A large pressure pulse decay method to simultaneously measure permeability and compressibility of tight rocks. J. Nat. Gas Sci. Eng. **98**, 104395 (2022)
11. Liu, M., Wu, J., Gan, Y., Hanaor, D., Chen, C.: Multiscale modeling of effective elastic properties of fluid-filled porous materials. Int. J. Solids Struct. **162**, 36–44 (2019)
12. Famà, A., Restuccia, L., Jou, D.: A simple model of porous media with elastic deformations and erosion or deposition. Zeitschrift für angewandte Mathematik und Physik **71**(4), 1–21 (2020)
13. Bazhenov, V., Zhestkov, M.: Computer Modeling Deformation of Porous Elastoplastic Materials and Identification their Characteristics Using the Principle of Three-dimensional Similarity. J. Siberian Federal Univ. Math. Phys. **14**(6), 746–755 (2021)
14. Epov, M.I., Shurina, E.P., Kutischeva, A.Y.: Computation of effective resistivity in materials with microinclusions by a heterogeneous multiscale finite element method. Phys. Mesomech. **20**(4), 407–416 (2017). https://doi.org/10.1134/S1029959917040051
15. Shurina, E.P., Epov, M.I., Kutischeva, A.Y.: Numerical simulation of the percolation threshold of the electric resistivity. Comput. Technol. **22**(3), 3–15 (2017)
16. Abdulle, A., Grote, M., Jecker, O.: Finite element heterogeneous multiscale method for elastic waves in heterogeneous media. Comput. Methods Appl. Mech. Eng. **335**(4), 1–23 (2018)
17. Babuska, I., Melenk, J.: The partition of unity finite element method. Int. J. Numer. Meth. Engng. **40**(4), 727–758 (1997)

Parallel Non-Conforming Finite Element Technique for Mathematical Simulation of Fluid Flow in Multiscale Porous Media

Sergey I. Markov[1,2]([✉]) [iD], Anastasia Yu. Kutishcheva[1,2] [iD], and Natalya B. Itkina[2,3] [iD]

[1] Trofimuk Institute of Petroleum Geology and Geophysics, SB RAS, 3 Koptug Ave., Novosibirsk 630090, Russia
`www.sim91@list.ru`
[2] Novosibirsk State Technical University, 20 Karl Marx Ave., Novosibirsk 630073, Russia
[3] Institute of Computational Technologies SB RAS, 6 Academician M.A. Lavrentiev Ave., Novosibirsk 630090, Russia

Abstract. We consider the three-dimensional steady-state incompressible fluid flow problem in a heterogeneous porous medium using the framework of the Newtonian rheology. The Stokes-Darcy equations are applied as a mathematical model of the mentioned physical process with the Biver-Joseph-Suffman interface conjugation conditions. For spatial approximation of the mathematical model, computational schemes of non-conforming finite element methods based on the discontinuous Galerkin technique are chosen. This approach allows one to use inconsistent macroscale mesh partitions and to satisfy the "inf-sup"-conditions when the Stokes-Darcy problem is solved. We proposed to apply the domain decomposition method based on the modified Schwartz procedure, which allows realizing a parallel computational algorithm. The effectiveness of the developed algorithms is shown by solving the fluid flow problem in the channels and pores of a geological rock. A three-dimensional geometric model of the geological rock is built using the results of computer tomography of cores.

Keywords: Heterogeneous medium · Porous medium · Seepage process · Stokes-darcy equations · Non-conforming finite element methods · Parallel calculations

1 Introduction

Direct mathematical simulation of hydrodynamic processes in heterogeneous porous media is required for solving many applied problems in oil and gas industry and exploration [1–3]. Due to sedimentation and geogenesis, geological rocks have a multiscale geometric structure, which essentially complicates the application of computational fluid dynamics methods [4, 5].

The development of modern multiscale finite-element methods in the approximation theory of hydrodynamic mathematical models allows applying full-scale discrete geometric models of heterogeneous media using the results of core computer tomography [6–8].

The fundamental mathematical model of hydrodynamic processes is based on the Navier-Stokes equations. Under the assumption of incompressibility and slow fluid flowability of the fluid, this mathematical model can be reduced to the Stokes system [9]. Further application of the averaging theory methods to the Stokes system allows writing down the phenomenological Darcy model for description of flows in homogeneous porous media [10].

Mathematical simulation of fluid flows in heterogeneous porous media requires using specialized multiscale mathematical models.

In this work we consider the system of Stokes-Darcy equations as a mathematical model describing the process of incompressible fluid seepage in a system of channels and voids surrounded by a homogeneous porous medium. Coupling of the mathematical models is performed by the Biver-Joseph-Suffman interface conditions [9–11]. We propose a modified non-conforming finite-element discretization of these mathematical models using the computational scheme of the discontinuous Galerkin method.

To build a parallel version of the algorithm for solving this problem, a modified technology of domain decomposition is applied [12–14]. The effectiveness of the developed computer program in comparing with the sequential version of the algorithm is shown by solving the incompressible fluid flow problem in a heterogeneous porous medium.

2 Problem Statement

A geological rock has a complex multiscale structure. Direct consideration of all structural features will lead to an explosive increase in the size of problem to be solved. To implement the procedure of mathematical simulation, idealized geometric models of geological media are used.

The idealized geometric model of the rock consists of a matrix and fluid-saturated inclusions. In our research, the rock sample has the cube shape with an edge length of 10 cm uniformly divided into cubic cells with an edge length of 1 cm (see Fig. 1). It is assumed that each discretization cell is characterized by its own set of transport properties.

We will simulate the incompressible fluid flow in the geological rock sample using its discrete geometric model and the domain decomposition principle. The incompressible fluid flow in the heterogeneous porous media is described by the Stokes-Darcy equations. To discretize this model, we apply the discontinuous Galerkin method which allows the solution to be broken on the interfragmentary boundary. The weighted distribution of local problems on computational threads is carried out at the level of subdomains. The algorithm for constructing the discrete geometric model of a rock core based on the results of computer tomography is described in [15].

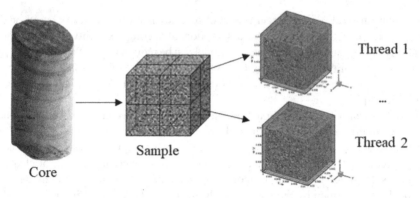

Fig. 1. Structure of the rock sample and computational domain decomposition.

The sample matrix consists of sandstone and contains inclusions (volume fraction 45%) through which water flows. The inclusion will be understood as a single-connected domain with a closed boundary inside the sample. The physical characteristics are given in Table 1.

Table 1. Physical properties

Material	Porosity, %	Permeability, m^2	Viscosity, Pa·sec	Density, kg/m^3
Sandstone	45	0.3	–	2500
Water	–	–	0.001	1000

Each macroscale discretization cell is independently tessellated by a microscale tetrahedral partition (see Fig. 2). Consistency of the microscale mesh at the interfragmentary boundaries is not required when using non-conforming finite-element approximations.

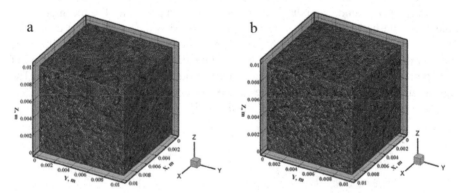

Fig. 2. Examples of microscale mesh partitioning: a – 5% porosity (1328950 tetrahedrons), b – 7% porosity (1659688 tetrahedrons).

3 Methods

3.1 Stokes-Darcy System

We denote the microporous medium as Ω^{matrix}, and macropores (voids) as Ω^{pores}. The steady-state slow flow of an incompressible fluid in a system of channels and pores is described by the following equations:

$$\nabla p_1 = \nabla \cdot \mu \left(\nabla v_1 + (\nabla v_1)^T \right) + \rho g \text{ in } \Omega^{\text{pores}},$$

$$\nabla \cdot v_1 = 0 \text{ in } \Omega^{\text{pores}}, \tag{1}$$

$$v_2 = -\frac{K}{\mu} \cdot (\nabla p_2 + \rho g) \text{ in } \Omega^{\text{matrix}},$$

$$\nabla \cdot v_2 = 0 \text{ in } \Omega^{\text{matrix}}, \tag{2}$$

ρ is fluid density [kg/m^3], μ is dynamic viscosity [Pa $\cdot$ sec], K is absolute permeability tensor [m^2], p_1 is fluid pressure in the macropores and voids [Pa], p_2 is fluid pressure in the microporous medium [Πa], v_1 is flow velocity in macropores [m/sec], v_2 is flow velocity in the microporous medium [m/sec], g is a gravitational acceleration vector.

The notation Γ_{12} means a boundary between the microporous medium and cavern. The ideal contact conditions for the normal component of the velocity and pressure should be satisfied [9]:

$$v_1 \cdot \mathbf{n}_{12}|_{\Gamma_{12}} = v_2 \cdot \mathbf{n}_{12}|_{\Gamma_{12}}, \tag{3}$$

$$\left[p_1 I - \mu \left((\nabla v_1)^T + \nabla v_1 \right) \right] \cdot \mathbf{n}_{12} \cdot \mathbf{n}_{12} \Big|_{\Gamma_{12}} = p_2|_{\Gamma_{12}}, \tag{4}$$

$\mathbf{n}_{12}$ is a unit external normal vector with respect to the channel wall and macropores.

The seepage boundary may not be smooth, then the friction condition is realized by using the Biver-Joseph-Suffman law [9]:

$$\mathbf{v}_1 \cdot \tau_{12}|_{\Gamma_{12}} = -m\mu \left((\nabla v_1)^T + \nabla v_1 \right) \cdot \mathbf{n}_{12} \cdot \tau_{12}|_{\Gamma_{12}}, \tag{5}$$

where τ_{12} is a unit tangential vector; m is a friction coefficient, which depends on the surface roughness and Reynolds number.

If the motion of the fluid is caused by a pressure difference on the S_{inlet} and S_{outlet} faces, the boundary conditions are formulated as:

$$\left[-p_1 I + \mu \left((\nabla v_1)^T + \nabla v_1 \right) \right] \cdot n \Big|_{S_{\text{inlet}}} = t_{\text{inlet}}, \tag{6}$$

$$\left[-p_1 I + \mu \left((\nabla v_1)^T + \nabla v_1 \right) \right] \cdot n \Big|_{S_{\text{outlet}}} = t_{\text{outlet}}, \tag{7}$$

$$p_2|_{S_{\text{inlet}}} = t_{\text{inlet}} \cdot n, \tag{8}$$

$$p_2|_{S_{\text{outlet}}} = t_{\text{outlet}} \cdot n, \tag{9}$$

n is the unit external normal vector with respect to the outer boundaries.

On the impermeable outer boundary S_{wall}, which does not intersect with the S_{inlet} and S_{outlet} faces, the boundary conditions are given:

$$v_1 \cdot n|_{S_{wall}} = 0, \quad v_2 \cdot n|_{S_{wall}} = 0. \tag{10}$$

Problems (1)–(10) can be solved in parallel on each decomposition cell. In accordance with the Schwarz method, an iterative procedure is implemented. Conditions (3)–(9) are formulated on the interface with respect to the solution obtained in an adjacent decomposition cell through an adjacent face. We assume that the macro-partition is consistent.

We denote the microporous medium in the l-th macroscale cell as Ω_l^{matrix}, and macropores (voids) as Ω_l^{pores}. The steady-state slow flow of an incompressible fluid in in the l-th macroscale cell is described by the following equations:

$$\nabla p_1 = \nabla \cdot \mu \left(\nabla v_1 + (\nabla v_1)^T \right) + \rho g \text{ in } \Omega_l^{pores},$$
$$\nabla \cdot v_1 = 0 \text{ in } \Omega_l^{pores}, \tag{11}$$

$$v_2 = -\frac{K}{\mu} \cdot (\nabla p_2 + \rho g) \text{ in } \Omega_l^{matrix},$$
$$\nabla \cdot v_2 = 0 \text{ in } \Omega_l^{matrix}, \tag{12}$$

On the boundary of the microporous medium and the cavern Γ_{12}, the conditions of ideal contact for the normal component of the velocity and the pressure should be satisfied as (3) and (4). The friction conditions are also realized in the form (5).

Let an iterative process be realized, and k be the number of the current iteration. On the boundaries between the cells of the macro-partition, the following conditions take place:

$$v_1^k \cdot n_{12}\Big|_{\Gamma_{12}} - v_2^k \cdot n_{12}\Big|_{\Gamma_{12}} = v_1^{k-1} \cdot n_{12}\Big|_{\Gamma_{12}} - v_2^{k-1} \cdot n_{12}\Big|_{\Gamma_{12}}, \tag{13}$$

$$\left[p_1^k \mathbf{I} - \mu \left(\left(\nabla v_1^k \right)^T + \nabla v_1^k \right) \right] \cdot n_{12} \cdot n_{12}\Big|_{\Gamma_{12}} - p_2^k\Big|_{\Gamma_{12}}$$
$$= \left[p_1^{k-1} \mathbf{I} - \mu \left(\left(\nabla v_1^{k-1} \right)^T + \nabla v_1^{k-1} \right) \right] \cdot n_{12} \cdot n_{12}\Big|_{\Gamma_{12}} - p_2^{k-1}\Big|_{\Gamma_{12}}, \tag{14}$$

$$v_1^k \cdot \tau_{12}\Big|_{\Gamma_{12}} = -m\mu \left(\left(\nabla v_1^{k-1} \right)^T + \nabla v_1^{k-1} \right) \cdot n_{12} \cdot \tau_{12}|_{\Gamma_{12}}. \tag{15}$$

The boundary conditions on the outer faces have the same form (6)–(10). The iterative process is repeated as long as there is inconsistency of the physical fields on the internal interfragmentary boundaries:

$$\frac{\left\| \mathbf{v}_1^k \cdot \mathbf{n}_{12}\big|_{\Gamma_{12}} - \mathbf{v}_1^{k-1} \cdot \mathbf{n}_{12}\big|_{\Gamma_{12}} \right\|_{\mathbf{H}(\mathrm{div})}}{\left\| \mathbf{v}_1^k \cdot \mathbf{n}_{12}\big|_{\Gamma_{12}} \right\|_{\mathbf{H}(\mathrm{div})}} > \varepsilon, \quad \frac{\left\| \mathbf{v}_2^k \cdot \mathbf{n}_{12}\big|_{\Gamma_{12}} - \mathbf{v}_2^{k-1} \cdot \mathbf{n}_{12}\big|_{\Gamma_{12}} \right\|_{\mathbf{H}(\mathrm{div})}}{\left\| \mathbf{v}_2^k \cdot \mathbf{n}_{12}\big|_{\Gamma_{12}} \right\|_{\mathbf{H}(\mathrm{div})}} > \varepsilon$$

$$(16)$$

In the paper, we use $\varepsilon = 10^{-5}$. Criterion (16) is sufficient, since the solution of the Stokes-Darcy equations is reduced to the solution of the saddle point problem and the definition of pairs (v_1, p_1) and (v_2, p_2).

3.2 Variational Formulation of the Discontinuous Galerkin Method for the Stokes-Darcy System

On a non-empty set $\Omega \subset R^3$, we consider a finite union of subsets $M_h(\Omega) = \bigcup_i K_i$, and $\mathfrak{I}_n(K_i)$ as a space of polynomials of degree n. On the union $M_h(\Omega)$, we introduce discrete function spaces for approximation of pressure and velocity:

$$P^h = \left\{ p^h | p^h \in L_0^2(\Omega) : p^h \in \mathfrak{I}_{n-1}(K_i) \forall K_i \in M_h(\Omega), \int_{K_i} p^h dK_i = 0 \right\}, \quad (17)$$

$$V^h = \left\{ v^h | v^h \in H_0(\mathrm{div}, \Omega) : v \in [\mathfrak{I}_n(K_i)]^3 \forall K_i \in M_h(\Omega), v \cdot n|_{\partial K_i} = 0 \right\}. \quad (18)$$

On the set $\Gamma = \bigcup_i \partial K_i$, a trace space is defined as

$$\mathrm{Tr}(\Gamma) = \prod_{\Omega_i \in M_h(\Omega)} L^2(\partial K_i). \quad (19)$$

Now, we denote by the $\Gamma_0 = \Gamma \backslash \partial \Omega$ a set of interior boundaries. It is convenient to express the traces of ambiguous functions on interfragmentary boundaries through the mean $\{\cdot\}$ and jump $[\cdot]$. For functions $\mathbf{v} \in [\mathrm{Tr}(\Gamma)]^3$, $p \in \mathrm{Tr}(\Gamma)$ and $T \in \mathrm{Tr}(\Gamma)$, the mean $\{\cdot\}$ and jump $[\cdot]$ on the outer boundaries $\partial \Omega$ are defined as:

$$\begin{aligned} \underline{[\mathbf{v}]}\big|_{\partial\Omega} &= \mathbf{v} \otimes \mathbf{n}, \quad [\mathbf{v}]|_{\partial\Omega} = \mathbf{v} \cdot \mathbf{n}, \quad \{\,\mathbf{v}\}|_{\partial\Omega} = \mathbf{v}, \\ [p]|_{\partial\Omega} &= p\mathbf{n}, \quad \{p\}|_{\partial\Omega} = p, \quad [T]|_{\partial\Omega} = T\mathbf{n}, \quad \{q\}|_{\partial\Omega} = T, \end{aligned} \quad (20)$$

on the outer faces $\Gamma_0 = \partial K_i \cap \partial K_j$ we have:

$$\begin{aligned} \underline{[\mathbf{v}]}\big|_{\Gamma_0} &= \mathbf{v}_i \otimes \mathbf{n}_i + \mathbf{v}_j \otimes \mathbf{n}_j, \quad [\mathbf{v}]|_{\Gamma_0} = \mathbf{v}_i \cdot \mathbf{n}_i + \mathbf{v}_j \cdot \mathbf{n}_j, \quad \{\,\mathbf{v}\}|_{\Gamma_0} = (\mathbf{v}_i + \mathbf{v}_j)/2, \\ [p]|_{\Gamma_0} &= p_i\mathbf{n}_i + p_j\mathbf{n}_j, \quad \{p\}|_{\Gamma_0} = (p_i + p_j)/2, \\ [T]|_{\Gamma_0} &= T_i\mathbf{n}_i + T_j\mathbf{n}_j, \quad \{T\}|_{\Gamma_0} = (T_i + T_j)/2. \end{aligned} \quad (21)$$

The variational formulation of the Stokes-Darcy problem based on the discontinuous Galerkin method has the form: to find $v_1^h \in V^h$, $p_1^h \in P^h$, $p_2^h \in P^h$, $\forall w_1^h \in V^h$, $q_1^h \in P^h$ and $q_2^h \in P^h$:

$$
\begin{aligned}
a_1\left(\mathbf{w}_1^h, \mathbf{v}_1^h\right) + b\left(\mathbf{w}_1^h, p_1^h\right) + \Lambda\left(\mathbf{w}_1^h, p_2^h\right) &= \left(\mathbf{w}_1^h, \mathbf{F}\right), \\
-b\left(\mathbf{v}_1^h, q_1^h\right) + d\left(q_1^h, p_1^h\right) &= 0, \\
a_2\left(p_2^h, q_2^h\right) + \Lambda\left(\mathbf{v}_1^h, q_2^h\right) &= \left(q_2^h, f\right),
\end{aligned}
\tag{22}
$$

$$
\begin{aligned}
a_1\left(\mathbf{w}_1^h, \mathbf{v}_1^h\right) = &\sum_K \int_{\Omega_K} \mu\left(\left(\nabla \mathbf{v}_1^h\right)^T + \nabla \mathbf{v}_1^h\right) : \nabla \mathbf{w}_1^h + \\
&+ \sum_{\Gamma_K \in \Gamma_{12}} \int_{\Gamma_K} \frac{\mu}{m}\left(\mathbf{v}_1^h \cdot 12\right)\left(\mathbf{w}_1^h \cdot 12\right) dS - \\
&- \sum_K \int_{\partial \Omega_K} \left\{\mu\left(\left(\nabla \mathbf{v}_1^h\right)^T + \nabla \mathbf{v}_1^h\right)\right\} : \left[\mathbf{w}_1^h\right] + \\
&+ \left\{\mu\left(\left(\nabla \mathbf{w}_1^h\right)^T + \nabla \mathbf{w}_1^h\right)\right\} : \left[\mathbf{v}_1^h\right] - \tau_1^{DG}\left[\mathbf{v}_1^h\right] : \left[\mathbf{w}_1^h\right] dS,
\end{aligned}
\tag{23}
$$

$$
b\left(\mathbf{v}_1^h, q_1^h\right) = -\sum_K \int_{\Omega_K} \nabla \cdot \mathbf{v}_1^h q_1^h d\Omega_K + \sum_K \int_{\partial \Omega_K} \left\{q_1^h\right\}\left[\mathbf{v}_1^h\right] dS,
\tag{24}
$$

$$
\Lambda\left(\mathbf{w}_1^h, p_2^h\right) = \sum_{\Gamma_K \in \Gamma_{12}} \int_{\Gamma_K} p_2^h\left(\mathbf{w}_1^h \cdot \mathbf{n}_{12}\right) dS,
\tag{25}
$$

$$
\left(\mathbf{w}^h, \mathbf{F}\right) = \sum_K \int_{\Omega_K} \rho \mathbf{g} \cdot \mathbf{w}^h d\Omega_K + \sum_{\Gamma_K \in S_{\text{inlet}}} \int_{\Gamma_K} \mathbf{t}_{\text{inlet}} \cdot \mathbf{w}^h dS + \sum_{\Gamma_K \in S_{\text{outlet}}} \int_{\Gamma_K} \mathbf{t}_{\text{outlet}} \cdot \mathbf{w}^h dS,
\tag{26}
$$

$$
d\left(q_1^h, p_1^h\right) = \sum_{\Gamma_K \in \Gamma_0} \tau^{DG} \int_{\Gamma_K} \left[q_1^h\right] \cdot \left[p_1^h\right],
\tag{27}
$$

$$
\begin{aligned}
a_2\left(p_2^h, q_2^h\right) = &\sum_K \int_{\Omega_K} \frac{\mathbf{K}}{\mu} \cdot \nabla p_2^h \cdot \nabla q_2^h d\Omega_K + \sum_{\Gamma_K \in \Gamma_{12}} \int_{\Gamma_K} \left(\mathbf{v}_2^h \cdot \mathbf{n}_{12}\right) q_2^h dS \\
&- \sum_K \int_{\partial \Omega_K} \left\{\frac{\mathbf{K}}{\mu} \cdot \nabla p_2^h\right\} \cdot \left[q_2^h\right] + \left\{\frac{\mathbf{K}}{\mu} \cdot \nabla q_2^h\right\} : \left[p_2^h\right] - \tau_2^{DG}\left[p_2^h\right] \cdot \left[q_2^h\right] dS,
\end{aligned}
\tag{28}
$$

$$
\begin{aligned}
\left(q_2^h, f\right) = &\sum_K \int_{K \in S_{\text{outlet}}} \mathbf{t}_{\text{outlet}} \cdot \mathbf{n} q_2^h + \frac{\mathbf{K}}{\mu} \cdot \mathbf{t}_{\text{outlet}} \nabla q_2^h + \tau_2^{DG} \mathbf{t}_{\text{outlet}} \cdot \mathbf{n} q_2^h dS \\
&+ \sum_K \int_{K \in S_{\text{inlet}}} \mathbf{t}_{\text{inlet}} \cdot \mathbf{n} q_2^h + \frac{\mathbf{K}}{\mu} \cdot \mathbf{t}_{\text{inlet}} \nabla q_2^h + \tau_2^{DG} \mathbf{t}_{\text{inlet}} \cdot \mathbf{n} q_2^h dS.
\end{aligned}
\tag{29}
$$

The uniqueness of solving the Stokes-Darcy problem is determined by a suitable choice of bases in the spatial approximation of the velocity and pressure. To approximate the velocity, we use the full second-order basis of the $\boldsymbol{H}(\text{div}, \Omega)$ -space. To approximate the pressure, we use the first-order hierarchical basis of the $H^1(\Omega)$ -space. Information about these bases can be found in [16]. In the matrix-vector form, the variational formulation (21) has the form:

$$\begin{pmatrix} A_1 & B & \Lambda^T \\ -B^T & D & 0 \\ \Lambda & 0 & A_2 \end{pmatrix} \begin{pmatrix} v_1 \\ p_1 \\ p_2 \end{pmatrix} = \begin{pmatrix} F \\ 0 \\ f \end{pmatrix} \tag{30}$$

To solve the system (21), we apply the Krylov methods (generalized minimal residual method) and preconditioner technologies described in [17].

4 Results

We consider the flow of incompressible fluid (water) at the pressure drop 1 Atm across the height of the sample. Figure 3 and Fig. 4 show the results of solving this problem in three cells of global domain decomposition.

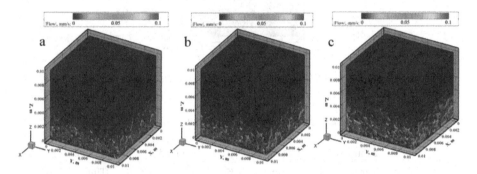

Fig. 3. Modulus of fluid flow velocity: a – 5% porosity, b – 7% porosity, c – 9% porosity.

Calculations were performed using AMD Ryzen 9 3950X 16-Core Processor 3.49 GHz, 64 GB RAM, OS Windows 10 (x64). Parallelization of the algorithm was performed using OpenMP technology. Table 2 summarizes the time required to solve a full-scale problem using different decomposition schemes and number of computational threads.

By the solution error δ we mean the next value:

$$\delta = \frac{\left\| \langle v \rangle - \langle v \rangle^* \right\|_{H(\mathrm{div},\Omega)}}{\left\| \langle v^* \rangle \right\|_{H(\mathrm{div},\Omega)}} \cdot 100\%, \tag{31}$$

where $\langle v \rangle$ is averaged flow velocity calculated using the domain decomposition method with regular coordinated mesh partition, $\langle v \rangle_v^*$ is averaged flow velocity calculated without domain decomposition (coordinated micro-scale tetrahedral partition was only used).

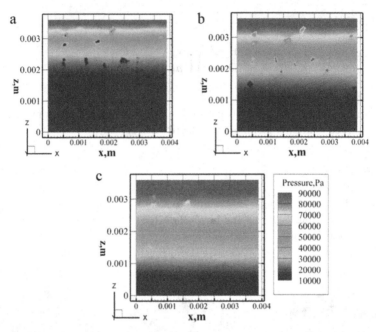

Fig. 4. Pressure in section Y = 0.002: a – 5% porosity, b – 7% porosity, c – 9% porosity.

Table 2. Cost of computational resources and errors

	2 threads		4 threads		8 threads		16 threads	
	Time, sec	δ, %	Time, sec	δ, %	Time, sec	δ, %	Time, sec	δ, %
8 macroelements	4320	9.3e10	2735	4.2e10	1938	1.9e10	691	8.9e10
64 macroelements	2308	7.6e9	1342	3.1e9	731	7.3e9	405	5.3e9
512 macroelements	1193	2.7e8	743	9.2e8	401	1.1e9	227	9.1e9

5   Conclusion

To solve the problem of incompressible fluid flow in multiscale incompressible heterogeneous porous media, a parallel version of a non-conforming finite-element discretization of the Stokes-Darcy problem was proposed using a variation of the domain decomposition technique.

Based on the results of computational experiments, it was found that the cost of time resources is reduced by a factor of 19. However, there is an increase in the error of solving the full-scale direct Stokes-Darcy problem due to the accumulation of computational error in implementing the iterative procedure of the domain decomposition method.

The magnitude of the computational error depends on the number of the macroscale partition cells. The full-scale problem is solved less accurately when more cells in decomposition are used.

Acknowledgements. The research was supported by RSF project No. 22–71-10037.

References

1. Li, J., Zhang, T., Sun, S., Yu, B.: Numerical investigation of the POD reduced-order model for fast predictions of two-phase flows in porous media. Int. J. Numer. Meth. Heat Fluid Flow **29**(11), 4167–4204 (2019)
2. Bosma, S., Klevtsov, S., Møynerc, O., Castelletto, N.: Enhanced multiscale restriction-smoothed basis (MsRSB) preconditioning with applications to porous media flow and geomechanics. J. Comput. Phys. **428**, 109934 (2021)
3. Cusini, M., White, J., Castelletto, N., Settgast, R.: Simulation of coupled multiphase flow and geomechanics in porous media with embedded discrete fractures. Int. J. for Num. and An. Methods in Geomechanics, 1–22 (2020). https://doi.org/10.1002/nag.3168
4. Hoffman, K., Chiang, S.: Computational Fluid Dynamics. Engineering Education System (2000). ISBN: 0962373109
5. Volker, J.: Finite Element Methods for Incompressible Flow Problems. Part of the Springer Series in Computational Mathematics book series (SSCM) 51 (2016)
6. Abgrall, R.: A residual distribution method using discontinuous elements for the computation of possibly non-smooth flows. Adv. Appl. Math. Mech. **2**(1), 32–44 (2010)
7. Droniou, J., Eymard, R., Gallouët, T., Herbin, R.: Non-conforming Finite Elements on Polytopal Meshes. In: Di Pietro, D.A., Formaggia, L., Masson, R. (eds.) Polyhedral Methods in Geosciences. SSSS, vol. 27, pp. 1–35. Springer, Cham (2021). https://doi.org/10.1007/978-3-030-69363-3_1
8. Pietro, D., Formaggia, L., Masson, R.: Polyhedral Methods in Geosciences. Part of the SEMA SIMAI Springer Series book series (SEMA SIMAI) 27, (2021)
9. Riviere, B.: Analysis of a discontinuous finite element method for the coupled stokes and darcy problems. J. Sci. Comput. **22**(23), 479–500 (2005)
10. Vassilev, D., Yotov, I.: Coupling stokes-darcy flow with transport. SIAM J. Sci. Comput. **31**(5), 3661–3684 (2009)
11. Ngondiep, E.: Unconditional stability over long time intervals of a two-level coupled Mac-Cormack/Crank–Nicolson method for evolutionary mixed Stokes-Darcy model. J. Comput. Appl. Math. **409**, 114148 (2022)
12. Ciarlet, P., Jamelot, E., Kpadonouc, F.: Domain decomposition methods for the diffusion equation with low-regularity solution. Comput. Math. Appl. **74**(10), 2364–2384 (2017)
13. Galvis, J., Efendiev, Y.: Domain decomposition preconditioners for multiscale flows in high contrast media: reduced dimension coarse spaces. Multiscale Model. Simul. **8**(5), 1621–1644 (2010)
14. Wang, C.: Domain decomposition methods for coupled Stokes-Darcy flows. 2016. Dissertation (Doctor of Philosophy): Dept. of Mathematics, University of Pittsburgh (2016)
15. Shurina, E., Dobrolubova, D., Shtanko, E.: Special techniques for objects with complex inner structure based on a CT image sequence. Cloud of Science **5**(1), 40–58 (2018)

16. Solin, P., Segeth, K., Dolezel, I.: Higher-order finite element methods. Chapman and Hall/CRC (2004)
17. He, Y., Li, J., Meng, L.: Three effective preconditioners for double saddle point problem. AIMS Mathematics **6**(7), 6933–6947 (2021)

Darcy Problem Solution in a Mixed Formulation for Geological Media of Various Structures

Ella P. Shurina[1,2] ⓘ, Natalya B. Itkina[2] ⓘ, and Svetlana A. Trofimova[1,2(✉)]

[1] Trofimuk Institute of Petroleum Geology and Geophysics, SB RAS, Koptug Ave. 3,
630090 Novosibirsk, Russia
svetik-missy@mail.ru

[2] Novosibirsk State Technical University, Karl Marx Ave. 20, 630073 Novosibirsk, Russia

Abstract. The possibility of using a mixed non-conformal finite element formulation to solve the fluid filtration problem in a porous geological medium with clay inclusions, as well as in a layered medium under pressure is analyzed. Within the framework of this work, for the construction of a computational scheme based on a mixed method for Darcy flow, the use of a set of hierarchic basis functions on rectangular finite elements for the velocity in the vector-valued function space H(div) is implemented for the first time. A software package in the C++ programming language that accepts as input a grid partition of the computational domain in the ".msh" format, obtained using a finite element mesh generator Gmsh, has been developed and verified. For further detailed analysis of the numerical pressure and velocity fields obtained as a result of solving the filtration problem, and also graphical display of the solution the Tecplot program can be used.

Keywords: Darcy law · Permeability field · Mixed finite element approximation · Discontinuous Galerkin method

1 Introduction

Filtration theory is a section of hydrodynamics and studies the flow (leakage) of liquids and gases through porous or fractured-porous media [1]. Mathematical modeling of seepage processes in a complex geological environment is an essential step in the development of efficient oil recovery methods. The fundamental point in the implementation of the mathematical modeling procedure is the use of modern apparatus of numerical methods, which makes it possible to take into account the multiscale nature of the geological environment.

When modeling processes associated with intensification (hydraulic fracturing of a rock formation) and the development of hydrocarbon deposits, a class of problems arises that do not provide for determining the explicit behavior of pressure at the boundary of the modeling area. But they specify the behavior of the normal velocity component. The need to solve such applied problems determines the use of specialized mixed variational formulations [2].

V. Jordan et al. (Eds.): HPCST 2022, CCIS 1733, pp. 83–96, 2022.
https://doi.org/10.1007/978-3-031-23744-7_7

Mixed finite element approximation of Darcy flow has the advantage of being able the ability to simultaneously determine both pressure and fluid flow velocity. Mixed variational principle may be discretized by seeking a critical point of the relevant functional over a finite dimensional subspace of the permissible trial functions. In fact this space is a direct sum of two corresponding subspaces [3, 4]. A key point, which is characteristic of the mixed approach, is that the solution of the filtration problem is not an extreme point of the functional, but it is a saddle point [5], i.e. the problem of finding the minimum of the functional in the general case does not have a unique solution and additional information is required to fulfill the correctness conditions. The most common approach is the introduction of Lagrange multipliers or the definition of a finite element basis from the space H(div), for which the Ladyzhenskaya-Babushka-Brezzi (LBB) conditions are satisfied [6]. The use of a special hierarchical basis makes it possible to determine a physically relevant solution to the problem.

In this paper, a mixed finite element discretization of the Darcy flow equation using the discontinuous Galerkin method (DG method) is considered [7]. This way has a number of advantages over the conformal finite element approximation when modelling filtration processes in complex geological areas. The DG method allows the solution to be independently approximated at each finite element, and in addition, special function trace operators – numerical fluxes – are used to determine the solution at interelement boundaries. The choice of the numerical fluxes is quite delicate, as it can affect the stability and the accuracy of the computational scheme.

The main feature of the discrete analogues of non-conformal mixed finite element formulations of Darcy's law with a tensor permeability coefficient in heterogeneous media is computational instability. An increase in the stability of such schemes is achieved by additional stabilization of the variational setting, choice of a basis [8] or a special solver. When constructing a discrete analogue, a high-order hierarchical basis system from the space H(div) for velocity was used, which includes divergently free edge-functions and also non-divergently free basis functions defined at the internal nodes of the finite element (bubble-functions) [9, 10]. The system of linear algebraic equations was solved using a combination of the Biconjugate Gradient method and the Generalized Minimum Residual method. This approach made it possible to use the advantages of both methods to accelerate the convergence of the solution [9].

2   Mixed Discontinuous Galerkin Formulation

Let's introduce a bounded n-dimensional domain $\Omega \subset R^n$ with boundary $\partial\Omega = \Gamma_D$. The Darcy's law can be represented as a system of equations for the velocity vector $\bar{v}$ and pressure p:

$$\begin{cases} \bar{v} = -K\nabla p \;\; in \;\; \Omega \\ \quad \nabla \cdot \bar{v} = f \;\; in \;\; \Omega \;, \\ \quad \bar{v} \cdot \bar{n} = g \;\; on \;\; \partial\Omega \end{cases} \tag{1}$$

where n is unit normal vector, K is a tensor permeability coefficient.

We consider the space $L_2(\Omega)$ of all square integrable vector fields with inner product

$$(\sigma_1, \sigma_2)_{L_2(\Omega)} = \int_\Omega \sigma_1 \sigma_2 d\Omega \text{ and norm} \|\sigma\|_{L_2(\Omega)} = (\sigma, \sigma)_{L_2(\Omega)}^{1/2}.$$

and the Hilbert spaces of scalar and vector functions:

$$H^1(\Omega) = \{\sigma \in L_2(\Omega), \; \partial\sigma/\partial x_i \in L_2(\Omega), \; 1 \leq i \leq n\}, \tag{2}$$

$$H^{div}(\Omega) = \{\overline{w} \in (L_2(\Omega))^n, \nabla \cdot \overline{w} \in L_2(\Omega)\}. \tag{3}$$

The mixed formulation for the Darcy problem has the form [11]: we find $(\overline{v}, p) \in H^{div}(\Omega) \times L_2(\Omega)$ such that:

$$\begin{cases} (K^{-1}\overline{v}, \overline{w})_{L_2(\Omega)} - (p, \nabla \cdot \overline{w})_{L_2(\Omega)} = 0 \; \forall \overline{w} \in H^{div}(\Omega) \\ (\nabla \cdot \overline{v}, \sigma)_{L_2(\Omega)} = (f, \sigma)_{L_2(\Omega)} \qquad \forall \sigma \in L_2(\Omega) \end{cases}. \tag{4}$$

Let $\Xi_h = \{R\}$ is set of finite elements R that are the rectangular fragmentation of the computational domain Ω. Thus

$$\Gamma = \bigcup_R \partial R$$

is the set of edges, $\Gamma_0 = \Gamma \backslash \partial\Omega$. We need to introduce finite element subspaces associated with the partition $\Xi_h = \{R\}$. We set:

$$\Sigma_h = \{\sigma \in L_2(\Omega) : \sigma|_R \in P_l(R) \forall R \in \Xi_h\}, \tag{5}$$

$$W_h = \{\overline{w} \in (L_2(\Omega))^n : \overline{w}|_R \in (P_l(R))^n \forall R \in \Xi_h\}, \tag{6}$$

where $P_l(R)$ is the space of polynomial functions of degree at most $l \geq 1$, which is defined on a finite element R.

In order to determine the numerical fluxes at the boundaries between finite elements, we define the average $\{\cdot\}$ and jump $[\cdot]$ operators of the scalar and vector functions. Let $e \in \Gamma_0$ be an interior edge separating the elements R_1 and R_2, for which the outward normals $\tilde{n}_{11}$ and $\tilde{n}_2$, respectively, are defined on the edge e. Then the average and jump operators of the scalar function

$$\sigma \in R(\Gamma) = \prod_{R \in \Xi_h} L_2(\Omega)$$

can be introduced as follows:

$$\{\sigma\} = \frac{1}{2}(\sigma_1 + \sigma_2), \tag{7}$$

$$[\sigma] = \sigma_1 \overline{n}_1 + \sigma_2 \overline{n}_2, \tag{8}$$

where $\sigma_i = \sigma|R_i$, $i = 1, 2$. Operators $\{\cdot\}$ and $[\cdot]$ for a vector function $\overline{w} \in [R(\Gamma)]^2$ take the form:

$$\{\overline{w}\} = \frac{1}{2}(\overline{w}_1 + \overline{w}_2), \tag{9}$$

$$[\overline{w}] = \overline{w}_1 \cdot \overline{n}_1 + \overline{w}_2 \cdot \overline{n}_2 \tag{10}$$

We choose as fluxes the pressure jump through the boundary of the finite element and the average value of the velocity at the boundary, i.e.:

$$\hat{p} = [p] \text{ on } \Gamma = \Gamma_0 \cup \partial\Omega, \tag{11}$$

$$\hat{v} = \{\overline{v}\} \text{ on } \Gamma = \Gamma_0 \cup \partial\Omega. \tag{12}$$

For additional stabilization of the mixed method, penalty terms for pressure p and velocity $\overline{v}$ jumps are introduced. In accordance with [4], the mixed discontinuous Galerkin formulation has the form: we find $(\overline{v}, p) \in H^{div}(\Omega) \times L_2(\Omega)$ such that:

$$\begin{cases} \int_\Omega K^{-1}\overline{v} \cdot \overline{w}d\Omega - \int_\Omega p\nabla \cdot \overline{w}d\Omega + \int_{\Gamma_0} [p]\{\overline{w}\}dS + \int_{\Gamma_D} (p\overline{n}) \cdot \overline{w}dS \\ \quad -\theta \int_\Omega \left(K^{-1}\overline{v} + \nabla p\right) \cdot \overline{w}d\Omega = 0 \\ -\int_\Omega \nabla \cdot \overline{v}\sigma d\Omega + \int_{\Gamma_0} \{\overline{v}\}[\sigma]dS + \int_{\Gamma_D} \overline{v} \cdot (\sigma\overline{n})dS + \delta\theta \int_\Omega (\overline{v} + K\nabla p) \cdot \nabla\sigma d\Omega \\ = \int_\Omega f\sigma d\Omega - \int_{\Gamma_D} g\sigma dS \end{cases} \tag{13}$$

where θ is a stabilization parameter, $\delta = \pm 1$, $[\cdot]$ is the jump operator, $\{\cdot\}$ is the average operator.

3 Numerical Experiments

In the development of oil fields, the most common system of areal well location in practice is a five-point scheme. An element of such system is a square, in the corners of which there are production wells, and in the center - an injection well (see Fig. 1).

Fig. 1. Five-point system for the location of production and injection wells

When modeling physical processes in such system, the computational domain is often represented as a quarter of the entire system, i.e. rectangular area with an injection well in the lower left corner and a production well in the upper right corner.

We consider the computational domain $\Omega = [0, 1] \times [0, 1]$, shown in Fig. 2, with the injection velocity vector set in the lower left corner, and the runoff velocity vector in the upper right corner, the medium permeability coefficient (corresponding to sandstone)

$$K = \begin{bmatrix} 50 & 0 \\ 0 & 50 \end{bmatrix},$$

the right side of the second equation of the system $f = 0$. Table 1 shows the relative errors in calculating pressure and velocity in the L_2-norm. Figure 3 shows the numerical pressure fields and velocity vectors.

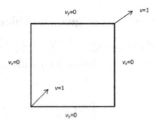

Fig. 2. Computational domain

Table 1. Relative errors in the L_2-norm

Mesh step	Number of iterations	Residual	$\dfrac{\|p_{h/2}-p_h\|_{L_2(\Omega)}}{\|p_{h/2}\|_{L_2(\Omega)}}$	$\dfrac{\|\bar{v}_{h/2}-\bar{v}_h\|_{L_2(\Omega)}}{\|\bar{v}_{h/2}\|_{L_2(\Omega)}}$
$h = 0.1$	144	$8.73652e^{-8}$	$1.49163e^{-2}$	$2.10535e^{-1}$
$h = 0.05$	139	$7.54663e^{-8}$		
			$3.48547e^{-3}$	$1.00726e^{-1}$
$h = 0.025$	175	$9.46792e^{-8}$		

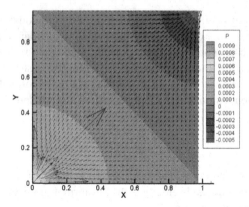

Fig. 3. Numerical pressure fields on the mesh $h = 0.05$.

3.1 Computational Domain with Rectangular Inclusions

We consider the computational domain $\Omega = [0, 1] \times [0, 1]$ with a different number of square inclusions, shown in Figs. 4, 6, 8, 10, the permeability coefficient of the medium (corresponds to sandstone).

$$K = \begin{bmatrix} 50 & 0 \\ 0 & 50 \end{bmatrix},$$

the permeability coefficient of inclusions (corresponds to clay)

$$K = \begin{bmatrix} 1 & 0 \\ 0 & 1 \end{bmatrix},$$

the right side of the second equation of the system $f = 0$. Table 2 shows the relative errors in calculating pressure and velocity in the L_2-norm. Figure 5 shows the numerical pressure fields and velocity vectors.

Fig. 4. Computational domain

Table 2. Relative errors in the L_2-norm

Mesh step	Number of iterations	Residual	$\dfrac{\|p_{h/2}-p_h\|_{L_2(\Omega)}}{\|p_{h/2}\|_{L_2(\Omega)}}$	$\dfrac{\|\bar{v}_{h/2}-\bar{v}_h\|_{L_2(\Omega)}}{\|\bar{v}_{h/2}\|_{L_2(\Omega)}}$
$h = 0.1$	183	9.410166e−8	1.35373e−2	2.03244e−1
$h = 0.05$	159	9.754978e−8		
			9.4679e−3	9.79569e−2
$h = 0.025$	182	8.115497e−8		

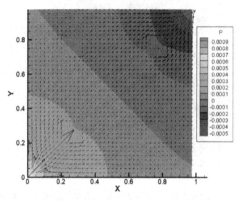

Fig. 5. Numerical pressure fields on the mesh $h = 0.05$

Table 3 shows the relative errors in calculating pressure and velocity in the L_2-norm. Figure 7 shows the numerical pressure fields and velocity vectors.

Fig. 6. Computational domain

Table 3. Relative errors in the L_2-norm

Mesh step	Number of iterations	Residual	$\dfrac{\|p_{h/2}-p_h\|_{L_2(\Omega)}}{\|p_{h/2}\|_{L_2(\Omega)}}$	$\dfrac{\|\bar{v}_{h/2}-\bar{v}_h\|_{L_2(\Omega)}}{\|\bar{v}_{h/2}\|_{L_2(\Omega)}}$
$h = 0.1$	197	7.712059e$-$8	1.30287e$-$2	2.0424e$-$1
$h = 0.05$	132	9.851416e$-$8		
			5.04539e$-$3	9.7946e$-$2
$h = 0.025$	276	3.729016e$-$8		

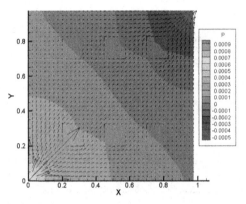

Fig. 7. Numerical pressure fields on the mesh $h = 0.05$

Table 4 shows the relative errors in calculating pressure and velocity in the L_2-norm. Figure 9 shows the numerical pressure fields and velocity vectors.

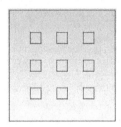

Fig. 8. Computational domain

Table 4. Relative errors in the L_2-norm

Mesh step	Number of iterations	Residual	$\dfrac{\|p_{h/2}-p_h\|_{L_2(\Omega)}}{\|p_{h/2}\|_{L_2(\Omega)}}$	$\dfrac{\|\bar{v}_{h/2}-\bar{v}_h\|_{L_2(\Omega)}}{\|\bar{v}_{h/2}\|_{L_2(\Omega)}}$
$h=0.1$	140	7.999886e−8	1.16483e−2	2.03446e−1
$h=0.05$	165	9.709769e−8		
			3.23100e−3	9.77259e−2
$h=0.025$	223	6.228467e−8		

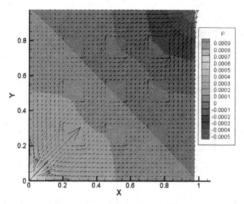

Fig. 9. Numerical pressure fields on the mesh $h=0.05$

Table 5 shows the relative errors in calculating pressure and velocity in the L_2-norm. Figure 11 shows the numerical pressure fields and velocity vectors.

Fig. 10. Computational domain

Table 5. Relative errors in the L_2-norm

Mesh step	Number of iterations	Residual	$\frac{\|p_{h/2}-p_h\|_{L_2(\Omega)}}{\|p_{h/2}\|_{L_2(\Omega)}}$	$\frac{\|\bar{v}_{h/2}-\bar{v}_h\|_{L_2(\Omega)}}{\|\bar{v}_{h/2}\|_{L_2(\Omega)}}$
$h = 0.1$	148	7.799563e−8	1.17462e−2	1.73646e−1
$h = 0.05$	175	9.375050e−8		
			3.40949e−3	9.43514e−2
$h = 0.025$	214	7.192013e−8		

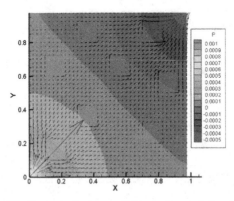

Fig. 11. Numerical pressure fields on the mesh $h = 0.05$

3.2 Layered Computational Domain

We consider a layered computational domain $\Omega = [0, 1] \times [0, 1]$ with different thick-nesses of the middle layer, shown in Figs. 12, 14, 16, the permeability coefficient of the medium (corresponding to sandstone).

$$K = \begin{bmatrix} 50 & 0 \\ 0 & 50 \end{bmatrix},$$

the permeability coefficient of the middle layer (corresponding to clay).

$$K = \begin{bmatrix} 1 & 0 \\ 0 & 1 \end{bmatrix},$$

the right side of the second equation of the system $f = 0$. Table 6 shows the relative errors in calculating pressure and velocity in the L_2-norm. Figure 13 shows the numerical pressure fields and velocity vectors.

Fig. 12. Computational domain

Table 6. Relative errors in the L_2-norm

Mesh step	Number of iterations	Residual	$\frac{\|p_{h/2}-p_h\|_{L_2(\Omega)}}{\|p_{h/2}\|_{L_2(\Omega)}}$	$\frac{\|\bar{v}_{h/2}-\bar{v}_h\|_{L_2(\Omega)}}{\|\bar{v}_{h/2}\|_{L_2(\Omega)}}$
$h = 0.1$	225	4.038231e−8	1.33729e−3	2.08169e−1
$h = 0.05$	274	1.978409e−8		
			3.09632e−4	9.95045e−2
$h = 0.025$	331	6.991224e−8		

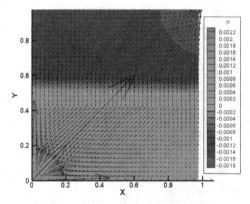

Fig. 13. Numerical pressure fields on the mesh $h = 0.05$

Table 7 shows the relative errors in calculating pressure and velocity in the L_2-norm. Figure 15 shows the numerical pressure fields and velocity vectors.

Fig. 14. Computational domain

Table 7. Relative errors in the L_2-norm

Mesh step	Number of iterations	Residual	$\dfrac{\|p_{h/2}-p_h\|_{L_2(\Omega)}}{\|p_{h/2}\|_{L_2(\Omega)}}$	$\dfrac{\|\bar{v}_{h/2}-\bar{v}_h\|_{L_2(\Omega)}}{\|\bar{v}_{h/2}\|_{L_2(\Omega)}}$
$h = 0.1$	229	9.572552e−8	1.88626e−3	1.84114e−1
$h = 0.05$	245	7.336942e−8		
			7.97283e−4	9.51487e−2
$h = 0.025$	319	9.196616e−8		

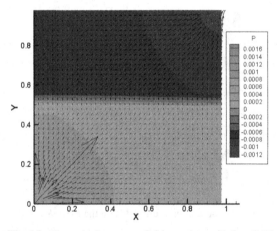

Fig. 15. Numerical pressure fields on the mesh $h = 0.05$

Table 8 shows the relative errors in calculating pressure and velocity in the L_2-norm. Figure 17 shows the numerical pressure fields and velocity vectors.

Fig. 16. Computational domain

Table 8. Relative errors in the L_2-norm

Mesh step	Number of iterations	Residual	$\frac{\|p_{h/2}-p_h\|_{L_2(\Omega)}}{\|p_{h/2}\|_{L_2(\Omega)}}$	$\frac{\|\bar{v}_{h/2}-\bar{v}_h\|_{L_2(\Omega)}}{\|\bar{v}_{h/2}\|_{L_2(\Omega)}}$
$h = 0.1$	226	3.510909e−8	3.34223e−3	1.88952e−1
$h = 0.05$	249	4.221870e−8		
			9.42187e−4	9.76676e−2
$h = 0.025$	319	7.166406e−8		

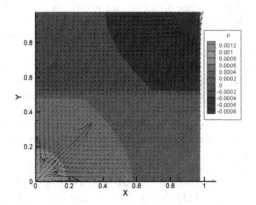

Fig. 17. Numerical pressure fields on the mesh $h = 0.05$

4 Conclusions

Within the framework of this work, a computational scheme based on a mixed discontinuous Galerkin method for solving the Darcy flow equations in porous media with a tensor permeability coefficient was developed, implemented and verified. For the first time, an approach using hierarchical basis systems in specialized functional spaces was used.

The authors have constructed and tested a software package in the C++ programming language.

Investigations on a class of model problems close to real ones have shown that for a computational domain with rectangular inclusions with a permeability coefficient corresponding to clay, the determination of the numerical fields of pressure and velocity is carried out with a relative error of 1e2 even on a coarse mesh. With a change in the location of inclusions and their number, the calculation accuracy is preserved. For a layered computational domain, when the mesh step is divided, the relative error in determining the numerical fields of pressure and velocity decreases by an order. It is observed that the accuracy of calculations is preserved when the thickness of the middle layer changes, the permeability coefficient of which corresponds to clay. In this regard, it can be concluded that the developed computational scheme can be used to calculate the pressure and velocity fields during forced injection of liquid in the area with rectangular clay inclusions and in a layered medium.

Acknowledgements. The research was supported by Fundamental Scientific Research Project No. 0266-2022-0030, Project No. 0266-2022-0025.

References

1. Leybenzon, L.S.: Podzemnaya gidrogazodinamika [Underground hydrodynamics]. AN SSSR Publ, Moscow (1953). (in Russian)
2. Arnold, D.N.: Mixed finite element methods for elliptic problems. Computer Methods in Applied Mechanics and Engineering **82**, 281–300 (1990)
3. Masud, A., Hughes, T.J.R.: A stabilized mixed finite element method for Darcy flow. Comput. Methods Appl. Mech. Engrg. **191**, 4341–4370 (2002)
4. Brezzi, F., Hughes, T.J.R., Marini, L.D., Masud, A.: Mixed discontinuous Galerkin methods for Darcy flow. Journal of Scientific Computing **22**(1), 119–225 (2005)
5. Brezzi, F.: On the existence, uniqueness, and approximation of saddle point problems arising from Lagrangian multipliers. RAIRO Anal. Numer. **8–32**, 129–151 (1974)
6. Ladyzhenskaya, O.A.: Matematicheskie voprosy dinamiki vyazkoy neszhimaemoy zhidkosti [Mathematical problems of viscous incompressible fluid dynamics]. Nauka Publ, Moscow (1970). (in Russian)
7. Arnold, D.N., Brezzi, F., Marini, L.D.: Unified analysis of discontinuous Galerkin methods for elliptic problems. SIAM J. Numer. Anal. **39**(5), 1749–1779 (2002)
8. Solin, P., Segeth, K., Dolezel, I.: High-order finite element methods. Chapman and Hall CRC (2004)
9. Trofimova, S.A., Itkina, N.B., Shurina, E.P.: Hierarchical basis in space for a mixed finite element formulation of the Darcy problem. Siberian Electronic Mathematical Reports **17**, 1741–1765 (2020)
10. Shurina, E.P., Itkina, N.B., Trofimova, S.A.: Multilevel Method Modifications for Discrete Analogues of Mixed Variational Formulations of the Filtration Problem. XIV International Scientific-Technical Conference on Actual Problems of Electronics Instrument Engineering (APEIE) (2018)
11. Brezzi, F., Marini, D.: A survey on mixed finite element approximations. IEEE Transactions on Magnetics **30**(5), 3547–3551 (1994)

Mathematical Modeling of the Electromagnetic Field from a Solenoidal Coil in the Frequency Domain

Dmitriy Arhipov[1] (ID), Nadezhda Shtabel[1,2](✉) (ID), and Ella P. Shurina[1,2] (ID)

[1] Trofimuk Institute of Petroleum Geology and Geophysics of Siberian Branch, Russian Academy of Sciences, Novosibirsk, Russia
orlovskayanv@ipgg.sbras.ru
[2] Novosibirsk State Technical University, Novosibirsk, Russia

Abstract. Numerical estimates of the measured electromotive force (EMF) for the electric field excited by a multi-turn solenoidal coil depend on the way the field source is approximated. The simple representation of the source as a point dipole is correct for measurements in the far field. For measurements in the vicinity of the field source, this method of approximation may give unreliable results. This paper proposes two ways of approximating a multi-turn solenoidal coil: surface approximation or single-coil approximation. Here we present the numerical simulation of the downhole logging procedure. As a result, we show the three-dimensional electromagnetic field, an electromotive force (EMF) induced in the receiver coils, and the phase difference of EMF depending on the method of source approximation. To calculate the three-dimensional electromagnetic field, the vector finite element method on the tetrahedral unstructured partition was used. The difference in the electromagnetic field configuration for different source approximation methods is shown. The effect of the number of cores on the time of solving a finite element system of linear algebraic equations is demonstrated. The use of a larger number of processor cores makes it possible to significantly reduce the time of solving the problem.

Keywords: Vector finite element method · SLAE · Electromagnetic field · Solenoidal source approximation · Helmholtz equation

1 Introduction

Electromagnetic field measurements are the basis for subsurface electrical prospecting methods used for mineral prospecting. Electrical prospecting methods can be divided into direct current methods, near- or far-field attenuation methods, and frequency-sounding methods. Electromagnetic measurements can be performed both at the surface and in the well. Methods of electrical exploration in the well are called well logging. Usually, well logging is used to distinguish oil-and-gas and water-saturated formations [1]. The idea behind the well logging method is the following: a device called a probe, which contains a generator and one or two receiver coils, is lowered into the well. The generator coil

© The Author(s), under exclusive license to Springer Nature Switzerland AG 2022
V. Jordan et al. (Eds.): HPCST 2022, CCIS 1733, pp. 97–111, 2022.
https://doi.org/10.1007/978-3-031-23744-7_8

with the current of a given frequency excites the electromagnetic field in the medium. Receiver coils located on the same axis with the generator coil measure induced EMF and phase difference of EMF in the case of two receiving coils [2, 3].

The medium around the well has a rather complex structure. To increase the reliability of data interpretation of measured EMF, a series of computational experiments for different models of the medium is carried out. The three-dimensional mathematical modeling based on the vector finite element method is a powerful tool for simulating the electromagnetic geophysical measurement procedure. In the paper, the curl-conforming basis from the Nedelec's space H (curl) is used for the approximation of vector electromagnetic fields [4].

There is a problem with approximating the field source such as a solenoidal coil. In the XX century, the theory of electromagnetic exploration was based on the works of Kaufman. Kaufman used a few assumptions and simplifications in his works to derive analytical and semi-analytical dependences between the geophysical model of the medium and the electromagnetic field [5]. The field sources were considered as vertical magnetic dipoles with momentum M for analytical calculations. In real geophysical devices, multi-turn coils with certain geometric sizes are used. In this paper, we show the dependence between the results of mathematical modeling of the electromagnetic field in the frequency domain on the choice of source approximation method.

2 Problem Statement

The behavior of the electromagnetic field in the frequency domain is described by the Helmholtz equation relative to the electric field in the domain Ω [4]

$$\text{curl}\mu^{-1}\text{curl}E + k^2E = -i\omega J,$$

where E – electric field [V/m], $\mu = \mu_r\mu_0$ – magnetic permeability of the medium [H/m], $\mu_0 = 4\pi \times 10^{-7}$, μ_r – relative magnetic permeability, $\varepsilon = \varepsilon_r\varepsilon_0$ – dielectric permeability of the medium [F/m], $\varepsilon_0 = 8.85 \times 10^{-12}$, ε_r – relative permittivity, σ – specific electrical conductivity of the medium [S/m], $k^2 = i\omega\sigma - \omega^2\varepsilon$ – wave number, $\omega = 2\pi f$ – cyclic frequency, f – source frequency [Hz], J - current in the source [A/m^2], which satisfies the charge conservation law $div J = 0$.

On the boundary $\partial\Omega$ of the computational domain Ω, homogeneous electrical boundary conditions are given

$$n \times E|_{\partial\Omega} = 0$$

where n is the external normal to the boundary of the computational domain.

At the boundaries Γ_{ij} between subdomains with different electrophysical characteristics the following continuity conditions are fulfilled.

$$[n \times E]|_{\Gamma_{ij}} = 0,$$

$$[n(\sigma + i\omega\varepsilon) \cdot E]|_{\Gamma_{ij}} = 0$$

The continuity conditions provide the charge conservation law in regions with discontinuous electrophysical characteristics. Consequently, the computational schemes for solving the Helmholtz equation should be designed so that the continuity conditions are fulfilled with a given accuracy. The vector finite element method allows one to fulfill these requirements if the vector basis functions are of full order.

3 Vector Variational Formulation

Let Ω be a three-dimensional region consisting of physically inhomogeneous subdomains with boundary $\partial\Omega$ - continuous by Lipschitz. Let us introduce the Hilbert space of vector complex functions as the following

$$H(curl; \Omega) = \left\{ v \in \left[L^2(\Omega)\right]^3 : curl\ v \in \left[L^2(\Omega)\right]^3 \right\},$$

$$H_0(curl; \Omega) = \{v \in H(curl; \Omega) : v \times n|_{\partial\Omega} = 0\}$$

with scalar product and norm

$$(u, v) = \int_\Omega u \cdot v \partial\Omega$$

Let us introduce the following variational formulation: For $J \in [L_2(\Omega)]^3$ find $E \in H_0(curl; \Omega)$ such that for $\forall v \in H_0(curl; \Omega)$ the next equation is fulfilled:

$$\left(\mu^{-1}curl\ E, curl\ v\right) + \left(k^2 E, v\right) = -i(\omega J, v).$$

The modeling domain Ω is partitioned into N non-overlapping finite elements Ω_k. The mesh T is defined as follows

$$\Omega = \bigcup_{k=1}^{N} \Omega_k, \forall\Omega_k, \Omega_j \in T, \Omega_k \cap \Omega_j = \emptyset, k \neq j.$$

It is necessary to introduce a finite-dimensional subspace $H_0(curl; \Omega)$ and define the basic functions for approximating **E** to formulate the discrete variational formulation.

Let us introduce a finite element space $H_0^h(curl; \Omega) \subset H_0(curl; \Omega)$ defining vector basis functions w_k^i on an element Ω_k and formulate a discrete variational formulation:
For $J \in [L_2(\Omega)]^3$ find $E^h \in H_0^h(curl; \Omega)$ such that $v^h \in H_0^h(curl; \Omega)$ is satisfied:

$$\left(\mu^{-1}curl E^h, curl v^h\right) + \left(k^2 E^h, v^h\right) = -i\left(\omega J, v^h\right).$$

The vector function $E^h \in H_0^h(rot; \Omega)$ can be represented by.

$$E^h = \sum_{i \in S} u_i w_i(\overline{x}),$$

where u_i are weights in the expansion, w_i are basis functions, S is a set of indices of degrees of freedom, $\overline{x} = (x, y, z)$-coordinates in the space R^3.

Substituting the expansion of the vector functions into the discrete variational problem, we obtain a system of linear algebraic equations (SLAE):

$$Au = f,$$

$$A_{ij} = \left(\mu^{-1}\text{curl}v_i, \text{curl}v_j\right)_\Omega + \left(k^2 v_i, v_j\right)_\Omega,$$

$$f_i = -i\omega(J, v_i).$$

The kernel of the curl operator, which leads to indefinite SLAE matrix A, and the lack of diagonal predominance are the reasons why the standard methods of solving SLAE are inapplicable. This leads to the stagnation of iterative methods for solving SLAEs.

To solve this problem, algebraic multilevel solvers are used. It allows us to overcome the peculiarities of the problem.

4 Source Approximation

Let us consider a plane coil of the area S with current I. The momentum of themagnetic dipole is used for analytical calculations of the electromagnetic field from such coil [5–7]:

$$M = I \cdot S \cdot n,$$

where n is the normal to the coil plane, I is the current in the coil proportional to the current in one turn multiplied by the number of turns in the coil (see Fig. 1).

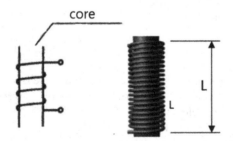

Fig. 1. Solenoidal coil with electric current.

The solenoidal coil approximation by a magnetic dipole is used when the field is investigated in the far zone from the source.

In well logging problems, the measurement of the induced EMF takes place in the same well, i.e. near the field source. In that case, the approximation of the coil by a magnetic dipole may not be correct.

Let us consider approximation approaches of the multi-turn coil regarding its geometrical sizes. A multi-turn solenoidal coil has a core with a diameter 2R, on which k turns of the conductor are tightly wound. The height of the winding is L.

A such coil can be represented as a thin single loop of radius R with current strength, $\tilde{I} = I \cdot k$, where I- the current in one coil turn [8].

Let's represent the source with δ-function then the current density is given by the formula

$$J(\overline{x}) = -i\omega\tilde{I}\delta(x - x_t, y - y_t, z - z_t)\tau,$$

where $\tilde{I}$ is the current in the circuit, τ is a unit vector co-directed with the current density vector. The points with coordinates (x_t, y_t, z_t) belong to the mesh edges approximating the generator coil loop.

Another way to approximate the coil is to represent the winding as a surface through which the current flows with linear density $J_S = (I \cdot k)/L$. Then the current density is given by the formula

$$J(\overline{x}) = -i\omega J_s\delta(x - x_t, y - y_t, z - z_t)\tau,$$

where τ is a unit vector co-directed with the current density vector. The points with coordinates (x_t, y_t, z_t) belong to the mesh faces approximating the surface of the generator coil.

5 Approximation of a Coil with a Current

Mathematical modeling of the electric field by the vector finite element method uses a tetrahedral mesh to approximate the modeling domain. Let us consider the problem of determining the EMF in receiver coils calculated for circular and square coils with current and circular and square receiver coils on a set of refining meshes over a weakly-conducting half-space. An important factor affecting the accuracy of the obtained results is the size of the tetrahedrons in the mesh. Since tetrahedral meshes are used, it is impossible to talk about the nestedness of the refining meshes.

We will compare the EMF obtained in the receiving coil with the theoretical values calculated by the approximate formula.

$$Re(\varepsilon) \approx \omega\mu_0\frac{M_rM_T}{4\pi r^3}\left(\frac{p^2}{4} - \frac{4p^3}{15\sqrt{2}} + O\left(p^5\right)\right),$$

$$Im(\varepsilon) \approx \omega\mu_0\frac{M_rM_T}{4\pi r^3}\left(-1 - \frac{4p^3}{15\sqrt{2}} + \frac{p^4}{8}O\left(p^5\right)\right),$$

where r is the distance between the generator and receiver coils, M_r, M_T are the momentum of the receiver and generator coils, $p = r\sqrt{\omega\mu_0\sigma}$.

Circular coil: generator coil has radius 0.1 m, receiver coil has radius 0.025 m. Square coil: generator coil side is 0.2 m, receiver coil side is 0.05 m (see Fig. 2). Distance between centers of generator and receiver coils 1.5 m. Current 1 A. Frequency 1 MHz. The conductivity of the half-space is 0.01 Sm/m.

The Fig. 3 shows plots of imaginary part of the EMF on the refining meshes. The partition size of the generator and receiver coils is shown in the Table 1.

Table 1. Characteristics of mesh partitions

N	h_c generator, m	h_c receiver, m	h_s generator, m	h_s receiver, m
1	0.01	0.005	0.1	0.01
2	0.005	0.005	0.05	0.01
3	0.002	0.005	0.025	0.01
4	0.001	0.005	0.02	0.01
5	0.001	0.001	0.01	0.01
6	0.0005	0.001	0.01	0.005

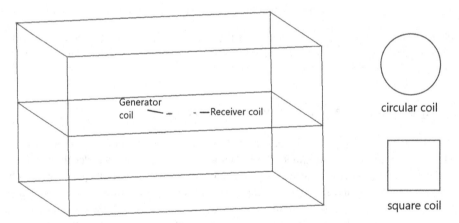

Fig. 2. Half space domain with generator and receiver coils.

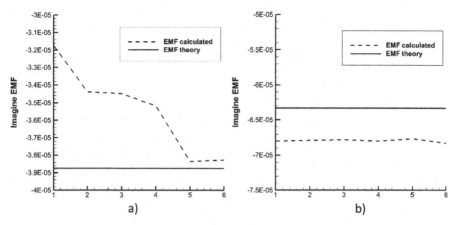

Fig. 3. Imaginary part of induced EMF in circular (a) and square (b) loops.

It can be seen that for a square coil, the number of mesh edges of the generator and receiver coils has no significant effect on the induced EMF. The relative error with the theoretical values is 7%. For circular coils, the choice of the characteristic size of the circle partition is decisive for the accuracy of the calculated EMF. On meshes 1–4 the generator coil partition is refined. The calculation accuracy of induced EMF thus increases from 82 to 91%. Additional refinement of the receiving coil allows us to obtain a difference between the theoretical and calculated data of 1%. Further refinement of the mesh in the source area does not improve the solution. The number of edges in the circular coil approximation affects the calculated electric field. In the coarse partition of the circle, the edges connecting the points of the circle outline a figure whose area is less than the original area of the circle. Thus, the momentum of the source is smaller, which is reflected in the electric field and the induced EMF in the receiver.

6   Calculation Results

Let us consider the problem of electrical logging in the well. The computational domain consists of a well with a solenoidal generator coil and receiver coils, the host medium, and the weakly-conducting formation (see Figs. 4 and 5).

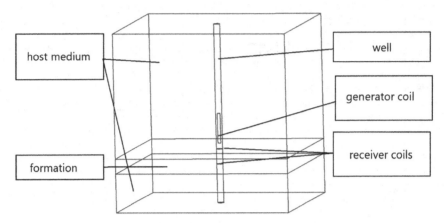

Fig. 4. Simulation area.

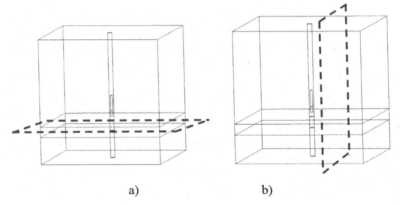

a)   b)

Fig. 5. Cross-section of the calculated area: a) $z = -155$ m; b) $x = 0.6$ m.

The following geometrical parameters of the medium were used for the calculations: well diameter is 0.2 m, generator coil diameter is 0.1 m, 60 turns of generator coil, the thickness of one turn is 1 mm, winding height is 60 mm, the diameter of receiving coils is 0.01 m. Receiving coils contain one turn. The current is 1 A, source frequency is 1 MHz. 4

Generator and receiver coils are coaxial, i.e. their centers are on one vertical axis. The first receiver coil is 0.5 m below the generator one, the second coil is 0.75 m below the generator. The system of generator and receivers will be shifted relative to the formation to investigate the induced EMF in the receiver coils. The thickness of a weakly conductive layer is 0.5 m, and the depth of the formation is from 0.9 to 1.4 m (Table 2).

Table 2. Electrophysical parameters of the medium

Domain	ε_r	μ_r	$\sigma\,[Sm/m]$
Drilling liquid	1	1	0.5
Host medium	1	1	0.1
Formation	1	1	0.01

We considered the distribution of the electromagnetic field for two approximation methods in the cross-section xOy ($z = -1.155$ m), passing through the center of the weakly conducting formation (see Fig. 3a). Figure 6 shows the distribution of components E_x and E_z of electric field for two methods of source approximation.

The real components E_x of electric field (see Fig. 6 a, b) are close in absolute value but differ in behavior. The imaginary components coincide in field amplitude. For source

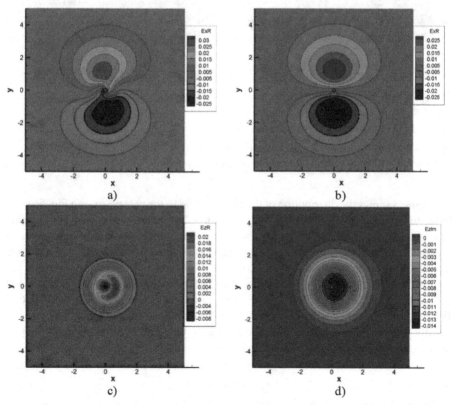

Fig. 6. Distribution of the real component E_x in the cross-section xOy ($z = -155$ m): a) the real component E_x at the source set as a surface, b) the real component E_x at the source set as a single loop, c) the real component E_z at the source set as a surface, d) the imaginary component E_z at the source set as a surface.

approximation, by surface, the lines of equal level have a small slope relative to the line $y = 0$. The real component E_z for the field at source approximation by a current surface has extremes near the well axis and distorted lines of equal level. In the case of source approximation by one loop, the field near the well is homogeneous. The component E_z at the source given by the loop is equal to 0. The component E_z at the source given by the surface is non-zero (see Fig. 6 c, d). The distribution of the E_z field component is also not symmetric with respect to the well, and there is a shift of the extremums to the right.

Figure 7 shows the distribution of the real component H_x in the formation plane for two ways of the source approximation. The field picture does not depend on the approximation method, but the maximum field values differ by two orders of magnitude.

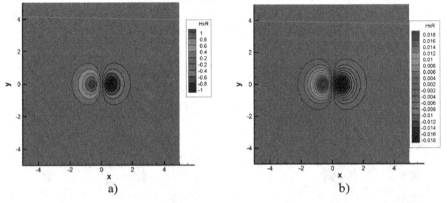

a)   b)

Fig. 7. Distribution of the real component H_x in the cross-section xOy ($z = -1.155$ m) with the source set as a surface (a) and as a single twisted loop (b).

Figure 8 shows the distribution of the electric field in the yOz plane (see Fig. 5b), the dotted lines indicate a weakly conducting formation. The behavior of the real components E_x of the electric field (see Fig. 8 a, b) are close in module, but differ in behavior, and imaginary components coincide. The electric field obtained from the source as a surface is concentrated in the well projection. Extremes of the field calculated from the loop are outside the well projection. The component E_z for the source given by the loop is equal to 0. The component E_z for the source given by the surface is non-zero (see Fig. 6 c, d). The picture of the distribution of the nonzero component of the E_z field allows us to identify the weakly-conducting formation in the vertical cross-section for both the real and imaginary components.

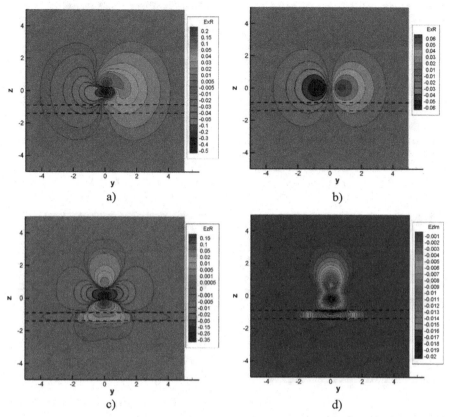

Fig. 8. Distribution of the electric field components in the cross-section yOz (0.5 m from the well axis): a) the real component E_x at the source set as a surface, b) the real component E_x at the source set as a single loop, c, d) the real and imaginary component E_z at the source set as a surface.

Figure 9 shows the imaginary components of the magnetic field in the vertical cross-section. The field distribution for both methods of source approximation is similar. The magnetic field responds to the presence of a weakly conductive formation by distorting the field lines. The amplitude of the magnetic field in the coil surface approximation is several tens of times greater.

The magnetic field is less responsive to the source approximation technique than the electric field in terms of equal-level line distortion. For the electric field in the formation plane, the amplitude of the field for different approximations is of the same order. In the vertical cross-section, the electric field near the source represented by the surface is three times greater than the field for a single loop in the same cross-section. Approximation of the coil by a surface generates a vertical component of the electric field, which does not appear in the case of a single loop. The magnetic field amplitude is more sensitive to the approach to the source approximation. The magnetic field components from the surface field source are two orders of magnitude greater than the corresponding magnetic field components from the loop field source.

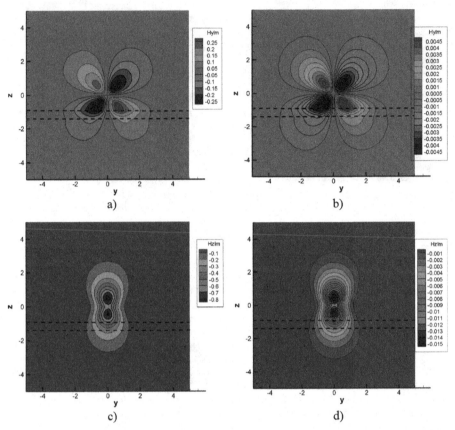

Fig. 9. Distribution of imaginary components of the magnetic field H_y, H_z in the cross-section yOz (0.5 m from the borehole axis) with the source set as a surface (a, c) and as a single loop (b, d).

Let us consider the change of induced EMF in the receiver coils when the generator-receiver system moves relative to the weakly conducting formation (see Figs. 10, 11 and 12).

As can be seen from Figs. 10 and 11, the imaginary part of induced EMF changes weakly depending on the position of the receiver coil and the weakly conductive formation. The real part of the induced EMF varies in a wider range and there is a slight increase in the value of the real component of EMF before the upper boundary of the formation to the approximation of the source through the conductive surface in the first (near) receiver coil. The second (far) receiver coil responds weaker to the presence of the formation and the magnitude of induced EMF in the far coil is about half as much as in the near coil. Both receiver coils show a decrease in EMF as the receiver coils pass inside the formation. The minimum of the real component of EMF is around the lower boundary of the formation. The lower boundary of the formation also distorts the curve of the imaginary component of EMF in the far receiving coil. Approximation of the source as a current loop gives a greater response of EMF for all measurements except for the real component in the first receiver coil near the upper boundary of the formation and when approaching the lower boundary of the region. Inside the weakly conductive formation, the EMF from the current loop has a larger amplitude than the EMF from the current surface.

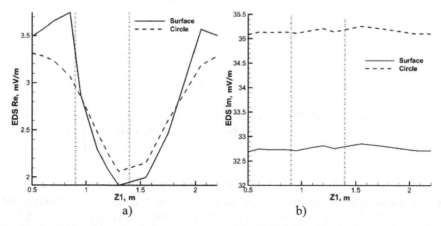

Fig. 10. Real (a) and imaginary (b) components of the EMF induced in the first receiver coil for two methods of source approximation. The vertical dotted line indicates the boundaries of the formation. z1 is the depth of the first receiver coil.

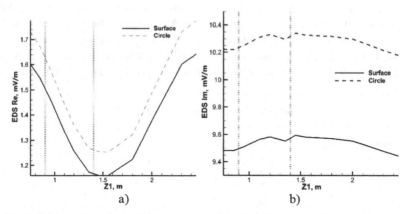

Fig. 11. Real (a) and imaginary (b) components of the EMF induced in the second receiver coil for two methods of source approximation. The vertical dotted line indicates the boundaries of the formation. z1 is the depth of the second receiver coil.

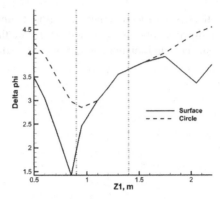

Fig. 12. Phase difference of EMF induced in the receiver coils for two methods of source approximation. The vertical dotted line indicates the boundaries of the formation. z1 is the depth of the first receiver coil.

Figure 12 shows the phase difference of the induced EMF in the receiver coils for the two source approximation methods. Inside the weakly-conducting formation, the phase difference for the two methods gives the same values when approaching the lower boundary of the formation. Near the first boundary of the formation, the phase difference of the EMF obtained from the surface approximation has a larger gradient than in the current loop approximation. This is due to a larger difference between the real parts of the EMF in the near and far receiving coils.

When using finite element methods, the most time-consuming procedure is the finite element system of linear algebraic equations (SLAE) solution. In this paper, OpenMP technology for matrix-vector multiplications is used to speed up this process. Table 3 shows the dependence of time to solve a system of linear algebraic equations (SLAE) on the number of CPU cores (see Table 3). The dimension of SLAE is 926016 unknowns.

Table 3. Solution time of SLAE [s].

Approximation methods	Core numbers				
	1	2	4	8	16
Loop	889	451	237	125	66
Surface	1304	665	342	178	93

The calculations were performed on the cluster "Information Computing Cluster of Novosibirsk State University" on nodes with 24 cores. Table 3 shows that the time decreases approximately twice with the increasing number of cores. This pattern decreases with the increasing number of cores, i.e. the dependence is not linear. This happens because only matrix-vector multiplication is parallelized, while other procedures are performed sequentially. Modeling using a surface field source takes longer since for this type of source the field has a more complex structure than for a single loop.

7 Conclusion

The paper proposes two ways of approximating the multi-turn generator coil used for field excitation in well-logging problems. It was proposed to approximate the multi-turn coil by one loop with full current and current surface, which preserves the height of the multi-coil winding. The mathematical modeling of the electric field in the well in the presence of weakly conductive formation showed that approximation by the current surface allows us to identify the formation by the field pattern in the vertical cross-section. The vertical component of the field appears through the formation. The magnetic field calculated for the source approximated as a surface has a much larger amplitude than when we use a single loop representation. The induced in the receiving coils EMF has a smaller amplitude for the current surface. The real component of EMF in the first receiving coil (closest to the generator) reacts to the presence of the formation medium boundary more than the second receiver coil.

Acknowledgements. The research was supported by Project No. FWZZ-2022–0025.

References

1. Epov, M.I., et al.: Modern algorithms and software for interpretation of resistivity logging data Geodynamics & Tectonophysics (2021). https://doi.org/10.5800/GT-2021-12-3c-054 6,669-682
2. Epov, M.I., Nikitenko, M.N., Glinskikh, V.N., Sukhorukova, K.V.: Numerical modeling and analysis of electromagnetic logging signals during drilling. NTV Logging **11**(245), 29–42 (2014). (in Russian)
3. Ratushnyak, A.N., Teplukhin, V.K.: Theoretical and experimental foundations of induction methods of well studies. Yekaterinburg, UrO RAS (2017). (in Russian)
4. Webb, J.P.: Edge elements and what they can do for you. IEEE Transaction on magnetic **2**, 1460–1465 (1993)
5. Kaufman, A.A., Morozova, G.M.: Theoretical foundations of the method of probing the formation of a field in the near zone. Nauka, Novosibirsk 125 (1970). (in Russian)
6. Casati, D., Hiptmair, R. and Smajic, J.: Coupling Finite Elements and Auxiliary Sources for Maxwell's Equations. Zurich Switzerland (Research Report No. 2018–13), 10 (2018)
7. Zhou, N.-N., Xue, G.-G., Wang, H.-Y.: Comparison of the time-domain electromagnetic field from an infinitesimal point charge and dipole source. Appl. Geophys. **10**(3), 349–356 (2013). https://doi.org/10.1007/s11770-013-0387-z
8. Chew, W.C., Kong, J.A.: Electromagnetic field of a dipole on a two-layer earth. Geophysics **46**(3), 309–315 (1981)

Computing Technologies in Data Analysis and Decision Making

Novel Planning Technique for ERP Systems Implementation Projects

Dmitry Yu. Stepanov$^{(\boxtimes)}$ (iD)

MIREA, Russian Technological University, Vernadskiy Avenue 78, 119454 Moscow, Russia
mail@stepanovd.com

Abstract. The article is devoted to the proposal of a new way to build a resource plan for ERP-system implementation projects, which has ease of use and speed of obtaining results, and this in turn allows you to flexibly emulate various situations. The method is applicable to corporate system implementation projects in which the software development is carried out. The proposed method uses the mechanism of the RICEF estimator, which is an expert assessment of the labor costs for preparing and developing programs for a pair of parameters «type of development – complexity». In the future, benchmarking is used, which determines the duration of the project phases in percentage terms. Based on the calculated efforts that define the stages of design and build, and the results of benchmarking, the timing of all remaining phases is calculated. The distribution of labor costs for the remaining stages of the project is carried out according to the pyramid principle, for which milestones of the analysis, integration testing and cutover are processed. The final plan of human resources is based on the calculated labor costs set for all stages of the project. Consequently, creation of the project schedule, determination of the scope of work for all phases, as well as finalization of the resource plan are carried out using only the planned labor costs of design and build stages.

Keywords: Resource plan · RICEF · Benchmarking · Pmbok · Corporate information systems · ERP-systems

1 Introduction

Having decided to link your life with a career as a consultant of corporate information systems implementation (hereinafter – ERP systems) [1], it does not matter American, German or Russian production, there may be a mistaken feeling that you should certainly be able to program and only your technical skills play a role. This is not quite true. The beginning of corporate information system implementation projects, which seems to be a very non-trivial task, is always preceded by a fairly large number of work, which is indirectly related to technology [2]. From the project management point of view, the beginning of the ERP-system implementation is almost final and fairly predictable activity. The stages of analysis, design, build, testing, cutover, as well as post go-live support are typical of the project of implementing corporate systems (Fig. 1). Usually, the analysis stage is preceded by the activity of creating a technical task justifying the feasibility of implementing an ERP-system, as well as the key requirements for it, which beginners may not know due to lack of experience [3]. Next, technical and commercial proposals are formed, which describe what stages the project will consist of, the duration of the stages, the amount of necessary human resources, as well as how the requirements will be covered [4]. The contract for the implementation of an ERP-system is concluded only in case of successful sales, followed by just those six stages of the project that we have described above.

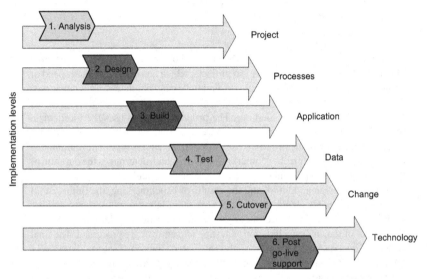

Fig. 1. Typical six stages of corporate information system implementation projects.

2 Problem Statement

Building a resource plan is an important issue in the preparation of technical and commercial proposals, it determines the necessary amount of human resources for the project. Project timeline, the amount of human resources, as well as the composition of the expected work are closely related, which can be set by the following simplified formula:

$$Project\ duration = Total\ efforts/Number\ of\ human\ resources. \qquad (1)$$

The preparation of resource plan should be carried out in accordance with the content of the project, since changing the content will lead to an adjustment of the duration, which is easy to see in (1). A natural question arises, how to build a resource plan? The project management body of knowledge (hereinafter – PMBoK) talks about critical path and chain methods that allow you to build a logical structure of the work performed, set their duration, assign human resources for tasks and fix time buffers [5]. PMBoK describes project work «from the bottom up», that is, the final value is the sum of many individual subtasks. The necessary amount of resources is determined by calculation of the duration and scope of work in (1), after which they are aligned and smoothed in accordance with the stages of the project. Both methods are time-consuming and, of course, not flexible, but do not require deep technical knowledge. The created project plan is not of great value, the value lies in its daily adaptation to the prevailing realities, it is idea that can be found in the work [6], devoted to Agile planning. Really, the method should perform calculations quickly in order to provide modeling of a variety of possible situations, one cannot disagree with it. In practice, the application of the methods described both in [5] and in other literary sources similar in content [7–9], which provide a general understanding of building a resource plan, is very laborious and problematic. The purpose of this paper is to propose a faster way of building a resource plan for ERP projects. Achieving the goal requires consideration of the following tasks:

- overview of ways to evaluate efforts for program development;
- analysis of the duration of stages for ERP-system implementation projects;
- quick creation of a resource plan for ERP projects.

3   Evaluation of Development Efforts

From the formula (1) it is easy to conclude that finding the number of human resources requires determining the timeline of the project and the amount of work performed. Let's rephrase: it means building a resource plan requires calculating the duration, as well as the labor costs of each stage of the project. Usually, the customer provides a potential contractor with a technical task even before the start of the ERP-system implementation project, which specifies a list of the user's initial requirements for the future information system. Each of the requirements is analyzed by the contractor's experts to be classified as Fit or Gap [10, 11], which is a mandatory step for preparing a preliminary ERP project plan. The requirements from the Fit area are considered already implemented by default in the basic package of ERP-system and do not require additional efforts, on the contrary, the Gap area serves as a signal about the need to refine or configure the information system (Fig. 2).

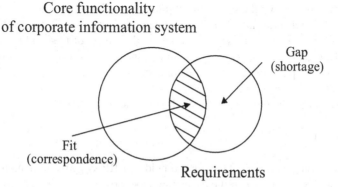

Fig. 2. Fit/Gap-analysis schema.

The type of development and complexity are determined for each Gap requirement. RICEF classification is used for it, where R is report, I – interface, C – dataprocessing program (i.e. convention), E – anextension of the standard functionality of ERP-system (i.e. enhancement) and F – printedform. The complexity of the development is characterized by such values as: low, medium, high, very high [12]. The matching of the two parameters «type of development – complexity» is performed for each Gap requirement. The contractor's company has efforts determined by experts, necessary for the preparation of project documentation and the development of a program for all possible values of these pairs. Thus, RICEF estimator (Fig. 3) is a matrix linking the type of development and its complexity, as well as the planned efforts to prepare technical documentation and software development.

RICEF type	Complexity	Efforts for documents preparation (man-days)	Efforts for building (man-days)
Report	Very high	7	15
Report	High	5	10
Report	Medium	4	7
Report	Low	3	5
Interface	Very high	10	25
Interface	High	7	20
Interface	Medium	5	15
Interface	Low	4	10

Fig. 3. Fragment of a RICEF estimator.

4   Benchmarking of the Stage Durations

Stages allow you to group the same type of project tasks into a logical sequence of work. In fact, the stages of corporate systems implementation are a set of phases inherited from the software development lifecycle, from which the stages of idea and termination are excluded. The content, quantity, and the name of the phases strongly depend on the type of ERP project, in particular, they are distinguished:

- implementation «from the scratch», including pilot projects;
- rollout;
- development.

If we summarize the knowledge in the field of ERP-systems implementation, then the typical stages, regardless of the project, will be those given in Fig. 1. Of course, the specifics of projects dictate their own features, for example, pilot projects often include a pilot operation stage, but this does not significantly change the course of the project, so the typical six stages remain virtually unchanged.

Literature sources devoted to ERP projects [1–4], unfortunately, give only a general estimate of the duration of such projects. The implementation of ERP-systems from one year or more is considered the norm, but it is problematic to find more detailed information on the duration of each phase. Therefore, as statistical information, we will use the Russian projects mentioned in [13, 14]. And based on them, we will try to formulate the recommended duration of the stages. This information is very important to us, without it we will not be able to build a human resource plan.

The RICEF estimator provides the calculation of efforts at the design and build stages. However, the duration of these two stages remains unknown. Let's take a closer look at all stages of ERP-system implementation projects. Statistics show that the sum of the design and build stages is at least 50% of the duration of the entire ERP project. Post go-live support stage is not included in the evaluation, since its duration is a constant agreed with the customer. Then it is possible to calculate the duration of all stages of the project in percentage terms using proposed benchmarking mechanism (Fig. 4).

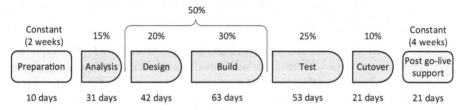

Fig. 4. Benchmarking of the stage durations for ERP-system implementation projects in percentage terms, as well as the approximate duration of the phases for a typical case.

So, the duration of the design and build stages is half of the project, and their percentage also varies. Usually the build stage lasts at least one and a half times longer than the design stage. The durations of the design and testing stages are comparable. The cutover stage is the shortest. Thus, you can find out the duration of the entire project by calculating the duration of at least one of the stages. The existing efforts for the preparation of project documentation and the development of programs calculated upon RICEF estimator will help us in this.

5   Building a Resource Plan

And now we will propose an algorithm for building a resource plan, knowing the efforts, as well as the preliminary duration of the stages of the ERP project. In our example, we will use the labor costs of the design and build stages, calculated based on the RICEF estimator, equal to 113 and 163 man-days, respectively. Then the procedure for building a resource plan will be as follows:

- it is required to set the initial duration of the design stage, considering the data in Fig. 4;

- then we impose the efforts of functional consultants found by RICEF estimator on the calendar grid. If the amount of labor costs exceeds the duration of the phase, you should add an additional human resource.The duration of the design stage and overall efforts of the functional consultants should coincide or differ by a minimum amount. Since calculation of the number of resources depends on the duration of the stage, we are guided by the rule that this number should be minimal. If it is not possible to select multiple resources, it is allowed to change the initial duration of the phase in greater or lesser directions.Selecting the optimal ratio of the phase duration and resource allocation, we find their final values;
- found duration of the design stage, equal in our example to 36 working days, allows us to find the duration of the entire project using a percentage of the benchmarking results (Fig. 4);
- we determine the number of developers, the same procedure is used as described in step 2, with the only exception that timing of the build stage has already been found on step 3;
- we will distribute the found number of consultants and developers to the remaining stages of the project, but exclude the analysis stage for developers, since they are not involved in it. We get a preliminary resource plan containing the total amount of effort in 1038 man-days (Fig. 5a).

Using the proposed procedure, we were able to build a resource plan for the corporate system implementation projects from the start date of the project. In this case, there is no limitation on the duration of the project. If the customer explicitly fixes the start and end dates of the project, the proposed scheme is also working, but you will need to reduce the duration of the design phase already at the 1^{st}step. After calculating the duration of all phases of the project, i.e. the duration of the entire project, it is compared with the start and end dates set by the customer. If necessary, the procedure is repeated many times, until the values of the phase durations satisfying the requirements are found. It is important to emphasize that the shorter project timeline, the more resources will be involved. The latter generates more likely risks of failure compared to a project of longer duration, but with less involvement of human resources.

The example of the plan in Fig. 5a seems redundant, in particular: the participation of not all functional consultants and developers is necessary at all stages of the project.

There is a hypothesis saying that the maximum amount of resources is required at the most critical stages of the project. In some literary sources [10, 15], this statement is called the pyramid principle. For us, these are the design, build and testing phases. Then we are able to propose the formula (2):

Number of resources at current stage $=$ RoundToInt (Number of resources at previous or subsequent stage/2)

$$(2)$$

which will reduce the number of people in functional groups, considering the number of resources involved in the previous or subsequent stages. The assessment is empirical in nature, based on such project activities as: analysis, integration testing and technical cutover.

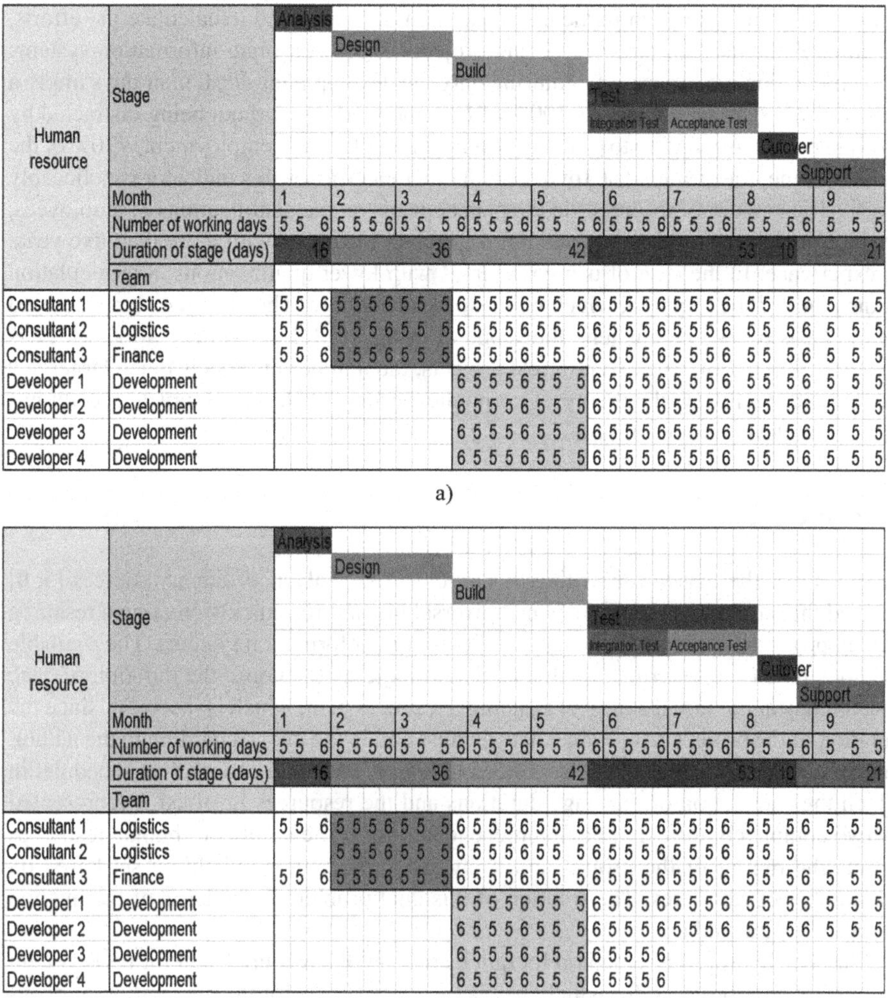

Fig. 5. Example of the human resources: a) preliminary redundant plan; b) final shortened plan.

We can apply (2) and clarify the number of.

- consultant resources at the analysis stage regarding the design phase;
- consultants at post go-live support stage regarding the cutover phase;
- developers in the acceptance testing, cutover, and post go-live support phases regarding to the build stage.

As a result, the volume of labor costs decreased by 15% and amounted to 887 man-days (Fig. 5b). Thus, it was possible to significantly reduce the involvement of team members in the ERP-system implementation project. The resource plan is based on a number of assumptions, in particular: ideal man-days are used to calculate the efforts, qualified personnel are considered, and projects of the corporate information systems implementation are processed, requiring only the development. Ideal man-days imply a situation where a person spends 100% of his time working without being distracted by any other tasks. According to [16], the best indicator of person employment is 70% of the working time, for example, at Toyota, at all other enterprises this indicator is noticeably worse. The amount of resources in the plan depends on the qualifications of employees, so the calculation is mainly carried out for human resources with more than five years of experience. In the case of using resources with lower qualifications, a recalculation of the plan will be required, which will lead to its increase. The resource plan is formed on the basis of two parameters: the scope of design and development work, which is its current limitation. However, in real projects, the number of tasks is much larger, for example, if we follow the theory of corporate systems [17], then a typical project has more than eleven groups of tasks.

6 Conclusions

So, what was the reason for writing this article? The analysis of literary sources [5, 6] showed that at the moment there is no universal method for quickly creating a resource plan in projects for the implementation of corporate information systems. The available methods based on the critical path and chain oblige us to choose the duration of each task already at the initial stages of building a resource plan, which is incorrect, since the duration of the project is the value that we must find in the end. According to the author, the creation of a resource plan is reduced to the construction of a project schedule, in which the calculation of the stage durations and the resources involved is represented by two completely different, but interdependent tasks. The number of human resources affects the timing of the project and, conversely, the timing should be set based on the number of resources. The solution of this dual problem is the basis of the method proposed by the author.

The article proposes a method for quick creation of a resource plan for ERP-system implementation projects, for which the planned labor costs for the preparation of documentation at the design stage, as well as the development of programs at the build stage are used as input parameters. Labor costs at the design and build stages are determined using the RICEF estimator, while benchmarking allows you to determine the duration of all remaining stages of the project, knowing the timing of the design phase. Thus, the method is two-parametric. The proposed method ensures the creation of a human resources plan from the start date of the project. To form a plan from the end date of the project, it is necessary to shorten the duration of the design stage and increase the number of its human resources. At the same time, it should be taken into account that the shorter duration of the phase, the higher risks of non-fulfillment of tasks, regardless of the increase in the number of resources [18].

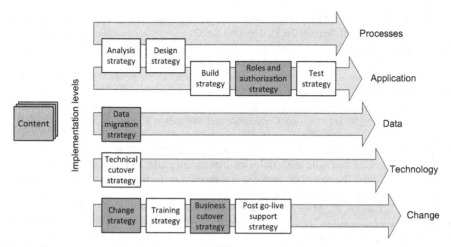

Fig. 6. Parameters determining the content of ERP-system implementation projects according to the theory of corporate information systems.

Further development of the method consists in refining and automating the algorithm for reducing resources (2), as well as including actions not related to design and build in the scope of calculation. In the first case, you can apply the concept of entropy from information theory [19], in the second, you should use the theory of corporate information systems [17], which will determine an additional set of parameters responsible for the content of the project, namely: data migration, business cutover, roles and authorization, as well as change management (Fig. 6). Other parameters shown in Fig. 6, have already been included in the two-parameters method. The use of extra parameters will make it possible to transform the method of constructing a resource plan into a more complex algorithm, in which a larger number of project tasks will be taken into consideration.

References

1. Samara, T.: ERP and Information Systems. Integration or Disintegration. Wiley, Hoboken (2018)
2. Habadi, A.: An introduction to ERP-systems: architecture, implementation and impacts. Int. J. Comput. Appl. **167**(9), 1–4 (2017)
3. Stepanov, D.: Analysis, design and development of corporate information systems: theory and practices. Russ. Technol. J. **8**(3), 227–238 (2015). (in Russian)
4. Ostroukh, A.V., Surkova, N.E.: Information system design. Lan (2019). (in Russian)
5. Shirenbek, H., Lister, M., Kirmse, S.H.: A guide to the project management body of knowledge: sixth edition. PMI (2017)
6. Cohn, M.: Agile Estimating and Planning. Pearson, London (2005)
7. Polkovnikov, A.V., Dubovik, M.F.: Project management. Full course MBA. Olymp-business (2018). (in Russian)
8. Heldman, K.: Project management jumpstart. Sybex (2018)
9. Tsiteladze, D.D.: Project management. Infra-M (2022). (in Russian)
10. Gvozdeva, T.V., Ballod, B.A.: Information systems design. Phoenix (2009). (in Russian)

11. Ali, M., Miller, L.: ERP system implementation at large enterprises – a systematic literature review. J. Enterp. Inf. Manag. **30**(1), 666–692 (2017)
12. Stepanov, D.: Preparing functional specification documents for corporate information system development on SAP ERP example (part 1). Corp. Inf. Syst. **3**(7), 29–52 (2019). (in Russian)
13. Kalyanov, G.: Consulting: from business strategy to corporate information system. Goryachya telecom line (2011). (in Russian)
14. Stepanov, D.Yu.: Using agile methodology in ERP-system implementation projects. In: Proceedings of the 35th International Conference on Information Technologies (InfoTech-2021), pp. 1–4 (2021)
15. Sudakov, V.A.: Corporate information systems. MAI (2016). (in Russian)
16. Blokdyk, G.: ERP and agile methodologies: a complete guide. 5STARCooks (2020)
17. Stepanov, D.Yu.: The theory of corporate information systems. In: Proceedings of International Russian automation conference (2022). (in press)
18. Laudon, J., Laudon, K.: Management information systems. Prentice hall (2002)
19. Mackay, D.: Information Theory, Inference and Learning Algorithms. Cambridge University Press, Cambridge (2003)

Development of an Automated Document Classification System with a Predefined Structure

Roman Semenov$^{(\boxtimes)}$ ⓘ and Alexey Sorokin

MIREA, Russian Technological University, Vernadsky Avenue 78, 119454 Moscow, Russia
9629790@gmail.com

Abstract. A variant of the algorithm of the software being developed is proposed, which automatically collects, analyzes and structures information into a convenient form for communication. The considered possibilities of conceptual representation and incoming information contribute to an increase in the speed of receiving a response. Due to the modeling based on the presented structure, the accuracy and unambiguity of the information will exceed human capabilities. Software has been created that provides the requested information on a specific subject area. The study of the possibilities of parallel analysis and modeling of the text of documents to increase the speed of the decision support system. Based on the achievements in the development of computing power, the methods of joint use of algorithms for processing text documents and algorithms for the conceptual representation of text modeling are described. The designed system is capable of processing a certain amount of information, comparatively, in less time than that in existing decision support systems. The article demonstrated the main stages of the analysis of the text of documents compatible with the conceptual structure for use in the processes of text modeling. A vector representation of the document information is considered, due to which it is possible to decompose the entire text into elements that will be analyzed. Calculations of the frequency characteristics of terms and the relative frequency of the document in question are considered in detail. The information obtained allows us to start creating an independent text processing system and bringing it to the concept. This design of the algorithms of the system will allow for the least short period of time to receive a response from the system that has just received the data. The designed system automatically selects the necessary ways to bring the text to the final form, through algorithmic structures of analysis and modeling. The decision support system, based on the algorithms presented in the article, conducts an appropriate analysis of terms and, based on the results obtained, assigns an index to the document to identify the main idea of the text and the relevance of the search query. Research in the field of the conceptual structure of the text allows you to more accurately determine the weight criterion of keywords and exclude from the list objects that less reflect the reliability of the information obtained from the document. Based on the data obtained, the system can be upgraded and receive constant support in the form of updates to the databases of relevance of search queries and calculation of criteria for weight characteristics. The designed decision support system can be implemented in large systems of Internet resources, used as an independent project that supports autonomous operation, through the introduction of dictionaries of words and rules with the possibility of updating, or for implementation in intelligent systems.

Keywords: Automatic text processing · Morphological analysis · Text classification · Conceptual structure · Neural network

1 Introduction

Every day, millions of requests on electronic resources are processed using various search, recognition and text processing algorithms. This allows you to find the necessary information among similar documents and those that do not contain it. The problem of the subject area of text processing and analysis is a huge flow of information, which is constantly increasing. To solve the problem of noisy text documents, new ways of identifying the main idea of texts and processing search queries are being developed. There are well-known algorithms that have been used since the discovery of this problem and are currently being used. The processing time of such algorithms is long and cannot be used to get a quick response. Now the capabilities allow you to use simultaneously a certain number of algorithms that are executed in parallel, which allows you to conduct a multi-sided analysis in the shortest period of time. It is necessary to identify the most effective methods that will interact with each other to increase efficiency. The basic idea of the model on which the algorithm used is based was proposed by Gerard Salton in 1973. In the vector representation of the information model, the meaning of the transmitted information is highlighted by a list of terms. By vector assignment of the number of terms to each analyzed document, such a list allows you to find the requested information. The image of the semantic representation of the document information for a search query depends on the frequency of use of the selected term and its relative weight. A simple model uses a binary vector, its values depend on whether a particular term belongs to a given context or not. The dimension of the vector D increases depending on the amount of textual information in the analyzed document. When processing a query, which is a text, the same analysis takes place on the terms of the search query and assignment of the vector q [1]. After the research, it is necessary to obtain a structure for comparing images of search queries and images of documents, which, after the algorithms presented below, will reveal the correspondence of the meaning of terms.

2 Representation of Document Information in Vector Form

It will take a lot of computing power and processing time to analyze a large volume of text contained in one or more documents. To simplify and accelerate the processing of the necessary material, an information flow is used. It is an array containing sets of documents. The matrix of the information array is shown in Fig. 1.

	Term 1	Term 2		Term j		Term D
Document 1	W_{11}	W_{12}	$\cdots$	W_{1j}	$\cdots$	W_{1D}
Document 2	W_{21}	W_{22}	$\cdots$	W_{2j}	$\cdots$	W_{2D}
	$\cdots$	$\cdots$	$\cdots$	$\cdots$	$\cdots$	$\cdots$
Document i	W_{i1}	W_{i2}	$\cdots$	W_{ij}	$\cdots$	W_{iD}
	$\cdots$	$\cdots$	$\cdots$	$\cdots$	$\cdots$	$\cdots$
Document N	W_{N1}	W_{N2}	$\cdots$	W_{Nj}	$\cdots$	W_{ND}

Fig. 1. Representation of the information flow in the form of a "document-term".

In order to fully define the vector model, it is necessary to identify how the weight of the term in the document will be determined in order to compile a complete image. Next, standard ways of setting the weighing function will be presented:

- The weight of Boolean algebra. The value "1" is assigned if the term is contained in the analyzed document; if it is missing, the value "0" is assigned.
- The frequency of the term (TF – term frequency). When determining the frequency of a term, its weight is determined. To do this, a variable is determined by the function, which depends on the number of appearances of the term in the analyzed document.
- The inverse frequency of the document (TF-IDF – term frequency – inverse document frequency). The value is calculated by the product of the function in which the number of uses of the term in the document is determined, and the function in which the inverse value of the number of documents in which the term is used is determined [2].

To create a vector image of the semantic content of the document, and further represent the concept of the analyzed text, it is necessary to correctly, quickly and unambiguously index the incoming information. The first stage of indexing is the selection of a range of terms that best characterize a particular selected document. This is due to the problem of clutteriness of the information flow. Now, a large number of documents containing a large amount of textual information are freely and publicly available. Each search query processes documents that contain the entered words and phrases, but in fact, the material found does not fully correspond to the one that was meant. Therefore, it is important to correctly select the terms of each analyzed document. This will bring us closer to solving the problem of the noisy information field on the Internet and will allow us to more quickly present the text for modeling a text response to a request.

3 Analysis of the Characteristics of the Frequency of the Term

To isolate indexed terms from a document, it is necessary to use patterns of frequency distribution of the occurrence of various words in statistical texts. For the decision support system and the display of indexed elements of interest to us, a hyperbolic pattern will be considered. The most well-known hyperbolic law, which relates to statistical text processing, is used in large-volume text samples [3]. The result of the analysis using this law is the definition of terms in an arbitrary document, the keywords are selected from the text for subsequent analysis. According to the law under consideration, the frequency of occurrence of a word will be determined by the expression:

$$(TF)_i = {}^{n_i}/_T$$

where n_i is the number of uses of a certain word t_i in the analyzed text.

T is the total number of words of the analyzed text.

The frequency of a term (TF) is the ratio of the number of occurrences of a particular word to the total number of words in a document. After identifying the words with the highest frequency, it is necessary to arrange the words of the text in descending order of their frequency of occurrence. This will allow you to immediately highlight the words most significant for an unambiguous description of the meaning in the document. The product of the frequency of the word $(TF)_i$ by the ordinal number of this word will be constant for any previously defined word t_i:

$$(TF)_i \cdot r_i = C,$$

where C is some constant;

r_i is the ordinal number of the frequency of the word.

The function $y = k/x$ is described by expression, the graph represents a hyperbola or a straight line in logarithmic coordinates, which are shown in Fig. 2.

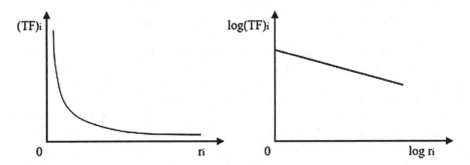

Fig. 2. Graphs of a function of the form $y = k/x$.

The words that are of the greatest interest for highlighting as keywords will be located in the middle part of the dependency graph, which is shown in Fig. 3.

Therefore, the most informative words will be those terms that occur not too often as the text progresses, but also not infrequently. This is explained by the fact that the most commonly used words will be prepositions and conjunctions, and rarely occurring words will be related to a specific section of the text and may not be related to the main idea of the narrative at all.

Rarely encountered terms give a small number of document and query matches, so it is impossible to consider them as keywords, since this will reduce the accuracy of document search. On the other hand, if a rare term really has to do with the general meaning of the document, then the document found by the keyword is highly likely

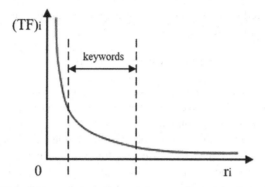

Fig. 3. Marking the boundaries of keywords according to the dependency graph.

to be relevant. Common terms have more weight and there will be a lot of relevant matches when comparing the terms of the request and the document. If we add to this an analyzer that excludes parts of speech that do not carry the main semantic load, then such a search will be even more accurate, which will be presented further when designing the concept of automatic text analysis. The words that are highlighted as keywords, namely the boundaries of the selections on the dependency graph, are an indicator of the accuracy of determining the meaning of the text and determine the quality of the search. In the existing methods of term allocation, a tool is used that excludes frequently occurring words that do not reflect the content of the document and are auxiliary and clarifying [4]. Such parts of speech are entered into a special dictionary of official and uninformative terms, called stop dictionaries, which obviously will not be indexed to improve the quality of indexing and keyword highlighting.

4   Relative Frequency of the Document

The frequency of occurrence of certain words in the text, called the absolute frequency, is effective with a relatively small amount of text, since when using a search query for one term, a huge number of documents containing the same word or phrase will be found. Other documents would also contain terms of absolute frequency and identical stop dictionaries, which would eventually lead to extremely low results of text analysis. Therefore, methods of reduction of relative frequency are used, which distinguish terms with a high frequency of use in this document using the same terms in the entire information array of the Internet. Models using relative frequency distinguish terms are more common in individual documents, and are used much less frequently outside of documents. This action allows you to display the content of a specific document and distinguish between different, analyzed, documents.

To calculate the document frequency in which the term t_i occurs, relative to the number of documents of the entire array, a weighing function is used, represented by the expression:

$$(IDF)_i = \log\left(N / (DF)_i\right)$$

where $(DF)_i$ is the frequency of occurrence of the term in the document, relative to the appearance of this term in the array of documents.

N is the number of documents that are in the information flow under consideration.

Using the above function, the largest values are assigned to those terms that appear only in a few documents. The more times the term occurs in the array documents, the lower the value of the inverse document frequency will be.

To specify the choice of the term in the analysis and assignment of the index, an assessment of the strength of the term is used. By force, we mean the individuality of the term, which characterizes documents as unlike as possible from each other. Keywords that only approximate documents that are difficult to distinguish from each other have the least power. The stronger the differences in the selected terms of individual documents, the easier it is to find certain documents. If the compared documents are represented by similar vectors of terms, then the indexing space is compressed, so it will be difficult to distinguish between relevant and irrelevant documents.

The significance of the strength of the term t_i is determined by its strength in the document $(DF)_i$. It is determined by the difference between the average values of the indicators of documents that are pairwise similar and when the term t_i is absent in the vectors of the array documents. It is also characterized by the average value of documents that are pairwise similar when the term t_i is present [5]. If the term in question is valuable for indexing, then having it in the vector makes the documents less similar to each other. This means that the average pairwise similarity of documents decreases, and the difference coefficient is positive. In a situation where similarity increases, the difference coefficient will be negative.

When indexing terms for each document, it is necessary to consider all the analyzed rules and algorithms. One of the main ways in which certain terms can be distinguished in a document is to distribute them according to the document frequency $(DF)_i$ and the frequency of occurrence of F_i. The occurrence of the term t_i in the array of documents is determined by the expression:

$$F_i = \sum_{k=1}^{N} (f_i)_k$$

The rules described below will determine the stages of the analysis of the automatic decision support system of text modeling when bringing the text information of the document to the concept. When indexing terms, more attention should be paid to such words that have the highest distinguishing power [6]. The terms with the average frequency of occurrence, denoted as F_i, and the document frequency, at which the total frequency in the information array is less than half of the frequency of occurrence of the term, will have the greatest strength. Less effective are terms with low documentary and total frequency, in which the strength will be close to zero. Terms with a negative force value are ineffective, which are determined by the high frequency of use in the information array. Their total frequency will be higher than the number of documents in the array. The conclusions are shown in Fig. 4.

The terms are arranged in order of increasing their document frequency, which is denoted as $(DF)_i$. The higher the document frequency, the larger the search query area for a particular term. The lower the frequency, the more accurately the term will determine the meaning of the indexed document [7]. Therefore, when analyzing such documents to bring them to an index division, it is necessary to choose as keywords terms with an average frequency that have term strength values above relative zero, denoted as DV.

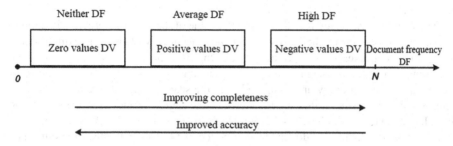

Fig. 4. Distribution of the document frequency of terms.

Figure 5 shows the different frequency distributions of terms depending on the amount of use of specific terms in the documents under consideration.

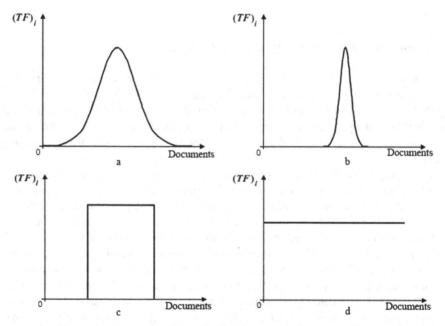

Fig. 5. Different representations of variants of terms depending on frequency.

The best terms for indexing are those with a smooth distribution, which is shown in Fig. 5a. The graph provides average values of search completeness and accuracy that best match the relevance of index assignment to terms for decision support systems. Graphical representations of the distribution of terms in Figs. 5b and 5c in one case increase the accuracy of the search, but reduce the area under consideration, and in the other, the search area is increased, but the accuracy is reduced. Figure 5d shows a uniform frequency distribution, which defines terms that do not have the necessary accuracy and area requirements [8]. Terms with this position prevail in the general array of information space and are not able to distinguish the meaning of a specific indexed document.

The next important step is to determine the value of the term of a previously defined frequency. The ability of the keyword that most accurately characterizes the content of the document determines its value for indexing [9]. The value is determined by the weighting factor that is assigned during the indexing process. The terms contained in the document search query are compared with the search image of the document and determined by how similar they are to determine the degree of relevance. Depending on the value of the term, which will correspond to the weight of the term of the search query image of the document, this document will be located higher in the list of results. The weighting of terms is necessary for the compilation of search images of text modeling documents. When analyzing the frequencies of terms, the described set of criteria allows for the selection of indexed terms [10]. The criteria for selecting terms for indexing are the weights of the terms. Keywords selected by frequency, but received the least weight, are not considered further, since the number of indexed terms for each document is limited.

5 Presentation of the Conceptual Structure

The text modeling decision support system uses a concept to calculate the weight criteria, according to the structure of which the most valuable terms used in the text are highlighted. A concept is a logical structure that allows evaluating information using morphological analysis. The conceptual structure presented in Fig. 6 demonstrates the interrelationships of elements to determine the greatest weight of each analyzed section of text.

Each element of the structure correlates with a set of words of specific parts of speech. Word sets combine parts of speech by morphological features, syntactic role in a sentence and a common grammatical meaning [1, 11]. A pair of characteristics < parameter name, parameter value> is called a morphological parameter. The parameter is characterized by gender, number, tense, declension and other signs of words accepted in the language used [12]. The parameter value is a specific value that this attribute can take. So, for example, gender can be average, masculine, feminine, several parameters can be compared in one word form [5]. After obtaining the normal forms of words with a set of selected parameters, the analysis of words in the text is carried out, which, after frequency analysis, turned out to be the most relevant [13]. This algorithm allows indexing terms in a document with the least number of errors and inaccuracies and preparing text components for deeper analysis for a text modeling system.

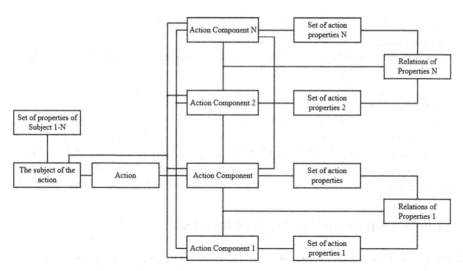

Fig. 6. Conceptual structure based on situational analysis.

6 Calculation of Weight Coefficients of Text Analysis Models

To compile a frequency model of terms of a single text, it is necessary to identify and calculate the weights. The frequency model of term weighting is used in the frequency indexing method [14]. The weight function given by expression will make it possible to calculate the weight characteristics of terms:

$$TG - IDF = W_i = (TF)_i \times (IDF)_i,$$

where W_i is the weight attributed to the term.

(TF)$_i$ is the frequency of the term in the document.

(IDF)$_i$ is the inverse document frequency.

After calculating the weights of terms, it is necessary to compare the values and identify the keywords with the highest value. The most relevant will be some terms, the number of which depends on the volume of the analyzed document; they will have values much higher than the rest.

Information that is more accurate can be obtained by following the algorithm for calculating the probabilistic model for estimating the weights of terms. This algorithm is based on the assessment of the probability that the document corresponds to the search image and is relevant. In comparison with the frequency weighting method, which can only be used to formally make an abstract calculation, without taking into account information needs [15], the probabilistic model compares terms with search images of documents and is represented by expression, which uses Bayes' theorem:

$$P(w_1|d) = \frac{P(d|w_1) \cdot P(w_1)}{P(d)}$$

where $P(w_1|d)$ is the probability of an event in which a certain document is relevant to the search query.

w_1 is a randomly selected document to match the search query.

d is the document being analyzed.

$P(w_1)$ is the probability of a randomly selected document being relevant to the search query.

$P(d)$ is the probability that the analyzed document is selected from the entire set of documents for consideration.

$P(d|w_1)$ is the probability that the analyzed document is selected from a set of relevant documents.

Based on the results of a certain probability, it is necessary to select the terms with the highest value [16]. The value indicates the probability of an event in which a document indexed by a specific term will be relevant for the corresponding search query.

7 Implementation of the Analysis Results

There are two strategies for applying pre-trained language representations for subsequent tasks: function-based, which contains separate, pre-trained representations for specific tasks, and point-based, which assumes a more accurate adjustment of the weights of the entire neural network to the required task. These two approaches have the same objective function during pre-training.

To implement the previously described structure, we use BERT (Bidirectional Encoder Representations from Transformers – bidirectional encoder). There are two stages: pre-training and fine-tuning, shown in Fig. 7. During pre-training, the model is trained on unlabeled data using various pre-training tasks. For fine tuning, the BERT model is first initialized with pre-trained parameters, and all parameters are fine-tuned using labeled data from subsequent tasks. Each subsequent task has separate customized models, even if they are initialized with the same pre-trained parameters.

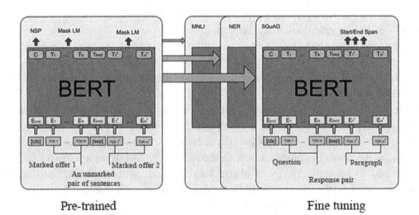

Fig. 7. Preliminary configuration of the executive structure.

The fine-tuning is based on the conceptual representation of the information that was discussed above. It allows BERT to model the results - regardless of whether they include single text or text pairs - by replacing the corresponding inputs and outputs with similar ones in a predefined structure.Most competing neural sequence transduction models have an encoder-decoder structure. Here, the encoder maps the input sequence of character representations $(x_1, \ldots, x_n)$ into a sequence of continuous representations $z = (z1, \ldots, zn)$. For a given z, the decoder then generates an output sequence $(y_1, \ldots, y_m)$ of characters, one element at a time. At each stage, the model is autoregressive, using previously generated symbols as additional input when creating the following.

The efficiency of the neural network is shown in Table, where 2 classes (Positive – interesting, Negative – uninteresting) of documents are presented. Based on the requested information, the analyzer is able to determine the importance of the term entered in the documents under consideration based on the predefined structure of the conceptual representation of each document under consideration (Table 1).

Table 1. Summary of network health check.

Class	Quantity	Percent
Positive	4340	93.94%
Negative	280	6.06%
Total	4620	100%

The result of the work carried out is successful and can be used in further developments. According to the information provided, 6% of the total number of documents does not correspond to the requested information after processing algorithms. Manual verification shows that the quality of text selection according to the conceptual structure is high, and the missing percentage is false positive positives. The same number of false negatives, which is from 1 to 3% of the total number of analyzed documents. The result of the work is a software capable of processing documents on the structures of the conceptual representation and identifying the most significant information. After structuring the documents, the request for the relevant information is classified according to the content of the documents and documents that do not correspond to the request are discarded. The presented solution will significantly accelerate the existing algorithms of document processing and is aimed at the subsequent creation of an autonomous system for analyzing the relevance of the requested information with the possibility of obtaining an instant response.

8   Conclusions

According to the results of the conducted research, the most effective methods of analyzing a text document have been identified. The most effective algorithms are demonstrated, which, according to the task, have the possibility of simultaneous data processing. This method will allow you to create an autonomous system that is able to simultaneously analyze, search for information and model text constructions. Based on the results obtained, software will be developed containing the analysis algorithms presented in the article, text modeling modules based on the conceptual structure model, and databases for compiling logically verified dictionaries. The task of text modeling for a decision support system is to choose a method for indexing terms and comparing such words with a conceptual structure. Each analyzed document is indexed by terms that are selected after certain algorithms are performed. For small-volume documents, the system will perform a general frequency analysis of words, then the data obtained after bringing the textual information to the concept will be correlated with the received terms and some will be excluded from the indexing list of the analyzed document, since they will not be relevant for further analysis. In cases where the text contains many complex sentences, the text modeling system conducts additional algorithms to search for relevant keywords. Constant support of the system is needed, which includes updating the databases of the rules of the language in which the analysis is carried out, updating algorithms for calculating the weights of terms and the relevance of search queries.

References

1. Ticher, S., Meyer, M., Vodak, R., Vetter, E.: Methods of text and discourse analysis, publishing house "Humanitarian Center" (2017). (in Russian)
2. Sorokin, A.B., Lobanov, D.A.: Conceptual design of intelligent systems. Inf. Technol. **1**(24), 3–10 (2018). (in Russian)
3. Chen, D.-Y., Zhao, H., Zhang, X.: Semantic mapping methods between expert view and ontology view. J. Softw. **31**(9), 2855–2882 (2020)
4. Krongauz, M.A.: Semantics: Textbook, Mjscow, Publishing center "Academy" (2005). (in Russian)
5. Sorokin, A.B., Smolyaninova, V.A.: Conceptual design of expert decision support systems. Theor. Appl. Sci. Tech. J. "Inf. Technol." **9** (23), 634–641 (2017). (in Russian)
6. Bolshakova, E.I.: Automatic text processing in natural language and computational linguistics: textbook. manual. Moscow, MIEM (2011). (in Russian)
7. Sosnina, E.P.: Introduction to applied linguistics: textbook. Ulyanovsk, UlSTU (2012). (in Russian)
8. Rodrigues da Silva, A.: Model-driven engineering: a survey supported by the unified conceptual model. Comput. Lang. Syst. Struct. **43**, 139–155 (2015)
9. Amaeva, L.A.: Comparative analysis of data mining methods. Innovative Sci. **2**(1), 27–29 (2018). (in Russian)
10. Feldman, S.E.: The answer machine. Synthesis lectures on information concepts, retrieval, and services. Morgan & Claypool Publishers, vol. **4**, pp. 1–137 (2012)
11. Andreychikov, A.V., Andreychikova, O.N.: Intelligent digital technologies of conceptual design of engineering solutions: textbook. Moscow. INFRA-M (2019). ISBN 978-5-16-014884-7. (in Russian)
12. Hoffman, M.D., Blei, D.M., Bach, F.R.: Online learning for latent Dirichlet allocation. In: NIPS. Curran Associates, Inc., pp. 856–864 (2010)
13. Nejdl, W., Krestel, R., Fankhauser, P.: Latent dirichlet allocation for tag recommendation. In: Proceedings of the Third ACM Conference on Recommender Systems. – ACM, pp. 61–68 (2009)
14. Beale, M.H., Hagan, M.T., Demuth, H.B.: Neural Network Toolbox. Math Works, Inc, User's Guide. Natick (2015)
15. Tsitulsky, A.M., Rogov, I.S, Ivannikov, A.V.: Intellectual analysis of the text, "StudNet" **6**, 476–483 (2020). (in Russian)
16. Anferov, M.A.: Genetic clustering algorithm. Russ. Technol. J. **6**(7), 134–150 (2019)

Ontology-Based Data Mining Platform for Diagnosing Sowing Quality of Wheat Seeds

Denis Baryshev$^{(\boxtimes)}$ ⓘ, Nadezhda Barysheva ⓘ, Ekaterina Avdeeva ⓘ, and Sergey Pronin ⓘ

Polzunov Altai State Technical University, Lenin Ave. 46, 656038 Barnaul, Russia
dennis.baryshev@gmail.com

Abstract. The paper presents the results of developing a unified data-mining platform using ontology to implement a seed sowing quality diagnostic system. The data mining ontology platform is based on a generalized ontological model, which includes the integration of several basic taxonomies. In the practical application of the model, the presented classes are characterized by hierarchical relationships and attributes and participate in the formation of decision-making results. For the data classification, the "Decision Tree" machine learning method is chosen, which allows the data to be divided into classes with high accuracy. The use of ontology-based platform allows solving such problems as the structuring of knowledge, its systematization and application for the seed sowing quality diagnostics. It also creates a versatile tool for storing, processing and analyzing data, with the ability to simultaneously use multidimensional descriptions of the concepts used in the subject area and minimize the impact of information uncertainty on the decision-making results.

Keywords: Ontology · Data mining · Diagnostics · Quality of wheat seeds

1 Introduction

The demand for knowledge related to the efficient development of agriculture has increased significantly all over the world [1]. Particular importance is given to the knowledge in the field of smart crop production [2]. To represent knowledge and information on agricultural development, ontological modeling can be used.

Today ontologies become important knowledge management tools providing a common conceptual model for expressing knowledge. In comparison with traditional methods of knowledge management, the use of ontologies makes it possible to meet modern requirements due to the ability to aggregate different knowledge from various sources [1], to analyze and implement it in practice for solving important application tasks [3]. Ontologies in information systems allow reusing subject domain knowledge and its implementation as a knowledge base for decision-making, as well as the ability to represent knowledge for the end user [4].

V. Jordan et al. (Eds.): HPCST 2022, CCIS 1733, pp. 137–146, 2022.
https://doi.org/10.1007/978-3-031-23744-7_11

2 Problem Statement

The development of ontology-based intelligent systems is currently gaining popularity, including in the agricultural sector. For example, in France, a vineyard monitoring system based on an ontological platform is used. The system allows the collection of data and provides useful knowledge for the prediction of grape diseases [5]. The issues of knowledge organization in the field of digital agriculture are widely presented in [5]. It has been established that knowledge about agricultural practices can be represented using ontologies, rule-based expert systems, or knowledge models built on the basis of data mining processes. The authors proposed a knowledge representation model called an ontology-based knowledge map [5].

In Bulgaria, a relevant solution for agriculture development by elaborating a knowledge base using ontologies to represent general knowledge in the subject area, a knowledge base for dynamic data from various measurements characteristics of evaluation and factors affecting the object, and the use of smart components are presented [2].

The presented work is devoted to the development of an ontological platform, an ontology-based knowledge representation model of the sowing quality of wheat seeds, which will allow the collecting of data from different sources, integrating data and obtained knowledge on seed quality alternative indicators into the subject area. It will also allow one to provide the storage of multi-year experimental research results on the quality indicators of seed material, and agronomic techniques for increasing yields, which can be used as input data for data mining to diagnose the wheat seeds sowing suitability and forecast the potential yield within the seed factor for management decisions.

The use of high-quality seeds is one of the most popular and affordable ways to increase yields [6, 7]. Ensuring the high quality of the seeds can be achieved by applying seed stock diagnostic and selection methods in the process of seed treatment [8], using agrotechnical separation techniques [9–11].

To solve this problem, an integrated approach to assessing the seeds quality and condition, which will allow giving adequate management decisions in a short time, is necessary. One of the alternative ways to increase yields is quality parameters diagnostics based on the study of the seeds electrophysical properties. This method allows a short time to assess the quality of seeds after sorting and immediately before sowing and to predict the potential yield. The method is based on the study of bioelectrical signals of wheat seeds before and after the process of their separation into fractions [12]. A bioelectric signal represents the change in voltage on the grain shell. Seeds with high sowing quality are found to be characterized by signal amplitude in the range of $160 \div 190$ mV, and low quality in the range of 240 mV and more [12–14].

The study of the electrophysical properties of seeds involves not only the procedures of seed preparation for laboratory analysis and their electrophysical parameters measurement but also data processing [13, 14]. There is a need for data mining in order to automate the process and increase the reliability of the seed quality assessment. Therefore, the development of a unified platform for data mining is an actual task today since it allows solving real problems of seed sowing quality diagnostics on the basis of accumulated knowledge and the results of experimental research.

This approach will allow one to take into account one of the data important characteristics – their variability – and make it possible to structure not only the electrophysical properties research results by the example of bioelectrical signals measurement, but the results of seed suitability indicators studies by standard methods as well.

In this regard, this work is aimed at the development of a platform, a universal tool for structuring, storing, and data mining using ontologies to implement a seed quality diagnostic system.

3 Materials and Methods

In ontological modeling, an ontological model was chosen in which the ontological model should be general, and reflect the general idea of solving the problem of diagnosing the quality of sowing wheat seeds in order to increase yields. On the other hand, the model should contain all the details necessary to solve the problem, including the results of seed quality studies using alternative methods.

The development of the ontology platform and its testing includes several main stages:

1. building an ontological platform as a knowledge base for the selected subject area and implementing data mining;
2. development of a model for representing and storing knowledge about the subject area, obtained as a result of multi-year experimental research, field testing, data mining results - data classification, final result forecasting;
3. practical testing of seed yield forecasting using datasets from the ontology platform.

Subject area – the seed factor in increasing the yield of grain, diagnosing the quality of seed material of cereal crops using the example of wheat seeds, and increasing productivity using agricultural practices.

The subject area is the seed factor in increasing grain yield, diagnosing the grain seed material quality on the example of wheat seeds, and increasing yields using agrotechnical methods.

For the implementation of the platform, a classification method with a types hierarchy was chosen, which is one of the most common ontology types [15].

A classification represents a tree-like hierarchy of types, a so-called taxonomy.

The data set used in this work was obtained as a result of numerous studies of seeds of soft spring wheat varieties, conducted over three years in collaboration with farms in Altai Krai [12–14]:

1. laboratory experimental studies of the electrophysical properties of wheat seeds of different varieties;
2. laboratory testing of wheat seed germination by ISTA methods, which involves 100 wheat seeds soaking in distilled water for 10–14 days at a temperature of 20–22 °C and counting the germinated seeds;
3. evaluation of the thousand-grain weight of analyzed varieties;
4. information on the results of application of purification technologies, wheat seeds fractionation, the results of electrophysical properties studies of wheat seeds divided into fractions;

5. field testing (trials)for the yield assessment as the final result.

   1. A bioelectrical signal measurement method was used to investigate the electrophysical properties of the seeds [12]. The method involves germinating the seeds for 12 h in distilled water at a temperature of 20 °C, measuring bioelectrical signals using steel electrodes, and entering the data into a computer.
   2. To divide the seeds by Petkus separator the fractionation method according to aerodynamic properties was applied. The separation speed ranged from 8 m/s to 11 m/s with a step of 1 m/s.
   3. The decision tree machine learning method was chosen as the classifier.

To assess the quality of the classification, the following metrics were selected: the fraction of correct responses of the algorithm, precision, recall and F-measure [16].

4 Results and Discussion

Today, the use of ontologies is gaining popularity for solving practical tasks related to data processing and data mining [17–19]. It is also not the exception to solving problems of assessing the wheat seeds quality.

The ontology-based platform is a methodological base [20] for the implementation of the design and management process of the wheat seed quality diagnostics system, identical to the knowledge base of an intelligent information system.

The implementation of the ontological model for solving the task of diagnosing the sowing quality of wheat seeds can be presented in the following general form:

$$O = \{C, R, A, F, Ax\} \tag{1}$$

where C is a finite set of classes describing concepts of the domain; R is a finite set of relations defined by classes (concepts); A is a finite set of attributes describing properties of classes C and relations R; F is a set of restrictions on the attributes values of concepts and relations; Ax is a set of axioms defining the semantics of classes and relations of ontology.

This model has been successfully tested for solving the problems of developing portals of scientific knowledge [20] and can be used to implement the problems of diagnosing.

This model has been successfully tested for solving the problems of developing scientific knowledge portals [21] and can be used to implement the tasks of diagnosing the seeds sowing quality.

The class structure is presented as follows:

$$C = \langle Name, (R_C C_{parent}), (A_1, ...A_{n(C)}\rangle, \tag{2}$$

where $Name$ is the class name; $R_C C_{parent}$ is classes of ontology C connected by the relationship of hierarchy R; $A_1, \ldots A_{n(C)}$ is class properties.

In this case, the structure of the property is defined as follows:

$$A_C = \langle Name_{A,C}, (F_{A1}, ...F_{Ak(A.C)}\rangle, \tag{3}$$

where $Name_{A,C}$ is the names of class properties C and relations R; $F_{A1}, ...F_{Ak(A.C)}$ is restrictions on the attributes values of concepts and relations.

The presented model contains a description of the data set, based on which the solution of the practical problem of seed quality diagnostics is performed.

The ontological platform based on the developed ontological model contains the following classes: "Agrotechnical methods", "Laboratory quality indicators", "Indicators of yield properties", "Bioelectric signal indicators", and "Control Methods" (Fig. 1).

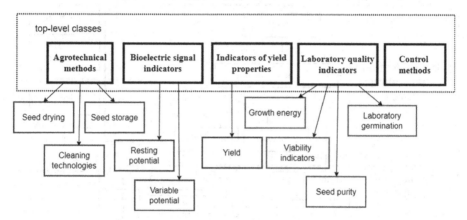

Fig. 1. Fragment of the ontological classification system.

The "Agrotechnical methods" class contains information about the technologies, properties, and parameters used. The "Laboratory quality indicators" class includes such laboratory quality indicators as laboratory germination, viability indicators, and growth energy.

The "Indicators of yield properties" class includes yield values, retrospective data from field studies, and thousand-grain weight and purity as well. The "Bioelectric signal indicators" class contains the results of studies of seeds electrophysical properties and includes the following derived classes: "Resting potential" and "Variable potential", and also informative features such as the time of signal change in the range of maximum value, the maximum value of potential and signal change rate.

The selection of informative features for implementing the classifier was performed according to the experimental studies results and the evaluation of correlation coefficients with the final result [13, 14].

For example, to diagnose the sowing quality of seeds before their fractionation (dividing seeds into fractions), three informative features are used to classify seeds and forecast the potential yield. After the seed dividing into fractions, informative features for each seed fraction are taken into account [13].

The main concept is to structure knowledge, systematize it and present it in the format of human-readable ontologies [16, 22, 23].

The ontological classification system (Fig. 2) consists of (1) preparation of reference data (pre-processed data on bioelectric signals of wheat grains, data on quality indicators and technological parameters), (2) selection of training and testing data, (3) - generation of classification rules using machine learning algorithms and, finally, (4) an ontology-based classification process.

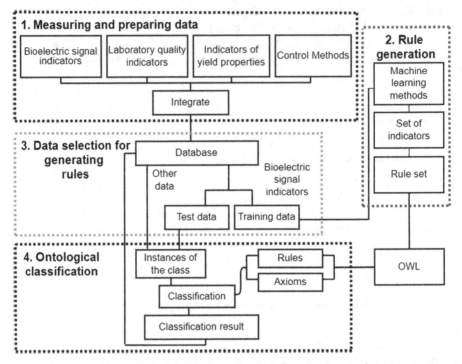

Fig. 2. Scheme of the ontology-based platform implementation.

The first block of the ontology-based model is data preparation. At this stage, the data on the quality of seeds are combined into a unified system, and all the data is stored in a database. The next stage is the selection of data on bioelectrical signals for machine learning method analysis in order to generate rules.

The data on bioelectric signals are divided into test data and training sets.

It should be noted that the data on bioelectrical signals are represented by the results of laboratory studies. Registration of the signal occurs after 12 h of seed preparation (the seeds are soaked in distilled water at a temperature of 20 °C for 12 h in a heat chamber). In total, 100 seeds of each variety and 100 grains of each seed fraction are selected for measurements according to the point sampling technology.

The measurement process is a piercing (puncture) into the germ area and 5 s signal recording. Then there is signal processing, and harmonic noise elimination (harmonic distortion suppression) in the recorded data using the moving average method with a window of 48 samples. The homogeneity of bioelectrical signals samples was checked using the Chow test.

Wheat seeds are a complex biological multilayer structure. Their morphological, physiological, and biochemical characteristics depend on genotype, variety, grain quality, infection with diseases, and belonging to a particular fraction, as shown by the results of the research [12]. In this regard, the application of machine learning methods will allow to implementation of the diagnosing seed sowing quality by their bioelectrical signals taking into account these features.

The decision tree method [24–26] was chosen for realization in the classification system. It allowed us to divide the data into classes with high accuracy (Fig. 3, 4). In the objective function, the data on grain yield was used. The results of two different experiments were demonstrated. The first experiment is of 2019, 4 classes total (Fig. 3). Objective function – yield of 16, 22, 35, and 39 q/ha.

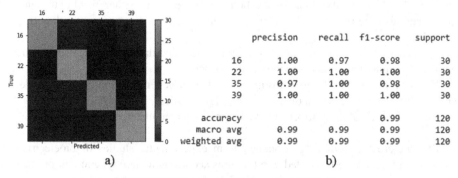

	precision	recall	f1-score	support
16	1.00	0.97	0.98	30
22	1.00	1.00	1.00	30
35	0.97	1.00	0.98	30
39	1.00	1.00	1.00	30
accuracy			0.99	120
macro avg	0.99	0.99	0.99	120
weighted avg	0.99	0.99	0.99	120

a)   b)

Fig. 3. Classification results based on 2019 data.

The results of the experiment showed that by the three informative features: time of signal change in the region of maximum value, maximum potential value, and signal change rate, the seeds can be classified with high accuracy into 4 groups according to the value of potential yield. This model allows for identifying seeds with higher sowing suitability by their electrophysical properties. The second experiment confirmed the effectiveness of the chosen method.

In the second experiment with the 2020 data, 6 classes were identified (Fig. 4). The objective function is the yield of 8, 16, 23, 28, 38 and 46 q/ha.

	precision	recall	f1-score	support		precision	recall	f1-score	support
8	0.97	1.00	0.98	30	8	0.97	1.00	0.98	30
16	1.00	0.97	0.98	30	16	1.00	0.97	0.98	30
23	0.94	0.97	0.95	30	23	1.00	1.00	1.00	30
28	1.00	0.97	0.98	30	28	1.00	1.00	1.00	30
38	1.00	1.00	1.00	30	38	1.00	1.00	1.00	30
46	0.97	0.97	0.97	30	46	1.00	1.00	1.00	30
accuracy			0.98	180	accuracy			0.99	180
macro avg	0.98	0.98	0.98	180	macro avg	0.99	0.99	0.99	180
weighted avg	0.98	0.98	0.98	180	weighted avg	0.99	0.99	0.99	180

a)   б)

Fig. 4. Classification results based on 2020 data: a - Data with noise emissions; b - Data after processing.

As a result of the experiment, it was found that the method coped with the task of classifying wheat seeds by yield indices without prior data processing, and without considering the wheat seed variety as an informative feature.

The proposed approach made it possible to integrate the knowledge obtained from numerous experimental studies of the grain electrophysical properties as alternative indicators of sowing qualities into the system of current knowledge, which together enhance the efficiency of grain quality assessment and contribute to increasing yields.

5   Conclusions

This paper presents a conceptual model of an ontology-based data mining platform for wheat seeds sowing quality diagnostics, built on the basis of a generalized ontological model, and also the results of its practical implementation on experimental data from 2019–2020.

The paper presents the results of developing a unified data mining platform using ontology to implement a seed-sowing quality system.

The data mining ontology platform is based on a generalized ontological model, which includes the integration of several basic taxonomies.

The application of the platform allowed for solving a number of important problems:

1. To structure the knowledge obtained from experimental studies of bioelectrical signals of wheat seeds divided into fractions according to aerodynamic properties.
2. To systematize the data and determine the form of presentation of seed quality diagnostics.
3. To generalize the process of data processing and analysis to solve practical problems. Including the seed quality indicators evaluations.
4. To automate the process of data collection and processing and to organize a repository of retrospective data stores.
5. To implement the possibility of simultaneous use of multidimensional descriptions and concepts of the subject area.
6. To minimize the impact of information uncertainty on the wheat seeds sowing quality diagnostics result which can be caused by the ambiguity of the system components due to randomness, or incomplete input initial data.

Thus, the main advantage of the developed platform is the possibility to use the accumulated experimental data, knowledge, and the results of multi-year research to solve not only the problems of seed quality diagnostics but also a variety of decision-making tasks in order to increase the yield of seed.

The proposed ontology-based platform can serve as an effective tool in planning the yield and the production process, as it allows taking into account all the elements associated with crop production, as well as the complexity of representing agricultural methods and restrictions not only in the selection of seed material but also in subsequent cycles of high-quality grain production.

References

1. Alfred, R. et al.: Ontology-based query expansion for supporting information retrieval in agriculture. In: Uden, L., Wang, L., Corchado Rodríguez, J., Yang, HC., Ting, I.H. (eds)

The 8th International Conference on Knowledge Management in Organizations. Springer Proceedings in Complexity. Springer, Dordrecht (2014). https://doi.org/10.1007/978-94-007-7287-8_24

2. Stoyanova-Doycheva, A., Ivanova, V., Doukovska, L., Tabakova, V., Radeva, I., Danailova, S.: Architecture of a knowledge base in smart crop production. In: 2021 International Conference Automatics and Informatics (ICAI), pp. 305–309 (2021)

3. Zheng, Y.-L., He, Q.-Y., Qian, P., Li, Z.: Construction of the ontology-based agricultural knowledge management system. J. Integr. Agric. **11**(5), 700–709 (2012)

4. Wilson, S.I., Goonetillake, J.S., Ginige, A., Walisadeera, A.I.: Towards a usable ontology: the identification of quality characteristics for an ontology-driven decision support system. IEEE Access **10**, 12889–12912 (2022)

5. Ngo, Q.H., Kechadi, T., Le-Khac, N.-A.: OAK: ontology-based knowledge map model for digital agriculture. In: Dang, T.K., Küng, J., Takizawa, M., Chung, T.M. (eds.) FDSE 2020. LNCS, vol. 12466, pp. 245–259. Springer, Cham (2020). https://doi.org/10.1007/978-3-030-63924-2_14

6. Zhang, T.: A reliable methodology for determining seed viability by using hyperspectral data from two sides of wheat seeds. Sensors **18**(3), 813 (2018)

7. Yang, L., Wen, B.: Seed quality Encyclopedia of Applied Plant Sciences, Second Edn, vol. **1**, pp. 553-563 (2017)

8. Anisur, R., Byoung-Kwan, C.: Assessment of seed quality using non-destructive measurement techniques: a review. Seed Sci. Res. **26**(4), 285–305 (2016)

9. Pasynkova, E.N., Zavalin, A.A., Pasynkov, A.V.: Change in quality parameters of hulled oats grain at fractionation Russ. Agricult. **44**, 409–413 (2018)

10. Lullien-Pellerin, V., Haraszi, R., Anderssen, R.S., Morris, C.F.: Understanding the mechanics of wheat grain fractionation and the impact of puroindolines on milling and product quality. In: Igrejas, G., Ikeda, T.M., Guzmán, C. (eds.) Wheat Quality For Improving Processing And Human Health, pp. 369–385. Springer International Publishing, Cham (2020). https://doi.org/10.1007/978-3-030-34163-3_16

11. Orobinsky, V., Gievsky, A., Baskhakov, I., Chernyshov. A.: Seed refinement in the harvesting and post-harvesting process In: International Scientific and Practical Conference "AgroSMART - Smart Solutions for Agriculture" (2018)

12. Barysheva, N.N., Pronin, S.P.: Method of determining seed germination by using membrane potential of wheat seeds. Eng. Technol. Syst. **29**(3), 443–455 (2019)

13. Barysheva, N.N., Guner, M.V., Baryshev, D.D., Pronin, S.P.: Analysis of seed quality indicators based on neural network. J. Phys: Conf. Ser. **1615**, 1–10 (2020)

14. Baryshev, D.D., Barysheva, N.N., Pronin, S.P.: Comparison of machine learning methods for solving the problem of wheat seeds classification by yield properties Russ. Agricult. Sci. **46**(4), 410–417 (2020)

15. Moran, N., Nieland, S., Tintrup, G.S., Kleinschmit, B.: Combining machine learning and ontological data handling for multi-source classification of nature conservation areas. Int. J. Appl. Earth Observ. Geoinf. **54**, 124–133 (2017). https://doi.org/10.1016/j.jag.2016.09.009

16. Velupillai, S., Dalianis, H., Hassel, M., Nilsson, G.H.: Developing a standard for de-identifying electronic patient records written in Swedish: precision, recall and F-measure in a manual and computerized annotation trial. Int. J. Med. Inf. **78**(12), e19–e26 (2009). https://doi.org/10.1016/j.ijmedinf.2009.04.005

17. Ingram, J., Gaskell, P.: Searching for meaning: Co-constructing ontologies with stakeholders for smarter search engines in agriculture. NJAS - Wageningen J. Life Sci., 90–91 (2019)

18. Dooley, D.M., Griffiths, E.J., Gosal, G.S.: FoodOn: a harmonized food ontology to increase global food traceability, quality control and data integration. npj Sci. Food **2**(23), 1–10 (2018)

19. Lytvyn, V., Vysotska, V., Dosyn, D., Lozynska, O., Oborska, O.: Methods of building intelligent decision support systems based on adaptive ontology. In: IEEE Second International Conference on Data Stream Mining & Processing (DSMP), pp. 145–150 (2018)

20. Zagorulko, Y.: Technology of scientific knowledge portals development. Softw. Prod. Syst. **4**, 25–29 (2009)

21. Nikolaychuk, O.A., Pavlov, A.I., Stolbov, A.B.: The software platform architecture for the component-oriented development of knowledge-based systems. In: 41st International Convention on Information and Communication Technology, Electronics and Microelectronics (MIPRO), pp. 1064–1069 (2018)

22. Davydenko, I.: Semantic models, method and tools of knowledge bases coordinated development based on reusable components. Open Semant. Technol. Design. Intell. Syst. **8**, 99–119 (2018)

23. Maksimov, N.V., Gavrilkina, A.S., Andronova, V.V., Tazieva, I.A.: Systematization and identification of semantic relations in ontologies for scientific and technical subject areas. Autom. Documentation Math. Linguist. **52**(6), 306–317 (2018). https://doi.org/10.3103/S00051055 1806002X

24. Wu, W., Li, A.-D., He, X.-H., Ma, R., Liu, H.-B., Lv, J.-K.: A comparison of support vector machines, artificial neural network and classification tree for identifying soil texture classes in southwest China. Comput. Electron. Agric. (Elsevier) **144**, 86–93 (2018)

25. Guo, P., Wu, W., Sheng, Q.: Prediction of soil organic matter using artificial neural network and topographic indicators in hilly areas. Nutr. Cycl. Agroecosyst. **95**, 333–344 (2013)

26. Das, H., Naik, B., Behera, H.S.: An experimental analysis of machine learning classification algorithms on biomedical data. In: Kundu, S., Acharya, U.S., De, C.K., Mukherjee, S. (eds.) Proceedings of the 2nd International Conference on Communication, Devices and Computing. LNEE, vol. 602, pp. 525–539. Springer, Singapore (2020). https://doi.org/10.1007/978-981-15-0829-5_51

Algorithm for the Classification of Coronary Heart Disease Based on the Use of Symptom Complexes in the Cardiovascular Environment

Akhram Kh. Nishanov[1] , Gulomjon P. Juraev[2], Malika A. Khasanova[3],
Fazilbek M. Zaripov[1(✉)] , and Saidqul X. Saparov[1]

[1] Tashkent University of Information Technologies (TUIT) named after Muhammad
al-Khwarizmi, Amir Temur Avenue 108, Tashkent 100084, Uzbekistan
`fazilbek_zaripov@gmail.com`
[2] Regional Center for Retraining and Advanced Training of Employees of Public Education of
Kashkadarya, Olimlar Street 2, Karshi City 180100, Uzbekistan
[3] Tashkent Medical Academy, Farabi Street 2, Tashkent 100109, Uzbekistan

Abstract. The article reveals a software tool "E-ischemic cardio diagnosis" based on an algorithm focused on the diagnosis (classification) of medical logos based on selected informative symptom complexes in coronary heart disease. This is the object of research, and it discusses the use of this software in the diagnosis of coronary heart disease (5 classes, so class x_1 – "Stenocardia tension", class x_2 – "Acute myocardial infarction", class x_3 – "Arrhythmic form", class x_4 – "Postinfarction cardiosclerosis", class x_5 – "Permanent form of atrial fibrillation"). The use of this software tool in the field of medicine is explained by the development of a modern recognition system that helps specialists in solving the problem of classifying medical logos. The electronic support system for medical workers developed as a result of the research allows specialists in this field to make significant decisions on such important issues as the primary processing of medical logos, the formation and evaluation of character sets, the classification of medical objects in selected symbols. As a result, the processes of medical diagnostics were automated, which made it possible to ensure the speed and reliability of diagnostics. In this case, a system has been created that helps the doctor to make a final diagnosis to the patient.

Keywords: Medical symbol · Primary treatment · Classification · Informative signs · Informative symptom complexes

1 Introduction

At the videoconference of the President of the Republic of Uzbekistan on November 9, 2021 dedicated to "Improving the system of prevention and treatment of cardiovascular and endocrine diseases", he touched upon problematic issues in the field of cardiology and noted the following. "Cardiovascular diseases still account for 53% of deaths among the population aged 30–70 years old. Over the past five years, the number of cardiovascular diseases has increased by 20%, even among young people. 20–25% of the population over the age of 25 have signs of hypertension.

In general, according to the results of the correspondence, about 4 million people in the republic are diagnosed with cardiovascular diseases and make up 12% of the total population" [1]. From the above it can be seen that with this type of disease, one of the most important tasks is a correct diagnosis and quick decision-making. This indicates the necessity of widespread implementation of advanced mathematical apparatuses, information and communication technologies in the field. It should be noted that in matters of primary processing of medical data, in particular, the identification of medical signs, it is important to select and classify the most important of the complex of signs describing the disease, that is, informative. A number of methods and algorithms have been proposed to solve these problems [2–24, 25]. The article describes in detail the principle of operation of the software tool "Eischemic cardio diagnosis", developed on the basis of the diagnostic algorithm (classification), based on the example of ischemic heart diseases, basing on informative symptom complexes.

2 Problem Description

This section discusses the principle of operation of the software "E-ischemic cardio diagnosis", used in the diagnosis of coronary heart diseases using the general structure of the software and its constituent modules. The general structure of the principle of operation of this software is expressed in Fig. 1.

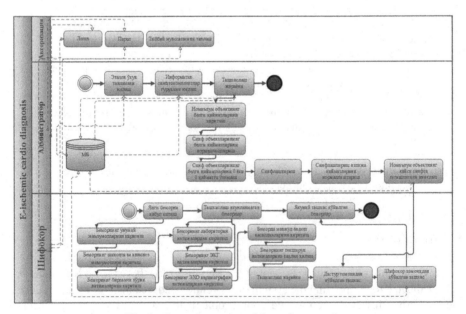

Fig. 1. General structure of the software tool.

The developed software tool is mainly divided into two categories depending on the category of operations, that is, the software tool containing the modules "Administrator" and "Doctor". To use the program, you need to log in to the browser and access the computer server (Fig. 2).

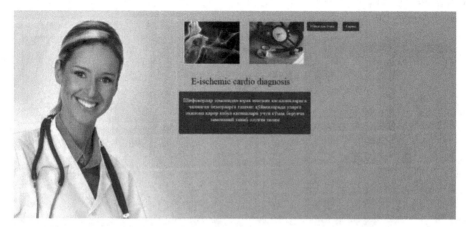

Fig. 2. Appearance of the main window of the software tool.

Initially, the user of the system can create his account through the "registration" item (Fig. 3).

Fig. 3. Appearance of the account creation window.

Each user has access to the system using their own created account when logging in. As a result, the authentication process is carried out in the system, that is, the login password is checked and the necessary module is loaded in accordance with your category.

3 Software Modules

3.1 Administrative Module

When the Administrator module of the software tool is launched, its general appearance will be as follows (Fig. 4).

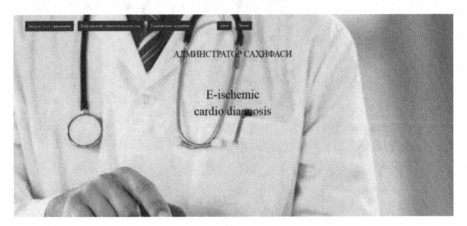

Fig. 4. Overview of the Administrator window.

This module "Administrator" also contains the following modules:

- selection of reference training;
- informative symptom complexes;
- the diagnostic process.

3.2 Reference Selection

The reference training sample, which is formulated for the class of coronary heart diseases, consists of 5 classes, 524 objects, 89 characters, one class is considered as one type of disease, so class x_1 *is* "Stenocardia tension", class x_2 *is* "Acute myocardial infarction", class x_3 *is* "Arrhythmic form", class x_4 *is* "Postinfarction cardiosclerosis", class x_5 *is* "Permanent form of atrial fibrillation".

At the same time, the space of signs characterizing each class (type of disease) was formed by specialists in this field and consisted of 89 signs characterizing each class (Table 1).

Table 1. Naming of signs characterizing ischemic heart diseases.

No.	Name of the sign	Code	No.	Name of the sign	Code
1	Pain in the heart area	x^1	46	Urea	x^{46}
2	Chest pain	x^2	47	Creatinine	x^{47}
3	Epigastric pain	x^3	48	The quantity of sugar in the blood	x^{48}
4	Pain in the left shoulder blade area	x^4	49	CPK (creatine phosphokinase)	x^{49}
5	Pain in the left shoulder and arm	x^5	50	Cholesterol	x^{50}
6	Lower jaw and neck pain	x^6	51	Atrialfibrillation	x^{51}
7	Duration of pain	x^7	52	Arrhythmia	x^{52}
8	Nitroglycerin helps	x^8	53	Throbbing arrhythmia	x^{53}
9	Pain during exercise	x^9	54	Blockade	x^{54}
10	Pain in a calm condition	x^{10}	55	Sinus rhythm	x^{55}
11	Rapid heartbeat	x^{11}	56	Deviation of the electrical axis of the heart to the left	x^{56}
12	Cold sweat	x^{12}	57	The electrical axis of the heart is in normal condition	x^{57}
13	Weakness	x^{13}	58	Ventricular extrasystole	x^{58}
14	Shortness of breath on exertion	x^{14}	59	Supraventricular extrasystole	x^{59}
15	Dyspnea at rest	x^{15}	60	S-T rise(mm)	x^{60}
16	Orthopnea position	x^{16}	61	S-T Lowered(mm)	x^{61}
17	Swelling on the legs	x^{17}	62	T rise	x^{62}
18	Feeling of lack of air	x^{18}	63	T Lowered	x^{63}
19	Increased blood pressure	x^{19}	64	Left atrium	x^{64}
20	Frequency of breathing duration	x^{20}	65	Interventricular septum	x^{65}
21	Heart frequency	x^{21}	66	Posterior wall of the left ventricle	x^{66}
22	Systolic blood pressure	x^{22}	67	Last diastolic measurement	x^{67}
23	Diastolic blood pressure	x^{23}	68	Last systolic measurement	x^{68}
24	Saturation (SPO2)	x^{24}	69	Last diastolic volume	x^{69}
25	Hemoglobin	x^{25}	70	Last systolic volume	x^{70}
26	Erythrocyte	x^{26}	71	Ejection fraction (EF%)	x^{71}
27	Color indicator	x^{27}	72	Stroke volume (UOml)	x^{72}

(*continued*)

Table 1. (*continued*)

No.	Name of the sign	Code	No.	Name of the sign	Code
28	Leukocyte	x^{28}	73	Hypertension	x^{73}
29	Segment core	x^{29}	74	Diabetes	x^{74}
30	Erythrocyte sedimentation rate(ESR)	x^{30}	75	Postinfarction cardiosclerosis	x^{75}
31	Thrombocyte	x^{31}	76	Circulatory insufficiency IIa	x^{76}
32	Haematocrit	x^{32}	77	Circulatory insufficiency IIb	x^{77}
33	Lymphocytes	x^{33}	78	COPD (Chronic obstructive pulmonary disease)	x^{78}
34	Monocytes	x^{34}	79	Chronic cholecystitis	x^{79}
35	Blood clotting time (beginning)	x^{35}	80	Total LV contractility	x^{80}
36	Blood clotting time (ending)	x^{36}	81	Reduction of regional contractility	x^{81}
37	PTI (Prothrombin)	x^{37}	82	Hypokinesis	x^{82}
38	Thrombotest	x^{38}	83	Leukocytosis	x^{83}
39	Fibrinogen	x^{39}	84	Muting heart tones	x^{84}
40	Fibrinolytic activity of blood	x^{40}	85	Violations of local contractility of the left ventricle of the heart	x^{85}
41	Plasma tolerance to heparin	x^{41}	86	Lung congestion	x^{86}
42	Hemostasis	x^{42}	87	LV hypertrophy (left ventricle)	x^{87}
43	AST	x^{43}	88	Liver enlargement	x^{88}
44	ATL	x^{44}	89	The appearance of the toothed Q	x^{89}
45	Bilirubin	x^{45}			

In the research work carried out, the values of the initial data were normalized during the initial processing of medical data, at the next stage, the values of the results obtained as a result of normalization of the initial data were transferred to a continuous quantitative form of 0 or 1 Type. Then the classification process was carried out, during which it was established that the objects of the class belong to their own class or another class. Then the classification process was carried out, during which it was established that the objects of the class belong to their own class or another class. This process was carried out in stages, at each stage objects that did not correspond to their class were excluded, and objects in the corresponding classes were completely performed, that is, up to 100% of the detection of their class, as a result of which a reference educational choice was formed [4, 7–18].

In this generated reference educational selection, objects of each class are entered into Excel in the context of path elements, and signs are entered in the format.xls or.xlsx. Then this file is loaded into the module "Selection of reference training" of the software tool "Adminstrator" (Fig. 5).

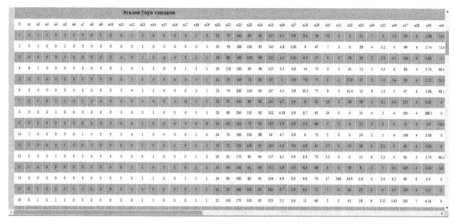

Fig. 5. Window view of the module "Selection of reference training".

3.3 Informative Symptom Complexes

As a result of the conducted scientific research on the formation of groups of informative symptom complexes for the class of ischemic heart diseases, groups of informative signs $l = 12$ to $l = 9$ were isolated from the groups of the complex of informative signs and groups of informative symptom complexes were formed [19, 20]. These groups of formed informative symptom complexes are presented below in Table 2.

The groups of information symptom complexes (Table 2) are included in the Excel program and are stored in the format.xls or.xlsx. Then the file is loaded into the "Informative symptom complexes" module of the "Administrator" module (Fig. 6).

Table 2. Groups of informative symptom complexes isolated from the groups of the complex of informative signs $l = 12$ to $l = 9$.

l	Number of options	Number of characters	Symptom complexes
12	91	58	$x^3, x^4, x^5, x^6, x^7, x^8, x^{10}, x^{11}, x^{12}, x^{14}, x^{15}, x^{17}, x^{19}, x^{20},$ $x^{21}, x^{22}, x^{24}, x^{25}, x^{26}, x^{31}, x^{32}, x^{33}, x^{35}, x^{36}, x^{37}, x^{40}, x^{41},$ $x^{42}, x^{45}, x^{48}, x^{50}, x^{51}, x^{52}, x^{53}, x^{55}, x^{58}, x^{59}, x^{60}, x^{62}, x^{63},$ $x^{64}, x^{65}, x^{66}, x^{68}, x^{69}, x^{71}, x^{72}, x^{75}, x^{76}, x^{77}, x^{79},$ $x^{80}, x^{81}, x^{85}, x^{86}, x^{87}, x^{88}, x^{89}.$
11	47	52	$x^1, x^3, x^5, x^6, x^7, x^8, x^{10}, x^{11}, x^{12}, x^{14}, x^{15}, x^{16}, x^{17}, x^{19},$ $x^{21}, x^{22}, x^{24}, x^{31}, x^{32}, x^{35}, x^{36}, x^{37}, x^{39}, x^{40}, x^{41}, x^{42}, x^{45},$ $x^{48}, x^{49}, x^{50}, x^{51}, x^{52}, x^{53}, x^{54}, x^{55}, x^{58}, x^{59}, x^{60}, x^{62}, x^{63},$ $x^{68}, x^{70}, x^{72}, x^{76}, x^{77}, x^{79}, x^{80}, x^{81}, x^{85}, x^{86}, x^{87}, x^{89}.$
10	23	32	$x^3, x^5, x^6, x^7, x^8, x^{10}, x^{11}, x^{12}, x^{14}, x^{17}, x^{20}, x^{37}, x^{40}, x^{41},$ $x^{42}, x^{48}, x^{50}, x^{51}, x^{52}, x^{53}, x^{54}, x^{59}, x^{60}, x^{62}, x^{63}, x^{66}, x^{68},$ $x^{77}, x^{80}, x^{81}, x^{85}, x^{89}.$
9	31	30	$x^3, x^5, x^8, x^{10}, x^{11}, x^{12}, x^{14}, x^{20}, x^{24}, x^{25}, x^{32}, x^{35}, x^{36}, x^{38},$ $x^{40}, x^{41}, x^{42}, x^{48}, x^{52}, x^{53}, x^{59}, x^{60}, x^{62}, x^{63}, x^{68}, x^{77}, x^{80},$ $x^{81}, x^{85}, x^{89}.$

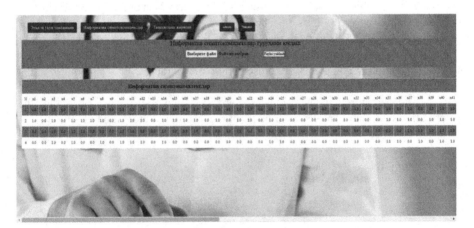

Fig. 6. Window view of the module "Informative symptom complexes".

3.4 Diagnosis Process

The work of this module is in the "doctor" module. The diagnostic process begins after the doctor accepts a new patient and enters general information about him into the system. One enters the values of the patient's complaints and anamnesis data, primary examination, laboratory examination, instrumental examination, concomitant diseases, analysis of the results of laboratory and instrumental examination (see Fig. 11, 12, 13, 14, 15, 16 and 17), and the diagnostic algorithm is performed.

3.5 Doctor Module

The doctor is registered in the "Registration" section of the program and the information about the doctor is confirmed by the administrator. Next, the doctor is logged in with a username and password, and the "Doctor" module is launched (Fig. 7).

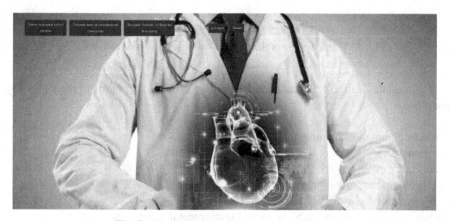

Fig. 7. Window view of the "Doctor" module.

At startup, the window of the "Doctor" module consists of such modules as "Admission of a new patient", "Patients with incomplete diagnosis", "Patients with a final diagnosis".

1. *Module "Admission of a new patient"*. This module is designed to enter general information about a patient at a doctor's appointment (Fig. 8).

The doctor enters general information about the new patient, and after clicking the "Next" button, general information about the patient is saved and transferred to the module "Patients with an incomplete diagnosis".

2. *"Patients with an incomplete diagnosis"*. When the module "Patients with an incomplete diagnosis" is launched, its window will look like this (Fig. 9).

Fig. 8. Window view of the module "Admission of new patients".

The module "Patients with an incomplete diagnosis" also consists of the following modules:

- patient complaints and anamnesis data;
- primary review;
- laboratory examination;
- instrumental examination;
- concomitant diseases;
- analysis of the results of laboratory and instrumental researches.

In the module "Patient complaints and anamnesis data", the doctor listens to the patient's complaint, anamnesis data is collected, if symptoms are shown in the window of this module, then 1 is entered, otherwise the digit 0 is used (Fig. 10).

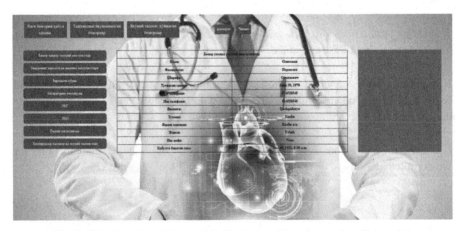

Fig. 9. Window view of the module "Patients with an incomplete diagnosis".

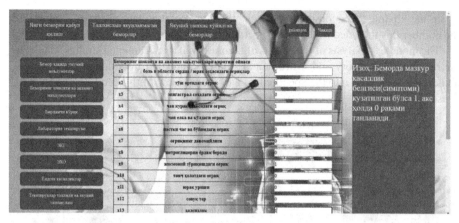

Fig. 10. Window view of the module "Patient complaints and anamnesis data".

In the "Primary medical examination" module, the doctor conducts a preliminary examination of the patient, the results of which are entered in the field with signs of the disease in the window of this module (Fig. 11).

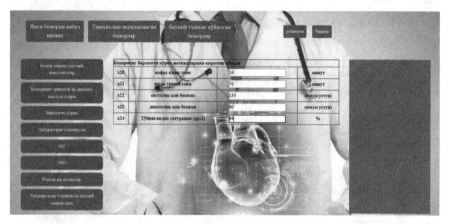

Fig. 11. View of the Preview module window.

In the module "Laboratory examination" in the field where the results of the laboratory examination of a patient with signs of the disease are shown, the value is entered in the digital form (Fig. 12).

The module "Instrumental examination" includes the results of cardiographic studies of ECG (Electrocardiography), ECHO of the patient (Fig. 13 and 14).

In the module "Concomitant diseases" there are concomitant diseases 1 otherwise the number 0 is included in the field specified in the window of this module (Fig. 15).

The module "Analysis of the results of laboratory and instrumental studies" analyzes the results of laboratory and instrumental studies conducted by a doctor on a patient, and digit 1 is entered into the specified area of the window of this module; otherwise 0 is used (Fig. 16).

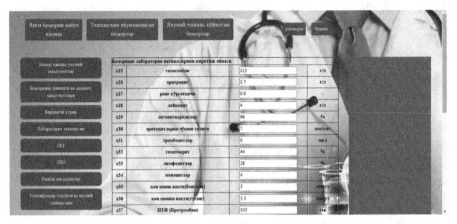

Fig. 12. Window view of the module "Laboratory tests".

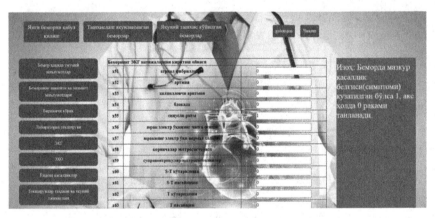

Fig. 13. Window for entering ECG results (Electrocardiography).

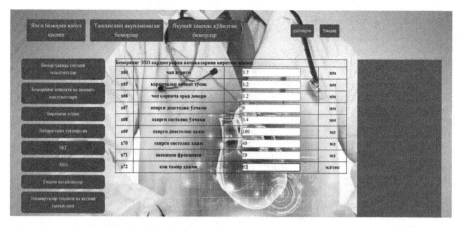

Fig. 14. Window for entering exocardiography results.

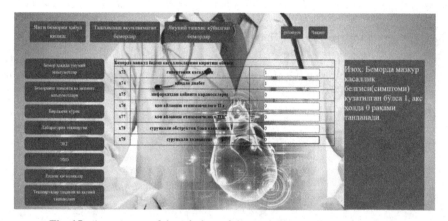

Fig. 15. Appearance of the window of the module "concomitant diseases".

After the doctor enters all the results of the examination, the diagnostic button is pressed, as a result of which the program diagnoses the patient.

The diagnosis established by the program is not the final diagnosis made to the patient, but the final diagnosis is made by the doctor and recorded in the section "diagnosis and recommendations identified by the doctor", and also stored in the system.

3. *Module "Patients with final diagnosis"*. This module stores a list of patients whose doctor diagnosed the disease, and their leaflets with a description of the disease in the doc format (Fig. 17).

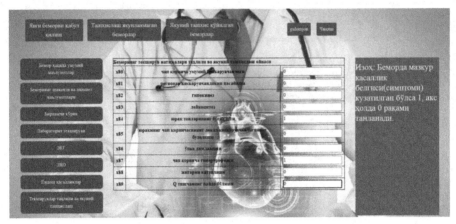

Fig. 16. The appearance of the module "analysis of the results of laboratory and instrumental examination"

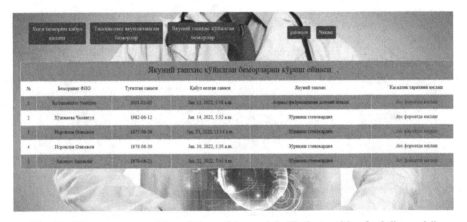

Fig. 17. The appearance of the window of the module "Patients with a final diagnosis"

In the process of working in the system, the "Administrator" or "Doctor" is carried out by pressing the "Exit" button to complete the work.

4 Algorithm of Formulation and Solution of the Issue

4.1 Inserting a Question

We suppose that the study sample is divided into classes and awarded as follows:

$$
K_1 = \begin{bmatrix} x_{11}^1 & x_{11}^2 & \cdots & x_{11}^N \\ x_{12}^1 & x_{12}^2 & \cdots & x_{12}^N \\ \vdots & \vdots & \vdots & \vdots \\ x_{1m_1}^1 & x_{1m_1}^2 & \cdots & x_{1m_1}^N \end{bmatrix} \cdots \quad K_r = \begin{bmatrix} x_{r1}^1 & x_{r1}^2 & \cdots & x_{r1}^N \\ x_{r2}^1 & x_{r2}^2 & \cdots & x_{r2}^N \\ \vdots & \vdots & \vdots & \vdots \\ x_{rm_r}^1 & x_{rm_r}^2 & \cdots & x_{rm_r}^N \end{bmatrix}
$$

In general terms, this can be expressed as follows:

$$K_p = \begin{bmatrix} x_{p1}^1 & x_{p1}^2 & \cdots & x_{p1}^N \\ x_{p2}^1 & x_{p2}^2 & \cdots & x_{p2}^N \\ \vdots & \vdots & \vdots & \vdots \\ x_{pm_p}^1 & x_{pm_p}^2 & \cdots & x_{pm_p}^N \end{bmatrix}$$

Here $p = \overline{1,r}$; and the training option $K = \bigcup_{p=1}^r K_p$ are expressed in appearance, regardless of whether they consist of classes that do not intersect, that is, the conditions $K_p \cap K_q = \emptyset$, $(p \neq q, p = \overline{1,r}; q = \overline{1,r};)$ are given.

Similarly, the components of the x_{pi} - object are x_{pi}^j - real numbers, this is read as follows: i - Patient, j - a sign of belonging to the p - class.

Here $p = \overline{1,r};, i = \overline{1,m_p};, j = \overline{1,N}$; and r is the total number of given classes, m_p - p is the total number of patients in the p-class and N is the total number of characters.

4.2 The Question

Construction of a decisive rule determines an unknown object in the diagnosis of a disease, that is, a person with a disease belonging to a class.

The algorithm for solving the question is below.

This software tool is built by a doctor during diagnosing a patient, that is, based on the decisive rule of determining an unknown object as an object of diagnosis in the classroom and works on this principle.

It should be noted that the research paper has developed an algorithm for medical diagnostics based on the data of informative symptoms, which is expressed as follows.

1. *Step 1. Enter the values of the symbols of the unknown object.*

At this stage, after the doctor accepts a new patient, one enters general information about him into the system, the patient's complaint and anamnesis data, initial examination, laboratory examination, instrumental examination, concomitant diseases, laboratory and instrumental examination enter the values of the analysis results (see Fig. 10, 11, 12, 13, 14, 15 and 16).

2. *Step 2. Normalization of the values of the signs of an unknown object.*

At this step, the values of the entered symbol of the unknown object (patient) are normalized, and this is done as follows:

The maximum values of symbols characterizing objects of class X_p the element $x_p^j(max)$ is detected, and this process is calculated and read as follows:

$$x_p^j(max) = \max_i x_{pi}^j, p = \overline{1,r}; j = \overline{1,N}; i = \overline{1,m_p}; \tag{1}$$

the maximum element of the j-sign column of features of objects belongs to the p-class.

The value of the symbol characterizing objects of class X_p is divided by the maximum value $x_{pi}^j(max)$, which is located in the column of the same symbol, and it is read as follows:

$$x_{pi}^j(Normal) = \frac{x_{pi}^j}{x_p^j(max)}, p = \overline{1, r}; j = \overline{1, N}; i = \overline{1, m_p};$$ (2)

the normalized state of an i-object belonging to class p by the j-sign. The program expresses the results of the normalization process (Fig. 18).

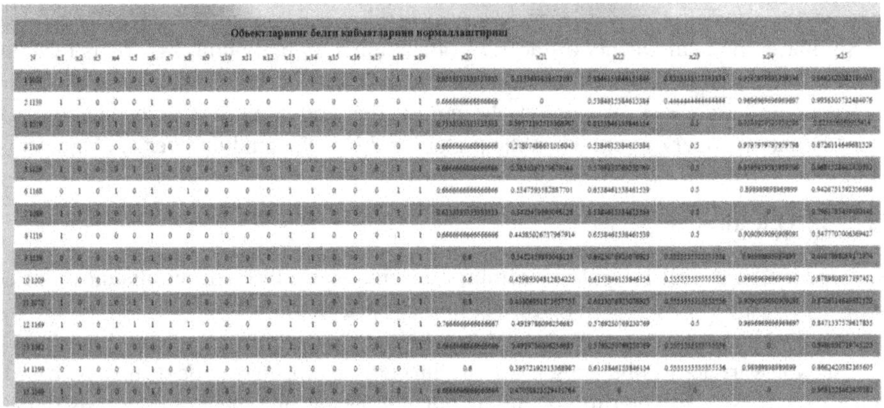

Fig. 18. Window for viewing normalization results.

3. *Step 3. Translation of normalized values of symbols of an unknown object to the value 0 or 1.*

The implementation of this step is as follows, that is, the characteristic of each object in each class is carried out by entering the following labels in the cross section of the entire class and all characters:

a. $\bar{x}_p = (\bar{x}_p^1, \bar{x}_p^2, \ldots, \bar{x}_p^N)$ vector, average representative objects of classes $X_p, p = \overline{1, r}$. Calculating its components using the following formula:

$$\bar{x}_p^j = \frac{1}{m_p} \sum_{i=1}^{m_p} x_{pi}^j, \quad p = \overline{1, r}; j = \overline{1, N}; i = \overline{1, m_p}.$$ (3)

b. Vectors $a_p = \left(a_p^1, a_p^2, \ldots, a_p^N\right)$ and $b_p = \left(b_p^1, b_p^2, \ldots, b_p^N\right)$ definition in the following form and calculation of its components by this formula:

$$d_p^j = \frac{1}{m_p} \sum_{i=1}^{m_p} (\bar{x}_p^j - x_{pi}^j)^2, \quad p = \overline{1, r}; j = \overline{1, N}.$$ (4)

$$b_{pi}^j = (\bar{x}_p^j - x_{pi}^j)^2, \quad p = \overline{1, r}; j = \overline{1, N}.$$ (5)

The actual number of components of the training selection supervised by X_p elements is transformed based on the following actions into the appearance of the boole

$$x_{pi}^{j} = \begin{cases} 1 \ \text{equals, if} \dfrac{b_{pi}^{j}}{a_{p}^{j}} \leq 1, \\ \text{otherwise equal to 0} \end{cases}. \tag{6}$$

The program describes the results obtained by converting the symbolic values of objects of 5 classes to the value 0 or 1 (Fig. 19).

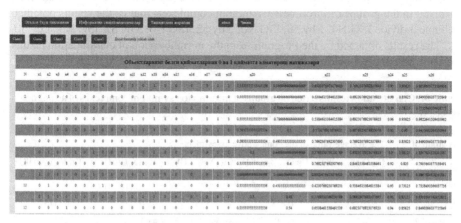

Fig. 19. A window for viewing the results converted to the value 0 or 1.

4. *Step 4. Classification based on selected informative symptom complexes.*

In Table 2 above, classification is carried out using selected groups of informative symptom complexes.

In the classification process, the proximity function between objects $\rho_i(x_{p1}, x_{p2})$ is introduced into this informative symbol space as follows:

$$\rho_i(x_{p1}, x_{p2}) = \begin{cases} 1, \ \text{if} \ \left(x_{p1}^{i} - x_{p2}^{i}\right) = 0, i = \overline{1, N}. \\ 0, \ \text{otherwise,} \end{cases} \tag{7}$$

If the first condition indicates the degree of similarity between two objects, then the second condition indicates the magnitude of their difference from each other, that is, it means that these components are not identical.

5. *Step 5. Assessment of the contribution of class objects to its class based on the classification results.*

The assessment of the contribution of class objects to their class at this 5-step is based on the following formula (8).

$$\Gamma_w\left(w, x_{pk}\right) = \sum_{\kappa=1}^{m_p} \sum_{i=1}^{N} \rho_i\left(w, x_{pk}\right), \ k = \overline{1, m_p}; \tag{8}$$

the results obtained by the program are shown in Table 3 below.

6. Step 6. Normalization of classification results.

At this step, the values reflected in Table 3 are divided by the number of characters involved in the groups of informative symptom complexes, that is, the results obtained in Table 3 divide ISG No. 1 by 58, ISG No. 2 by 52, ISG No. 3 by 32, ISG No. 4 by 30.

The results obtained by the Program from the normalization of classification results are shown in Table 4 below.

Table 3. Classification results based on groups of information symptom complexes of an unknown object W.

A group of information sign complexes (ISG)	The average value of the comparative assessment				
	$\overline{\Gamma_W}(S_{Class1}(W))$	$\overline{\Gamma_W}(S_{Class2}(W))$	$\overline{\Gamma_W}(S_{Class3}(W))$	$\overline{\Gamma_W}(S_{Class4}(W))$	$\overline{\Gamma_W}(S_{Class5}(W))$
ISG No. 1	47.77095	45.31818	43.25806	45.01111	42.65575
ISG No. 2	43.7654	40.6591	38.5968	41.1889	37.93452
ISG No. 3	26.7877	24.2727	24.5645	25.7222	24.49184
ISG No. 4	25.9385	22.4167	22.8226	24.4778	22.6557

Table 4. The values obtained as a result of the normalization process s of the results presented in Table 3

Groups of the complex of information signs (ISG)	The normal state of the average value of the comparative assessment				
	$\overline{\Gamma_W}(S_{Class1}(W))$ normal	$\overline{\Gamma_W}(S_{Class2}(W))$ normal	$\overline{\Gamma_W}(S_{Class3}(W))$ normal	$\overline{\Gamma_W}(S_{Class4}(W))$ normal	$\overline{\Gamma_W}(S_{Class5}(W))$ normal
ISG No. 1	0.823637	0.781348	0.745829	0.776054	0.735444
ISG No. 2	0.84164	0.78191	0.74225	0.79209	0.72951
ISG No. 3	0.83712	0.75852	0.76764	0.80382	0.76537
ISG No. 4	0.86462	0.74722	0.76075	0.81593	0.75519
Total	**3.36701**	**3.069**	**3.01647**	**3.18789**	**2.98551**

7. Step 7. Building a decisive rule.

At this 7th step, the total sum of the results obtained from normalization at the 6th step in the cross section of classes is calculated, and then, depending on the largest value of the results obtained in the cross section of these classes, that is, if

$$\Gamma_w(w, x_{pi}) > \Gamma_w(w, x_{pj})$$ (9)

the inequality is fulfilled. Then $W = (w^1, w^2, \ldots, w^N)$, the object is higher than others i-the degree of belonging to the diagnostic object.

The diagnosis made by the program, that is, an unknown object W (the patient was diagnosed with stress angina by the program), indicates belonging to the 1st class, the reason W as an unknown object was chosen as the second object of the 1st class.

5 Results

1. In the first column, in accordance with Table 2, the number of l - signs participating in the set of informative signs used by a medical professional for diagnosis is indicated.
2. The second column is the number of doctors who, based on their experience, diagnose using only their own, self-confident l - sign.
3. In the third column, this is expressed through the association of l - informative character sets used by doctors.

For example, in the first line, the number of characters participating in the set of informative characters is 12. So, each doctor makes a diagnosis based on only 12 characters. And the number of doctors is 91. The number of characters used by all doctors is 58, we call this set of symbols simtokomplexes. Thus, 91 doctors achieve a 100% diagnostic result using only 58 symptom complexes out of 89 characters. Each doctor uses 12 out of 58 symptom complexes and so on. So, the set of information signs 91 of $l = 12$ consists of 58 characters. Researches shows that when choosing 12 characters out of 89 characters, 91 informative complexes with informative character complexes were obtained, each of which gave a 100% diagnostic result. For example, for the case when $l = 12$, we denote the set of informative features as A_i^l. Here l is the number of characters, and i is the doctor. A_i^l reads like this: l characters used by the i-doctor.

$$A_1^{12} = \left\{x^3, x^8, x^{10}, x^{11}, x^{15}, x^{40}, x^{41}, x^{52}, x^{53}, x^{80}, x^{81}, x^{85}\right\};$$
$$A_2^{12} = \left\{x^3, x^6, x^8, x^{12}, x^{24}, x^{33}, x^{42}, x^{52}, x^{53}, x^{55}, x^{62}, x^{80}\right\};$$
$$A_3^{12} = \left\{x^{10}, x^{11}, x^{32}, x^{40}, x^{42}, x^{52}, x^{53}, x^{59}, x^{60}, x^{62}, x^{77}, x^{80}\right\};$$
$$\cdots\cdots\cdots\cdots\cdots\cdots\cdots\cdots\cdots\cdots\cdots\cdots;$$
$$A_{91}^{12} = \left\{x^3, x^5, x^8, x^{11}, x^{41}, x^{51}, x^{52}, x^{53}, x^{62}, x^{68}, x^{80}, x^{89}\right\}$$

Here, each set contains 12 characters, and the number of sets is 91. That is, each complex has a 100% diagnosis. The number of sets with a total of 12 characters is $C_{89}^{12} = \frac{89!}{(89-12)!12!}$, but only 91 of them had a 100% result. But only 91 of them had a 100% result.

Similarly, the symptom complex:

$$S^{12} = \bigcup_{i-1}^{91} A_i^{12} = \left\{ \begin{array}{c} x^3, x^4, x^5, x^6, x^7, x^8, x^{10}, x^{11}, x^{12}, x^{14}, x^{15}, x^{17}, x^{19}, x^{20}. \\ x^{21}, x^{22}, x^{24}, x^{25}, x^{26}, x^{31}, x^{32}, x^{33}, x^{35}, x^{36}, x^{37}, x^{40}, x^{41} \\ x^{42}, x^{45}, x^{48}, x^{50}, x^{51}, x^{52}, x^{53}, x^{55}, x^{58}, x^{59}, x^{60}, x^{62}, x^{63} \\ x^{64}, x^{65}, x^{66}, x^{68}, x^{69}, x^{71}, x^{72}, x^{75}, x^{76}, x^{77}, x^{79} \\ x^{80}, x^{81}, x^{85}, x^{86}, x^{87}, x^{88}, x^{89} \end{array} \right\}$$

$$\left| S^{12} \right| = card\left(S^{12} \right) = 58.$$

The power of the S^{12} symptom complex is 58, i.e. 58 signs are involved in this complex. In total, the experts used 58 characters out of 89. Each expert used 12 out of 58 characters and so on.

6 Conclusions

Thus, as a result of using the created software "Electronic ischemic cardio diagnostics", the processes of medical diagnostics were automated, it became possible to ensure the speed and reliability of diagnostics. As a result, the doctor created a system to assist the patient in the final diagnosis of the disease.

References

1. Mirzoyev, Sh.: In Uzbekistan, more than 50% of people die from cardiovascular diseases. In: News Portal Repost.uz. https://repost.uz/ne-korona. Accessed 04 Nov 2022
2. Adylova, Z.T., Umarova, D.M.: Methods of constructing an informative feature space for managing communication networks in conflict situations. Uzbek J. Probl. Inform. Energy **2**, 3–9 (1992)
3. Bykova, V.V., Kataeva, A.V.: Methods and means of analyzing the informative value of signs in the processing of medical data. Softw. Prod. Syst. Softw. Syst. **2**(114), 172–178 (2016)
4. Schulte, R.V., Prinsen, E.C., Hermens, H.J., Buurke, J.H.: Genetic algorithm for feature selection in lower limb pattern recognition. Front. Robot. AI **8**, 1–12 (2021)
5. Zagoruiko, N.G., Kutnenko, O.A., Borisova, I.A., Dyubanov, V.V., Levanov, D.A., Zyryanov, O.A.: The choice of informative signs for the diagnosis of diseases based on genetic data. Vavilovsky J. Genet. Breeding **18**(4/2), 898–903 (2014)
6. Mahajan, P.: Applications of pattern recognition algorithm in health and medicine. Int. J. Eng. Comput. Sci. **5**, 16580–16583 (2016)
7. Shehab, M., et al.: Machine learning in medical applications: a review of state-of-the-art methods. Comput. Biol. Med. **145**, 105458 (2022)
8. Shi, B., et al.: An evolutionary machine learning for pulmonary hypertension animal model from arterial blood gas analysis. Comput. Biol. Med. **146**, 105529 (2022)
9. Xia, J., et al.: Performance optimization of support vector machine with oppositional grasshopper optimization for acute appendicitis diagnosis. Comput. Biol. Med. **143**, 105206 (2022)
10. Ashok, B., Aruna, P.: Comparison of feature selection methods for diagnosis of cervical cancer using SVM classifier. J. Eng. Res. Appl. **6**(1), 94–99 (2016)

11. Eklund, P., Karlsson, J., Rauch, J., Šimůnek, M.: On the logic of medical decision support. In: de Swart, H., Orłowska, E., Schmidt, G., Roubens, M. (eds.) Theory and Applications of Relational Structures as Knowledge Instruments II. LNCS (LNAI), vol. 4342, pp. 50–59. Springer, Heidelberg (2006). https://doi.org/10.1007/11964810_3

12. Soldaini, L., Cohan, A., Yates, A., Goharian, N., Frieder, O.: Retrieving medical literature for clinical decision support. In: Hanbury, A., Kazai, G., Rauber, A., Fuhr, N. (eds.) ECIR 2015. LNCS, vol. 9022, pp. 538–549. Springer, Cham (2015). https://doi.org/10.1007/978-3-319-16354-3_59

13. Portela, F., Santos, M.F., Machado, J., Abelha, A., Silva, Á., Rua, F.: Pervasive and intelligent decision support in intensive medicine – the complete picture. In: Bursa, M., Sami Khuri, M., Renda, E. (eds.) ITBAM 2014. LNCS, vol. 8649, pp. 87–102. Springer, Cham (2014). https://doi.org/10.1007/978-3-319-10265-8_9

14. Nishanov, A.K., et al.: Algorithm for the selection of informative symptoms in the classification of Medical Data. In: Developments of Artificial Intelligence Technologies in Computation and Robotics, pp. 647–658 (2020)

15. Pusztová, Ľ., Babič, F., Paralič, J.: Semi-automatic adaptation of diagnostic rules in the case-based reasoning process. Appl. Sci. **11**, 292 (2020)

16. Nishanov, A.K., Akbaraliev, B.B., Samandarov, B.S., Akhmedov, O.K., Tajibaev, S.K.: An algorithm for classification, localization and selection of informative features in the space of polytypic data. Webology **17**(1), 341–364 (2020)

17. Moses, J.C., Adibi, S., Shariful Islam, S.M., Wickramasinghe, N., Nguyen, L.: Application of smartphone technologies in disease monitoring: a systematic review. Healthcare **9**, 889 (2021)

18. Gurazada, S.G., Gao, S.C., Burstein, F., Buntine, P.: Predicting patient length of stay in Australian emergency departments using Data Mining. Sensors **22**, 4968 (2022)

19. Nishanov, A.Kh., Turakulov, Kh.A., Turakhanov, Kh.V.: A decision rule for identification of eye pathologies. Biomed. Eng. **33**(4), 178–179 (1999)

20. Nishanov, A.Kh., Turakulov, Kh.A.,Turakhanov, Kh.V.: A decisive rule in classifying diseases of the visual system. Meditsinskaia tekhnika **4**, 16–18 (1999)

21. Kamilov, M., Fazilov, S., Mirzaeva, G., Gulyamova, D., Mirzaev, N.: Building a model of recognizing operators based on the definition of basic reference objects. In: Journal of Physics: Conference Series, vol. 1441, p. 012142 (2020)

22. Fazilov, S., Khamdamov, R., Mirzaeva, G., Gulyamova, D., Mirzaev, N.: Models of recognition algorithms based on linear threshold functions. In: Journal of Physics: Conference Series, vol. 1441, no. 1, p. 012138 (2020)

23. Fazilov, Sh.Kh., Mirzaev, N.M., Radjabov, S.S., Mirzaeva, G.R.: Hybrid algorithms of the person identification by face image. In: Journal of Physics: Conference Series, vol. 1333, no. 3, p. 032016 (2019)

24. Fazilov, S.K., Mirzaev, N.M., Radjabov, S.S., Mirzaeva, G.R.: Determination of representative features when building an extreme recognition algorithm. In: Journal of Physics: Conference Series, vol. 1260, no. 10, p. 102003 (2019)

Business Process Designing of the Institutional Ranking System of Higher Education of the Kyrgyz Republic

Bibigul B. Koshoeva(✉) ⓘ, Mirlan K. Chynybaev ⓘ, Aigiuzel T. Bakalova ⓘ, and Asel R. Abdyldaeva ⓘ

I. Razzakov Kyrgyz State Technical University, Aitmatov Ave. 66, 720044 Bishkek, Kyrgyz Republic
koshoeva@kstu.kg

Abstract. The paper considers the "Methodology for determining the national ranking of universities in the Kyrgyz Republic". The mechanism for calculating rating indicators is described. A business process diagram of an automated system for generating a national ranking of universities in the Kyrgyz Republic was designed. There was also a designed conceptual model of the ASFNR of universities in the Kyrgyz Republic in the form of a Use Case diagram and Activity diagram using MS Visio tools. The result of the design is an automated system for calculating the national ranking of higher educational institutions in Kyrgyzstan.

Keywords: Automated information system · University rankings · MVT architecture · Conceptual model · IDEF0 · UML

1 Introduction

The system of higher education concentrates a great potential of scientific, technical, managerial and intellectual resources. Therefore, in the relevant regulatory and program documents and scientific research, the priorities of directions and targets of universities are sufficiently formed and given; indicators and criteria for evaluating performance are formalized.

At the same time, the use of a rating system is recommended as a tool for achieving the targets.

Currently, there are more than 50 national and 7 global university rankings in the world practice. The lists of rankings were originally compiled taking into account the national level, the comparison of universities or the results of activities in the relevant areas of universities took place within the framework of clearly defined political, scientific and educational activities.

Rating of universities is an integral assessment of their activities based on established criteria with the presentation of ranking results. The ranking of universities is designed to provide the most complete information to potential applicants, students, parents, professional communities and employers about the potential of universities.

V. Jordan et al. (Eds.): HPCST 2022, CCIS 1733, pp. 168–182, 2022.
https://doi.org/10.1007/978-3-031-23744-7_13

The program for the development of education in the Kyrgyz Republic for 2021–2040, approved by the Government of the Kyrgyz Republic dated May 4, 2021 No. 200, provides for the creation of a National University Ranking Model as a launching pad for preparing universities for participation in international rankings.

Currently, in the higher education system of Kyrgyzstan, the rating of educational programs is used only on a fee basis, developed by IAAR (Independent Agency for Accreditation and Rating, Republic of Kazakhstan). Due to the fact that the budget of not all universities allows allocating funds, data on programs are entered partially or not entered at all, because of this, the formation of the rating is not sufficiently objective. Therefore, it is not possible to obtain an integrated overall assessment of the activities of universities, as well as a real picture of the assessment of profile and specialized universities.

In connection with the foregoing, the relevance of designing and developing an automated information system has arisen, which will allow collecting data from universities by questioning them, as well as forming a rating using the given data. This system will serve as a tool for assessing the competitiveness of universities. This work is a state order of the Ministry of Education and Science of the Kyrgyz Republic (MES KR) on the topic "Development and implementation of a Model of the National Rating System of Higher Education".

The ranking of universities is an assessment procedure for the quality of educational services, determined by the result of the components - education, scientific research work, material and technical base, educational and methodological work, international integration, social work, university development dynamics [1].

The institutional ranking methodology makes it possible to compare universities and stimulates the desire to develop integration into the international educational space.

The formation of a university ranking involves the substantiation and selection of a representative number of indicators and criteria for evaluating the activities of universities with appropriate weight values that make up the database of the pilot study. The collection and processing of data, the implementation of the calculation according to the methodology, are a complex and multifaceted process that requires a lot of time, material and human resources. Under these conditions, automation of the processes of determining the institutional ranking of universities based on information systems will be the most appropriate means of making a decision on the issue under consideration.

The relevance of the study is justified by the growing interest of consumers of educational services in ratings as tools for assessing the competitiveness of universities. Universities are interested in improving their positions in national and international rankings, and therefore, they are focused on studying and choosing ranking methods, as well as providing appropriate indicators.

The beginning of the ranking is considered to be in 1966, when a book by Alan Carter from the American Council on Education was published in the United States, in which the ranking of 106 universities was presented. Around this time, university rankings began to be used more widely in the world. In 1982, an Assessment of Research Doctorate Program was launched in the United States, using 16 indicators for evaluating scientific research and doctoral programs and covering 200 universities (2700 relevant academic programs were analyzed) [5].

National, regional and international ratings are compiled today in Canada, Poland, Germany, France, Ukraine, and a number of Asian countries. It continued most actively in the UK, where in 1971 the Times Higher Education rating was published [5].

The methodology of formation and results in the institutional rankings of universities are currently relevant both for consumers of educational services, various professional and academic communities, and for the universities themselves. To develop an institutional rating system of universities, various methods are used that allow taking into account systems and assessments of assessment indicators [8–10], general approaches to building a rating system model [8].

2 Methodology of National Institutional Ranking

In this work, it was assumed that a noisy audio signal can be represented in the time-frequency domain as follows. Our research team conducted a study of university rankings, key indicators and evaluation criteria [2–5]. Based on this analysis, the "Methodology for determining the national ranking of universities in Kyrgyzstan" was developed [6], which includes 4 criteria and a certain number of indicators:

- criterion "Conditions for obtaining quality education" - 27 indicators;
- criterion "Level of demand for graduates by employers" - 4 indicators;
- criterion "Level of research activity" - 17 indicators;
- criterion "Brand of the university" - 4 indicators.

The rating will be compiled according to different criteria and indicators. Since the system will be web-based, data collection and authenticity can be done in a short time. At the moment, there are no analogues of this system in the Kyrgyz Republic.

The main objects of the rating research are the universities of the Kyrgyz Republic by levels (BA, MA, PhD), and the subject is quantitative and qualitative indicators on key parameters of indicators and rating criteria [7].

The ranking is calculated according to the 100-point system according to the methodology for determining the national ranking of universities in Kyrgyzstan [6]. At the same time, 40 points are given for the conditions for obtaining a quality education, 20 points for the level of demand for graduates by employers, 30 points for the level of research activity and 10 points for the brand of the university. At the same time, if for some of the indicators the university gains the maximum value among others, then it is assigned the highest score equal to 1. The rest of the universities are assigned points as a percentage.

Example:
In criterion 1 "CONDITIONS FOR OBTAINING QUALITY EDUCATION" of the group "LEVEL OF TEACHING" of subgroup 1.7, the number R_i of full-time teaching staff per 100 students (hereinafter, the share values related to teaching staff are given to the full rate) is calculated by the following formula:

$$R_i = \frac{R_t}{R_{st\_sum}} * 100 \; stud,$$

where R_t is the number of full time teaching staff, R_{st_sum} is the total number of students. The weight for this indicator is $h = 1.5$ [6].

In the same criterion "CONDITIONS FOR RECEIVING QUALITY EDUCATION" of the "INTERNATIONAL INTEGRATION" group of subgroup 1.18, the academic external mobility of the teaching staff (for the last year), the number of R_i (at least 5 working days) (per 100 full-time teaching staff) will be calculated according to the following formula:

$$R_i = \frac{R_{t\_mobil\_contry}}{R_t} * 100 \ teachingstaff,$$

where $R_{t_{mobi}*l_{contry}}$ is the number of teaching staff who completed internships at other universities within the country, R_t is the number of full time teaching staff. The weight for this indicator is $h = 0.5$ [6].

Criterion 2 "LEVEL OF DEMAND FOR GRADUATES BY EMPLOYERS" of the group "QUALITY OF GRADUATES' CAREER" of subgroup 2.2. Availability of structures and their effectiveness in working with the labor market (alumni association, career center, etc.) (the number of contracts, content and forms of interaction with the market are also given) labor, work plan, information about activities on the university website) points are calculated according to the following Table 1.

Table 1. Distribution of scores by sections.

Alumni association	Career center	Agreements	Work plan	Information about activities on the site	Sum
0.4	0.4	0.4	0.4	0.4	2

The weight for this indicator is $h = 2$ [6].

In criterion 3 "LEVEL OF SCIENTIFIC RESEARCH ACTIVITY" of the group "SCIENTIFIC ACHIEVEMENTS" of subgroup 3.6, the number of publications for the last 3 years in scientific journals indexed in foreign databases per 1 scientific and pedagogical worker (SPW) is calculated by the following formula:

$$R_i = R_{t\_article}/R_t,$$

where $R_{t\_article}$ is the number of publications, R_t is the number of scientific and pedagogical workers. The weight for this indicator is $h = 3$ [6].

Criterion 4 "BRAND OF THE UNIVERSITY" of group 4.3 "ATTENDANCE TO THE OFFICIAL WEBSITE OF THE UNIVERSITY" indicates R_i as the average attendance per day based on the data of the last academic year. The weight for this indicator is $h = 3$ [6].

Each indicator has its own weight. The ranking is based on four criteria, 52 indicators, which are aggregated into integrated and global data that characterize all the basic areas of activity of universities.

3 Business Process Designing of the Institutional Rating System of Higher Education of the Kyrgyz Republic

In the course of the study, a business process for generating a rating (AS-IS model) was designed, which is presented in the form of an IDEF0 diagram in Figs. 1 and 2 [11]. The IDEF0 diagram represents the organization of logical relationships between processes as a set of modules consisting of:

- processes – the functions to be performed by the organization in question;
- input data - the parameters necessary for the implementation of the process;
- controls being the rules or requirements that must be followed in carrying out a process;
- mechanism, usually something by which the process is perturbed, and
- output representing the result of the process.

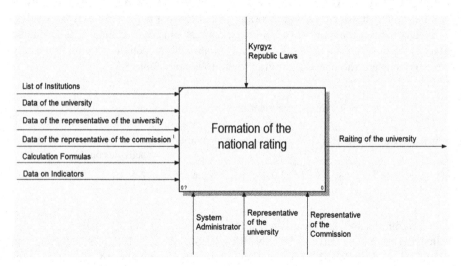

Fig. 1. AS-IS model context diagram – ranking formation.

By the example of the presented diagrams, you can see exactly what information is needed for the presented processes, who will be involved in the processes of forming the institutional ranking of the university, and that as a result of all processes, the former university ranking is obtained.

Based on the problems identified in the description of the business process, there is a need to develop a web-based system that will free universities from paying money for educational programs, provide a rating according to various criteria to all interested parties.

The developed automated system should perform the following functions:

- Functions available without authorization:

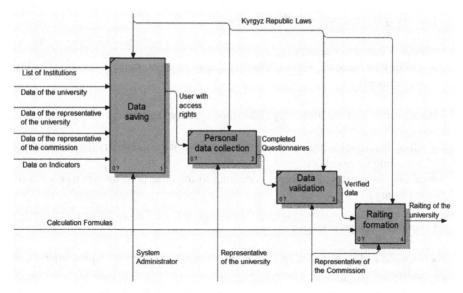

Fig. 2. Decomposition of the ranking formation process.

- rating output for all criteria;
- rating output according to the selected criterion;
- filling out the questionnaire by employers;
- filling out the questionnaire by university graduates;
- filling out the questionnaire by students (by definition of the index of digital skills of students).

- The system will require authorization for the following roles:

  - administrator;
  - university representative (Administrator);
  - representative of the university (Education Quality Department);
  - commission representative.

- Functions of each role:

  - Administrator:

    - creation of credentials for representatives of universities, representatives of the commission;
    - viewing the list of universities and editing the data of universities;
    - viewing the list of questionnaires from representatives of universities and editing these questionnaires;
    - viewing the list of criteria, groups, levels and editing data (setting points for the criterion).

– University representative (Administrator):

  • editing university data;
  • adding new accounts with the role "University Representative (Education Quality Department)".

– Representative of the university (Education Quality Department):

  • filling out a questionnaire from a representative of the university (entering values according to criteria);
  • adding supporting documents according to the criteria in the questionnaire (if necessary).

– Commission Representative:

  • viewing and verification of supporting documents added by representatives of universities.

– Non-functional requirements that the developed system should have:

  • reliability requirements - it is necessary to provide for the blocking of incorrect user actions when working with the program.
  • interface requirements - the interface must be graphical; messages should be as informative as possible.

A conceptual model was also developed that demonstrates the semantic structure of the designed system and describes the interaction of users with the system [11]. The conceptual model is presented in Fig. 3 as a Use Case diagram described in the Unified Modeling Language (UML) tool provided by MS Visio.

The developed conceptual model clearly demonstrates the main functionality of the automated system for the formation of an institutional rating and the interaction of its end users with the presented functionality. As a rule, users in the Use Case diagram are called actors (or actants), and the functions performed in the system are called use cases of this system. Thus, in the presented conceptual model, the actors are the system administrator, representatives of universities, members of the accreditation commission, employers, students, graduates, and users interested in the ranking of universities. Main use cases are authorization in the system, viewing the list of universities, filling out questionnaires, confirming data, forming a university ranking, viewing the ranking of universities, and adding, editing and deleting data.

Next, Activity Diagrams (Figs. 4 and 5) were designed for the main processes/use cases of the formation of a university ranking system.

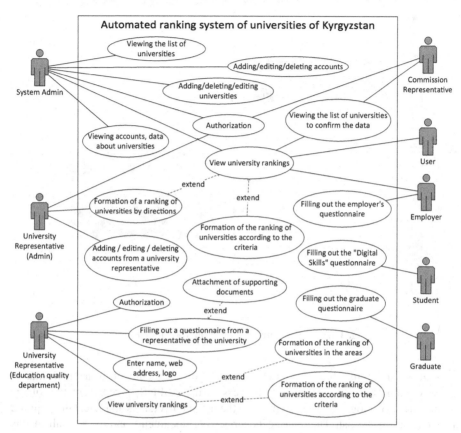

Fig. 3. Use Case diagram.

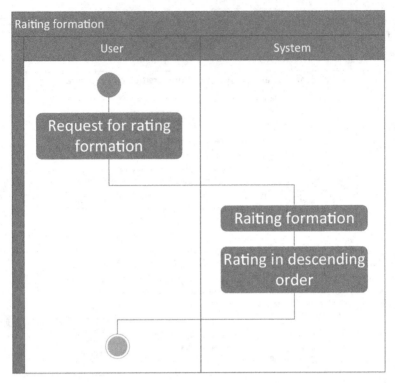

Fig. 4. Activity diagram of the lating formation process.

Figure 5 shows the algorithm for checking supporting documents by a representative of the commission. To perform this process in the ASFNR, a representative of the commission must log in to the system, open the list of universities and select the required university from the list. They check its supporting documents and, as a result of this process, either confirm the documents of the university or leave their comments about the inconsistency of documents in the form of a comment.

The commission module executed in the system, in turn, checks the authorization of the commission representative (user) in the system and all the rights granted to him, displays a list of universities for him and opens a list of supporting documents from the representative of the university. At the end it saves all the changes made.

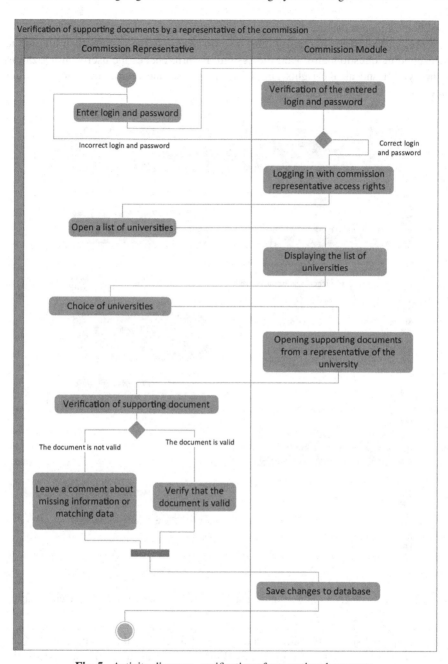

Fig. 5. Activity diagram - verification of supporting documents.

When filling out the questionnaire by a representative of the university (see Fig. 6), the representative of the university must also first log in to the system, then fill out the corresponding questionnaire. And the system, in turn, checks the user's authorization in the system and all the rights granted to him, and then returns the corresponding questionnaire to him to fill out.

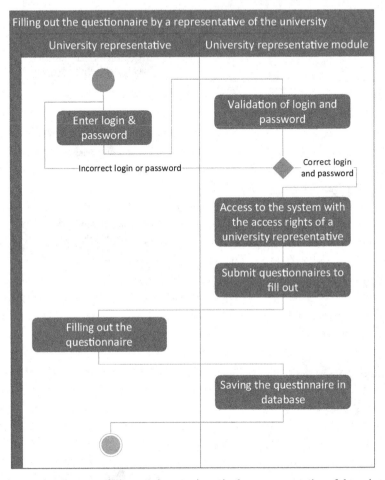

Fig. 6. Activity diagram - filling out the questionnaire by a representative of the university.

Viewing the list of universities is a function performed by the Administrator of the ranking formation system of universities (Fig. 7.). Like all users of the system, the administrator must log in to it, i.e. enter his username and password. Then, he requests the system to display a list of universities that are available in the ASFNR database, on which the administrator can perform operations of adding new, modifying or deleting existing data and request the system to save the changes made.

The administrator module in the system performs the following actions: checks the authorization of the system administrator, checks the rights granted to him to perform the requested operation, and saves the user's changes.

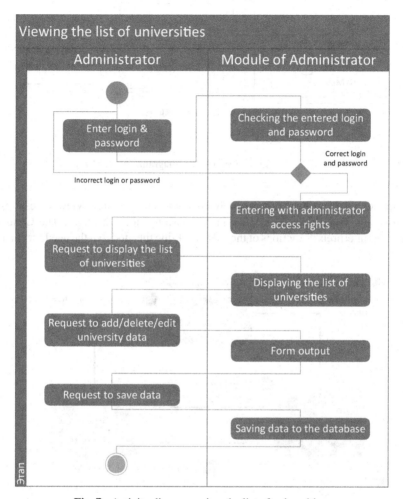

Fig. 7. Activity diagram - view the list of universities.

A UML Deployment Diagram has been built that displays the physical deployment of artifacts on nodes, i.e. what hardware and software components run on each node and how the various parts of this complex connect to each other. There are two types of nodes: Devicenode, Runtimehost. In our example, a runtime node is presented - a software-computing resource that runs inside an external node and is a service that executes other executable software elements (Fig. 8).

For this system, the MVC (Model-View-Controller) architecture was chosen.

The MVC architecture (Fig. 9) allows you to divide the application code into 3 parts: Model (Model), View (View) and Controller (Controller). Model (Model) provides data

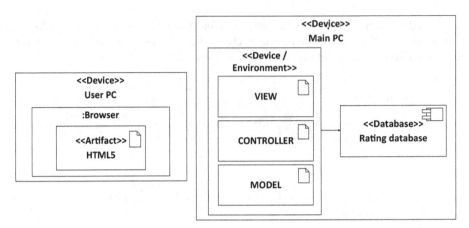

Fig. 8. Deployment diagram.

and responds to controller commands by changing its state. The View is responsible for displaying model data to the user in response to model changes. The Controller (Controller) interprets the actions of the user, notifying the model of the need for changes.

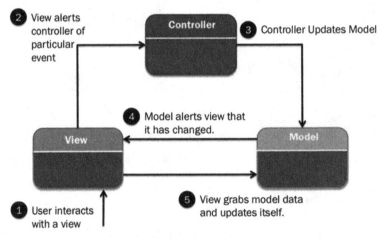

Fig. 9. MVC architecture.

The database structure consists of 13 tables: University_Questionnaire_Data, QuestionnaireCriteriaValue, Universities, University_questionnaire_questions_Mapping, University_questionnaire_questions, Levels, Groups, Criteria, Users, Positions, User-Roles, Student_Questionnaire_Data,Employee_Questionnaire_Data (Fig. 10).

For the development of an automated rating system, the following were selected:

- Web application - HTML + JavaScript for the implementation of the client side and Framework - ASP.NET MVC Framework,

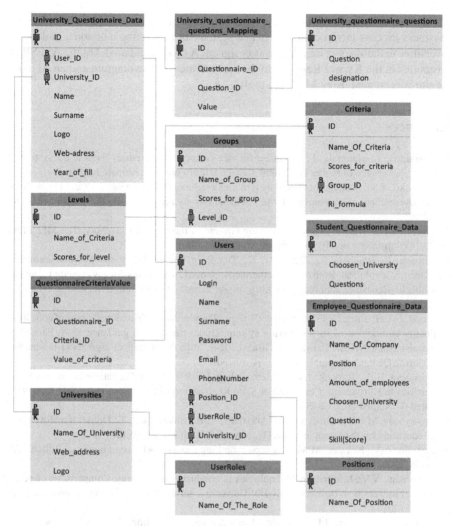

Fig. 10. Structure of the ASFNR database.

- Database - MS SQL Server for the implementation of the server part of the ASFNR.

4   Conclusion

The developing Web Application of the Automated System for the Formation of the National Ranking of universities will be a database of basic information on the universities of Kyrgyzstan, which automates the process of ranking formation and will identify the rating of educational institutions according to various indicators of the criteria and their individual groups.

In the framework of the grant of the Ministry of Education and Science of the Kyrgyz Republic for implementing research and development work on the topic "Development

and implementation of a Model of the National Ranking System of Higher Education",
a Business Process model of an automated system for generating a national ranking of
universities in the Kyrgyz Republic was designed. A conceptual model of the ASFSR of
universities in the Kyrgyz Republic in the form of a Use Case diagram and an Activity
diagram was created using MS Visio tools.

References

1. Navodnov, V.: Materials of the seminar for vice-rectors of universities "New in the system of quality assurance of higher education" (2007). http://www/nica.ru/main.ru/phtml
2. Gladkikh, Y.N.: Analysis of university rankings as a basis for determining target indicators for the development of innovative activities of the university PSNRU. Alley Sci. **10**(26), 181–189 (2018)
3. Losh, E.: The War on Learning: Gaining Ground in the Digital University. The MIT Press (2014). https://doi.org/10.7551/mitpress/9861.001.0001
4. Altbach, P., Reisberg, L., Rumbley, L.: Trends in global higher education: tracking an academic revolution. Rotterdam, Netherlands. Sense (2010). https://www.cep.edu.rs/public/Altbach,_Reisberg,_Rumbley_Tracking_an_Academic_Revolution,_UNESCO_2009.pdf. Accessed 15 Oct 2020
5. Wood, J., Carpenter, J.: (Education for Change): Development of a national model of education quality and design of its monitoring system. Under the general guidance of A. Parsi, Yu. Narolskaya, UNICEF Uzbekistan Office (2014). https://www.unicef.org/uzbekistan/media/3656/file/QEF_Conceptual_Framework_rus.pdf
6. Koshoeva, B.B., Torobekov, B.T.: Methodology for determining the national ranking of universities in Kyrgyzstan. Certificate of State Registration of the object. Copyright No. 3916, dated 07 August 2020 (2020)
7. Chynybaev, M.K., Koshoeva, B.B., Arzybaev, A.M., Bakalova, A.T.: Digital transformation of education by the example of KSTU. News KSTU **4**(52), 88–95 (2019)
8. Kozulin, A.V., Kovalev, M.M.: University ranking models. belarusian banking. Bulletin **23**, 18–26 (2001)
9. Grinshkun, V.V.: Information technologies in conducting monitoring comparisons of pedagogical universities. Bulletin of RUDN University, series "Informatization of education" **3**, 19–23 (2014)
10. Ivanov, S.S., Volkova, I.E.: Conceptual foundations of the rating of universities and educational programs of the Eurasian Union. Eurasian Union of Scientists (ESU). Pedagogical Sci. **11**(20), 28–31 (2015)
11. Chynybaev, M.K., Koshoeva, B.B., Bakalova, A.T., Abdyldaeva, A.R.: Designing an automated system for the formation of a national ranking of universities in the Kyrgyz Republic. High-Perform. Comput. Syst. Technol. **6**(1) (2022)

Evaluation of the Efficiency of the Vaccination Against Chickenpox Based on the Results of Simulation Modeling

Anton V. Kulshin[1] , Denis Yu. Kozlov[1] , Ekaterina A. Peredelskaya[2] ,
Lyubov A. Khvorova[1 (✉)] , and Tatyana V. Safyanova[2]

[1] Altai State University, Lenin Avenue 61, 656049 Barnaul, Russia
KhvorovaLA@gmail.com
[2] Altai State Medical University, Papanintsev Street 126, 656031 Barnaul, Russia
katrin_05_07_1995@mail.ru

Abstract. Chickenpox is one of the most widespread diseases in the world. Since the domestic vaccine has not been developed at present, and the purchase of an imported vaccine for routine vaccination is an expensive investment of the state, there is a need to assess the epidemiological and socio-economic effectiveness of vaccination based on the results of simulation modeling. This article describes the implementation of a simulation model of the spread of chickenpox in the population as a system for managing and strategic planning of vaccination against the Varicella Zoster virus. An algorithm for computer simulation of the spread of chicken pox with the features of various types of vaccination is given. A comparative analysis of the results obtained in the process of constructing the model with real medical data was carried out. The software implementation of the model was carried out in the high-level programming language Python. To determine the parameters of the simulation model, statistical data on the incidence of chickenpox in the Altai Territory for 2008–2018 were used. Such a system of vaccine prevention management allows, without long-term medical examinations and expensive procedures, to identify and, in particular, determine the best strategy for vaccination in a fairly short period of time. As a result of modeling, it was possible to identify the most optimal strategy for vaccination against the varicella-zoster virus in the Altai Territory, build graphs of the dynamics of the spread of the virus, reflecting the number of cases during one of the types of vaccination and without such procedures. It allowed one to evaluate the epidemiological and socio-economic efficiency of vaccination programs against chickenpox for children in the Altai Territory. Thus, computer technologies and computer modeling, a comprehensive computational study of the dynamics of the development of various diseases can help medical workers determine the optimal strategy in the treatment of diseases, as well as predict its consequences.

Keywords: Simulation model · Varicella zoster virus · Chickenpox · Python · Control system · Vaccination · Monte-Carlo simulation · Altai krai

V. Jordan et al. (Eds.): HPCST 2022, CCIS 1733, pp. 183–195, 2022.
https://doi.org/10.1007/978-3-031-23744-7_14

1 Introduction

Among the medical measures implemented in the fight against infectious diseases, one of the leading places is given to vaccination. All over the world it is recognized as the most effective, economical and affordable tool in the fight against infections. In our country, vaccination has been raised to the rank of state policy capable of preventing, limiting the spread and eliminating infectious diseases.

Chickenpox is an acute systemic, usually childhood infection caused by the varicella zoster virus (varicella zoster). It is an extremely contagious and widespread worldwide disease with a number of specific characteristics: it spreads by airborne droplets, in which the virus is inoculated onto the mucous membrane, or through direct contact with the carrier of the virus; has a high contagiousness [1], susceptibility to infection among susceptible people reaches 95–100% [2, 3]; characterized by fever, moderate intoxication, and widespread vesicular rash [4, 5]. Children of preschool age who attend children's educational institutions are mainly susceptible to infection. They account for an average of 94–96% of the total number of cases, while the peak of the disease occurs at the age of 3–4 years. After the disease, 97% of those who have been ill develop lifelong immunity [3, 6].

Based on the experience of organizing vaccine prophylaxis against varicella zoster virus abroad, on the territory of the Russian Federation, including on the territory of the Altai Territory, various strategies and methods of organizing vaccine prophylaxis are currently being considered [7–9]. Approximate cost calculations indicate that at least 5.1 billion rubles were spent from the country's budget in 2021 on the treatment of patients and anti-epidemic measures for "controlled" infections. At the same time, it is known that the cost of vaccination for any infection, the epidemiological effectiveness of which has been proven, is about 10 times less than the cost of treating the disease. In these conditions, the medical and economic importance of vaccine prophylaxis becomes obvious, the organization and conduct of which carries elements of increased responsibility for the epidemiological well-being of the country's population. In addition, the relevance and practical significance of the study are related to the needs of doctors in a system capable of managing and planning the timing and methods of vaccine prophylaxis, when determining the timing of reaching the peak of the spread of the varicella zoster virus and the maximum decrease in the incidence rate with reaching a minimum plateau in the level of disease transmission.

The purpose of the study is the use of computer technologies, simulation modeling and computational technologies to study the dynamics of the spread of chicken pox, as well as to develop a control system and predict the spread of the disease among the population of the Altai Territory, the timing of reaching the peak and maximum reduction in the incidence rate with reaching a minimum plateau in the level of disease transmission to assess the effectiveness of planned vaccination on the general epidemic situation.

As information support for the simulation model of the spread of chicken pox and to determine the parameters of the model, data on the incidence of chickenpox from statistical reporting forms No. 2 of the Federal State Statistical Observation "Information on infectious and parasitic diseases" in the Altai Territory for 2001–2019 and data on the population of the Federal Agency for State Statistics in the Altai Territory were used.

2 Development and Implementation of Simulation

Simulation is often used in medicine. This is due to the fact that it is often impossible to experiment on a real object, but there is a need to replicate the behavior of the system under study based on the analysis of the most significant relationships between its elements [10]. The simulation model can be considered as a set of rules that determine which state the system will go to in the future from a given current state.

Many researchers use the classic SIR-model [11], in which the entire population is divided into three groups: disease-susceptible (Suspect), infected (Infected) and Recovered/Removed (Recovered/Removed). The model assumes that those who have been ill cannot be infected again. The model is described by a system of 3 differential Eqs. (1) and initial condition (2):

$$\frac{dS}{dt} = -r \cdot S \cdot I, \ \frac{dI}{dt} = r \cdot S \cdot I - v \cdot I, \ \frac{dR}{dt} = v \cdot I, \tag{1}$$

$$S(t_0) = S_0, \ I(t_0) = I_0, \ R(t_0) = R_0. \tag{2}$$

Here $S(t)$ – the number of people susceptible to the disease, $I(t)$ – the number of infected, $R(t)$ – the numbers of recovering individuals at time t; r is the rate of infection transmission; v is the rate of recovery of infected individuals.

It can be seen from the system of Eq. (1) that for the total population M

$$M(t) = S(t) + I(t) + R(t) = const$$

the conservation law is satisfied:

$$\frac{dS}{dt} + \frac{dI}{dt} + \frac{dR}{dt} = 0.$$

The number of cases in the model is determined by the formula:

$$Z(t) = M - S(t) = I(t) + R(t).$$

We used a modified SEIR simulation model of the spread of the epidemic. Here, one more state is added to the three states in the SIR model – Exposed (infected, in the incubation period). In addition, the study takes into account the level of risk depending on the age of the patient, the number of social connections, the likelihood of infection, the presence of a vaccine in the body, and others.

One of the simulation methods is Monte Carlo simulation. The constructed simulation models are based on the Monte Carlo simulation method [12, 13], the essence of which is to build mathematical models using a random number generator. The Monte Carlo method refers to simulation modeling, in which, when calculating a system, the behavior of all its components is reproduced and studied [13].

Such models help to assess the rate of infection spread, characteristics of vulnerable groups of the population, the optimal age of vaccination and other social and economic factors associated with the disease.

The construction of simulation models by the Monte Carlo method is implemented in the following steps:

- the system is decomposed into simpler parts-blocks;
- laws and "plausible" hypotheses are formulated regarding the behavior of both the system as a whole and its individual parts;
- depending on the questions posed to the researcher, the so-called system time is introduced, which simulates the course of time in a real system;
- the necessary phenomenological properties of the system and its individual parts are specified in a formalized way;
- random parameters appearing in the model are associated with some of their implementations, which are constant for one or more cycles of system time.

```
#-----------------Human simulation--------------------

class Person():
  def __init__(self, age = 0):
    self.age = age #Age in days
    self.immunity = False #Resistance
    self.infected = False #Infected or not infected
    self.isolation = False #Self-isolation
    self.live = True #Alive/not alive
    self.contagiousDays = 0 #How many days sick
    self.EndDisease = 0 #When the sickness ends
    self.friends = 0 #Number of friends
    self.probability = 0 #Chance of Infection
    self.vac = 1 #Availability of a vaccine
    self.startDisease = 0 #Day of infection
```

Fig. 1. Model parameters.

The software implementation of the model was carried out in the high-level programming language Python [14–17]. For the software implementation of the simulation model, the following most significant parameters were selected for each individually special person: age, the presence of developed immunity, the duration of the course of the disease, the number of social connections, the likelihood of infection, the presence of a vaccine in the body, and others. A complete list of model parameters is shown in Fig. 1.

Under natural conditions, the virus infects a person by entering the body through airborne droplets or through direct contact with a sick person [3]. Based on this, the simulation model for several virtual days simulates social meetings in various age groups, within which they may be infected, by generating a random number. An analysis of the spread and severity of the course of the disease suggests the need to divide the entire study population into age categories: 0–2 years, 3–6 years, 7–17 years, 18 and older.

For 2005–2018 according to the Altai Territory, the age group 0–2 years old included 4% of the total population, which averaged 2.5 million people, the group 3–6 years old – 4%, the group 7–17 years old – 11%, the group 18 years and older – 81%. The initial data are presented in Table 1.

Table 1. The population of the Altai Territory by age for 2005–2018.

Year	Children 0–2 years old (total)	Children 3–6 years old (total)	School children 7–17 years old	Adult AT	Total population
2005	82917	105285	279387	2071841	2539430
2006	78355	102999	276080	2046077	2503510
2007	80749	102532	272081	2017662	2473024
2008	79118	104002	278770	2046588	2508478
2009	83611	105600	266992	2040573	2496776
2010	81659	103135	260759	1974202	2419755
2011	78891	103033	263188	1972246	2417358
2012	86511	102601	254135	1963983	2407230
2013	88992	108454	240846	1960459	2398751
2014	92487	114748	263652	1919751	2390638
2015	93156	116629	268026	1907001	2384812
2016	93102	118351	272544	1892777	2376774
2017	91935	119714	277480	1876551	2365680
2018	87546	118554	284157	1859823	2350080
Total, %	**4%**	**4%**	**11%**	**81%**	**100%**

The probability of human infection as a result of contact with the patient was calculated as the average value of the ratio of the number of cases to the total number of people on a monthly basis. This is due to the peaks of the spread of the disease, which occur in the winter-spring period (from December to April). Empirical probability of infection by month for 2005–2018 was: January – 14%, February – 12%, March – 10%, April – 11%, May – 9%, June – 7%, July – 5%, August – 2%, September – 3%, October – 6% %, November – 10% and December – 11%.

Figure 2 shows a fragment of the software implementation of a virtual meeting of the simulated population. If the generated random number exceeds the probability of infection, then after contact with an infected person, the person will be considered infected.

The parameters related to the duration of the disease, as well as the incubation period and the period of self-isolation, were selected based on the observations of doctors. They found that from the moment the infection enters the body and until the infected person fully recovers, an average of 24 days pass. In this acse, the incubation period usually lasts 12 days, 1–2 days infected people are contagious and are considered carriers of the disease. And with the manifestations of the first symptoms, such as rash and hyperthermia, they go to self-isolation until full recovery and development of immunity [1, 2]. These features were also taken into account in the developed system.

```
#We will choose a random number of his half of friends,
#if there are any.
if peopleMeetToday > 0 and person.isolation == False \
  and person.contagiousDays > daysContagious//2 \
  and tms != 3 and tms !=4:
  peopleMetToday = random.randint(0,peopleMeetToday)
else:
  peopleMetToday = 0

#With a probability of 80%, a person will prefer to communicate
#with a person from their age group
for _ in range(0, peopleMetToday):
  if random.randint(0,100) <= 80:
    friendsMeet = peopleDic[random.randint(0,len(peopleDic)-1)]
    while Ages(friendsMeet.age) != Ages(person.age):
      friendsMeet = peopleDic[random.randint(0,len(peopleDic)-1)]
  else:
    friendsMeet = peopleDic[random.randint(0,len(peopleDic)-1)]

  #If the "friend" does not have immunity to the disease at the meeting
  #and he is not infected, then when he meets the patient,
  #we play the probability of the disease.
  if friendsMeet.infected == False and friendsMeet.immunity == False:
    if random.random()*friendsMeet.vac > (1-season[ind]):
      friendsMeet.infected = True
      friendsMeet.EndDisease = random.randint(daysContagious-1,
        daysContagious+1)
      friendsMeet.startDisease = tm
      infPeople += 1
```

Fig. 2. Fragment of software implementation of the model

In addition to the likelihood of a person becoming infected through contact with a sick person, the simulation model takes into account the number of social contacts, among which there may be a sick person. In the age group 0–2 years, their number is usually small, so the value was chosen from the interval of 1–5 people. This is due to the fact that newborn children most often contact with their parents and possible brothers, sisters, grandparents, that is, with a limited circle of people up to a certain point in time. Starting from the age of 3, when the child begins attending preschool institutions, the number of his social contacts increases dramatically. Therefore, for a group of 3–6 years, up to 20 relationships with other people were established. Similarly, for a group of 7–17 years old – up to 20 contacts per day. For the adult part of the population, that is, for the group of 18 years and older, most often working in offices, the number of social connections is about 10 people.

The above number of social connections was played as a normally distributed random number separately for each person in the simulation model in accordance with his age group.

One of the main objectives of the presented study was to model the consequences of three different types of vaccination [18, 19]:

- type 1 vaccination, carried out in two doses – at the age of 1 year, after which the probability of disease after the production of antibodies is 20%, and at 6 years – the probability of disease after the production of antibodies is only 5%;

- type 2 vaccination, carried out in two doses – at 6 years old – the probability of disease after the production of antibodies is 20%, and at 6 years 30 days after the first vaccination – the probability of disease after the production of antibodies is 5%;
- type 3 vaccination, carried out in one step – at 6 years old – the probability of disease after the production of antibodies is 20%.

3   Analysis of the Simulation Results

Based on the results of the developed management system and strategic planning of vaccination based on a simulation model, it was possible to find the optimal plan for vaccinating the population of the Altai Territory. To compare the effectiveness of different types of vaccination, the result of simulation modeling without vaccination was obtained (Fig. 3). So, according to Fig. 4, it can be determined that the first type of vaccine prophylaxis has a stronger effect than the others do.

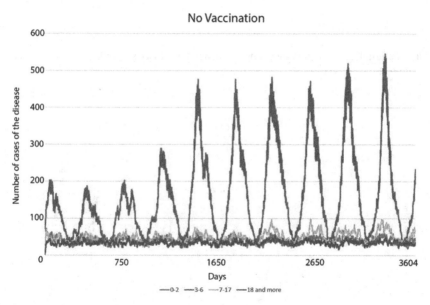

Fig. 3. The result of simulation modeling is the number of cases in different age groups without vaccination.

The graph clearly shows the seasonality of the disease and the gradual increase in infected children, especially in the age group from 3 to 6 years old.

Figure 4 shows the effect of the first vaccination strategy, where the first dose is given with parental consent at 1 year and the second at 6 years old. It can be seen that the number of sick children at the age of 3–4 years is significantly reduced.

This nature of the spread of the disease tells us that vaccinated children at the age of 1 year get sick less often after reaching 3–6 years. Strategies of types 2 and 3 do not have such significant effect. Figures 5 and 6 show that the number of cases has decreased by approximately 10%.

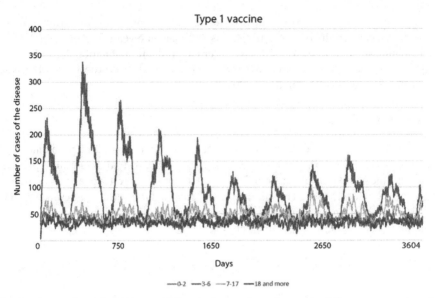

Fig. 4. The result of simulation modeling is the number of cases in different age groups after vaccination of the 1st type.

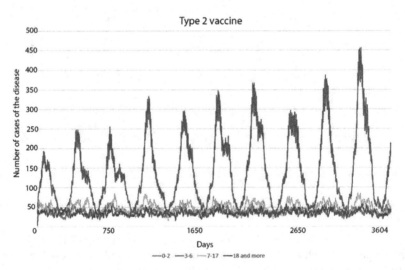

Fig. 5. The result of simulation modeling is the number of cases in different age groups with type 2 vaccination.

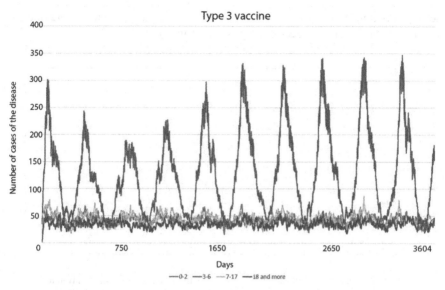

Fig. 6. The result of simulation modeling is the number of cases in different age groups with type 3 vaccination.

4   Evaluation of Vaccination Economic Efficiency for Altai Krai

The economic evaluation was carried out on the basis of guidelines 3.3.1878–04 dated 03/04/2004 "Economic efficiency of vaccine prevention", which outlines the basic principles and methods for performing calculations to assess the economic indicators of vaccine prevention of infectious diseases in order to select its optimal strategy. The guidelines use individual elements of the methodology implemented in the world practice. This involves the experience of employees of the laboratory of epidemiological analysis of the Central Research Institute of Epidemiology of the Ministry of Health of Russia, accumulated in the course of many years of work on the calculation and analysis of the economic aspects of vaccine prevention of infectious diseases.

When calculating the cost (damage) of one case of chickenpox, three main factors were taken into account:

1. damage from temporary disability;
2. hospitalization and outpatient doctor visits;
3. expenses of parents' own funds for transportation, treatment, diagnostics and other expenses.

The cost of a dose for vaccination of one child was taken equal to 2325 rubles. According to the recommendations of the World Health Organization, vaccination coverage was 95% [19]. According to the literature data, with a single vaccination, the efficiency is 80% [18]. When taking into account the costs of vaccination, the costs of treating ill children aged 0–6 years were taken into account, the average cost of treatment was 16.84 million rubles. The results of the calculations showed that the cost of treating chickenpox is almost twice the cost of the vaccination program, which indicates the high economic efficiency of this vaccination program (Fig. 7).

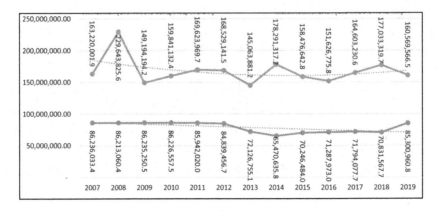

——◆—∟ the cost of vaccination at 6; ·········· – the linear cost of vaccination at 6;
——◆—– the cost of treatment; ········· – the polynomial cost of treatment.

Fig. 7. Dynamics of economic costs for the treatment and vaccination of chickenpox in the population of Altai Krai for 2007–2019 with trend lines [9]:

Taking into account the data on the incidence of chickenpox in the Altai Territory, on average for 2001–2019, the number of cases of diseases that should be expected in the analyzed cohort as it grows older was the calculated ones (Fig. 8).

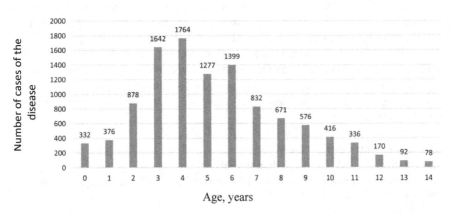

Fig. 8. Expected number of varicella cases in the analyzed cohort as they grow older (absolute number).

The number of varicella cases averted is related to the number of expected varicella cases in the analyzed cohort and differs from it by an amount proportional to vaccine effectiveness and vaccination coverage. With two vaccinations at the age of 1 year, an efficiency of 80% after the first dose [20] and 95% after the second dose [19] was taken at 95% vaccination coverage (Fig. 9).

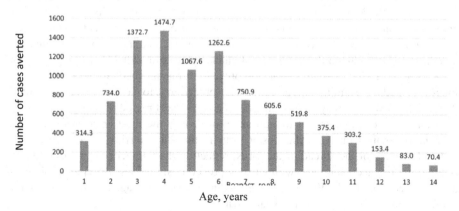

Fig. 9. Number of cases of chickenpox averted in the analyzed cohort as it matures (absolute number).

To determine the economic efficiency of vaccinal prevention against chicken pox and self-sufficiency of investments, we analyzed the ratio of the costs of vaccination and the amount of damage prevented.

The self-sufficiency of vaccination against chickenpox, carried out in 2022, for children aged 6 years, will be achieved only by 2032. The economic benefit will be about 7.52 million rubles. Effectiveness from the implementation of the vaccination program will be observed in five years. At the same time, the incidence rate may decrease by 2–2.5 times. The average rate of decline in the incidence of chickenpox among adults was 0.5%. There was no difference between the number of cases without vaccination and during the implementation of the program. The reason is the short period of observation, the insufficiency of the immune layer and the redistribution of the incidence among the population. However, modeling results show that the vaccination program leads to the earliest reduction in the incidence of adolescents, and most likely to the same effect in adults.

5   Conclusions

Information systems and software and computer technologies began to penetrate more and more into medicine. Information systems and software and computer technologies are transforming management methods and the efficiency of any institution. Medical informatics and software and computer technologies consider issues not only of healthcare informatization. First of all, medical information systems are tools designed for the daily work of a doctor and medical personnel, monitoring the quality of medical care, analyzing current medical and preventive work, monitoring health indicators, making managerial decisions and economic forecasting of the functioning of the healthcare system.

This article describes the implementation of a simulation model of the spread of chickenpox in a population as a system for managing and strategic planning of vaccination against the Varicella Zoster virus. An algorithm for computer simulation of the spread of chicken pox with the features of various types of vaccination is given. A comparative analysis of the results obtained in the process of constructing the model with real medical data was carried out. To determine the parameters of the simulation model, statistical data on the incidence of chickenpox in the Altai Territory for 2008–2018 were used. Such system of vaccine prevention management allows one, without long-term medical examinations and expensive procedures, to identify and, in particular, determine the best strategy for vaccination in a fairly short period of time.

As a result of modeling, it was possible to identify the most optimal strategy for vaccination against the Varicella Zoster virus in the Altai Territory. We built graphs of the dynamics of the spread of the virus, reflecting the number of cases during one of the types of vaccination and without such procedures, to assess the epidemiological and socio-economic effectiveness of the vaccination program against chicken pox for children using the example of the Altai Territory.

Conclusions are based on the simulation results:

1. The number of sick children aged 3–4 years after vaccination is significantly reduced. Taking into account the results of modeling and data from literature sources, it is advisable to implement a two-time immunization program with the introduction of the first dose of vaccination at 12 months and the second dose at the age of 6 years. However, given the specifics of some regions of Russia, in particular the Altai Territory, where varicella vaccination procedures have not previously been carried out, it is recommended to vaccinate children who have reached 6 years of age under a single vaccination program as initial measures to combat the "growing up" of the infection.

2. Vaccination to combat chickenpox in terms of its monetary costs will be more profitable than the usual treatment of a viral infection. The cost of therapy exceeds the cost of vaccination almost twice. It should be noted separately that the self-sufficiency of immunization of children who have reached the age of 6 years will be achieved in 2032, provided that the program is implemented in 2022. If these conditions are met, the economic benefit will amount to 7.52 million rubles, and the prevented damage will exceed the costs by 1.2 times. The effectiveness of vaccination in epidemiological terms will be traced in five years from the moment of its implementation – a reduction in the incidence of 2–2.5 times.

3. At the moment, the developed system allows planning and testing various vaccination strategies in a simulation experiment [9]. The use of simulation modeling in medicine makes it possible with a high degree of probability to conclude which vaccine prophylaxis method is the most justified.

The findings of such studies can be used by public health organizations and authorities to successfully combat the spread of infection.

Acknowledgments. This work was supported in the framework of "Priority-2030" Program by Altai State University.

References

1. Afonina, N.M., Miheeva, I.V.: Socio-economic significance of infectious pathology caused by the Varicella zoster virus. Materials of the XXI Congress of Pediatricians of Russia with international participation "Topical problems of pediatrics". Moscow (2019). (in Russian)
2. Balikin, V.F., Filosofova, M.S.: Expansion of clinical polymorphism and increasing severity of Varicella zoster infection in children. Materials of the XIII Congress of Pediatric Infectious Diseases of Russia "Topical issues of infectious pathology and vaccination". Moscow (2014). (in Russian)
3. Streng, A., Grote, V., Rack-Hoch, A., et al.: Decline of neurologic varicella complications in children during the first 7 years after introduction of universal varicella vaccination in Germany, 2005–2011. Pediatr. Infect Dis. J. **36**(1), 79–86 (2017)
4. Zryachkin, N.I., Buchkova, T.N., Chebotareva, G.I.: Complications of varicella (literature review). J. Infectology **9**(3), 117–128 (2017). (in Russian)
5. Chickenpox – MayoClinic Reference. https://www.mayoclinic.org/diseases-conditions/chickenpox/symptoms-causes/syc-2035128
6. Timchenko, V.N.: Infectious diseases in children. St. Petersburg, SpetsLit, 218–224 (2012). (in Russian)
7. Zryachkin, N.I., Buchkova, T.N., Elizarova, T.V., Chebotareva, G.I.: Pharmacoeconomical justification for the inclusion of vaccination against chickenpox in the regional calendar of preventive vaccinations on the example of the Penza region. Epidemiol. Infect. Dis. **22**(6), 288–294 (2017). (in Russian)
8. Order of the Ministry of Health of the Russian Federation No. 1122n dated 06.12.2021 "On approval of the national calendar of preventive vaccinations, the calendar of preventive vaccinations for epidemic indications and the procedure for preventive vaccinations" (in Russian)
9. Peredelskaya, E.A., Safyanova, T.V., Kozlov, D., Kulshin, A.V., Khvorova, L.A.: Epidemiological and socio-economic evaluation of the effectiveness of a single vaccination program against chicken pox for children 6 years old on the example of the Altai Territory. Medicine **4**, 66–79 (2021). (in Russian)
10. Taha, H.A.: Operations Research: An Introduction. Williams, Moscow, 697–737 (2007). (in Russian)
11. Kermack, W.O., McKendrick, A.G.: A contribution to the mathematical theory of epidemics. Proc. Royal Soc. London Series A, Containing Papers Math. Phys. Charact. **115**(772), 700–721 (1927)
12. Barton, P., Bryan, S., Robinson, S.: Modelling in the economic evaluation of health care: selecting the appropriate approach. J Health Serv Res Policy **9**(2), 110–118 (2004)
13. Briggs, A., Claxton, K., Sculpher, M.: Decision Modelling for Health Economic Evaluation. Oxford University Press, London (2007)
14. Vander, P.: Python for Complex tasks: Data Science and Machine Learning. Peter, St. Petersburg (2018). (in Russian)
15. Rashka, S.: Python and Machine Learning: A textbook. DMK Press, Moscow (2017)
16. Downey, A.B.: Studying Complex Systems Using Python. DMK Press, Moscow (2019)
17. Downey, A.B.: Modeling and Simulation in Python, Green Tea Press (2017)
18. Lee, Y.H., Choe, Y.J., Cho, S.I., et al.: Effects of one-dose varicella vaccination on disease severity in children during outbreaks in Seoul Korea. J. Korean Med. Sci. **34**(10), e83 (2019)
19. Perella, D., Wang, C., Civen, R., et al.: Varicella vaccine effectiveness in preventing community transmission in the 2-dose era. Pediatrics **137**(4), e20152802 (2016)
20. Marin, M., Marti, M., Kambhampati, A., et al.: Global varicella vaccine effectiveness: a meta-analysis. Pediatrics **137**(3), e20153741 (2016)

Hierarchical Volume Mesh Model of Heterogeneous Media Based on Non-Destructive Imaging Data

Daria Dobroliubova[1,2](✉) 🄳 and Ekaterina Shtanko[1] 🄳

[1] Trofimuk Institute of Petroleum Geology and Geophysics, SB RAS, Koptug Ave. 3, 630090 Novosibirsk, Russia
`mik_kat@ngs.ru`
[2] Novosibirsk State Technical University, Karl Marx Ave. 20, 630073 Novosibirsk, Russia

Abstract. In this paper, we have proposed an algorithm for building a hierarchical 3D mesh model of heterogeneous media, native or human-made, described by a stack of cross-section 2D images obtained via digital imaging methods, such as magnetic resonance imaging (MRI), computed tomography (CT), etc. The algorithm yields a multi-level hierarchical mesh accurately capturing the complex internal geometry of the medium at the micro-level. The meshes are oriented towards the modern nonconforming multiscale finite element methods. We have provided a validation of the algorithm in a test medium containing microinclusions of various geometric shapes. To simulate the output data of CT non-destructive imaging, we have generated the stack of slice images of the test media concerned. We have obtained the accuracy estimations for the volume fraction of inclusions and the interface surface area. We also have shown the scalability of the algorithm proposed using OpenMP technology.

Keywords: Discrete geometric model · Digital core · Unstructured mesh · Hierarchical mesh · Multiscale finite elements

1 Introduction

Imaging technologies, such as computed tomography (CT), nuclear magnetic resonance (NMR) spectroscopy, etc., are used to digitize the internal structure of the complex objects as a set of two-dimensional digital images of the cross-sections, without disrupting its integrity. Non-destructive imaging methods are widely applied in medical and biological applications, materials science, geophysics, and petrophysics. In medical applications, imaging techniques and reconstruction of the object's internal structure are used in the diagnostics of the diseases, development and insertion of various medical implants [1]. In materials science, imaging methods are employed in composite materials, meta- and nano-materials manufacturing [2], since they facilitate quality control of the internal structure without physical invasion into the studied object [3]. In geophysics investigations, they are widely used to study the internal structure of rock samples (cores)

[4]. As non-destructive imaging techniques are rapidly advancing, their wide application in various fields of science and engineering led to the progress in development of the methods for creating virtual counterparts to the physical heterogeneous objects (so-called "digital twins") based on the imaging data. When constructing a digital twin of the complex media, the aim is to capture, as accurately as possible, its internal structure and micro characteristics.

In a certain number of research fields, it is enough to navigate inside a reconstructed three-dimensional medium or to highlight specific anomalous zones (for example, cracks in geophysics, areas of structural changes in medicine, material defects in material science, etc.). However, digital twins also provide vast opportunities to study the effects of various physical impacts, microstructural properties, and features of heterogeneous media on their macroscopic characteristics via mathematical modeling. In fact, numerical simulation in this case can benefit or even replace the physical laboratory experiment.

2 Methods for Reconstructing the Internal Structure

The conversion from a stack of slice images to some three-dimensional structural model is a pre-modeling step. The format of the model strongly depends on the physical experiment that is numerically reproduced, and the numerical method to be implemented.

In some cases, researchers can use an idealized reduced model. For example, in the pore-network approach [5, 6] the porous medium is represented as a system of nodes and channels, neglecting their exact geometry. These models are widely used for such physical and chemical processes as non-Newtonian fluid displacement, interfacial exchange, simulation of turbulent flows, etc. The method is implemented in the freeware package OpenPNM for modeling single- and multi-phase transport processes in porous materials.

When using digital cores, the accuracy of numerical experiments largely depends on the accuracy of capturing the microstructural features of the physical rock core samples [7]. In this application, it is preferable to use methods that are aimed at constructing three-dimensional discrete models representing the internal geometry of the medium. Such discrete geometric models should capture all the internal features of the object under study as accurately as possible. Currently, there is no unified standard for constructing discrete models of the complex media based on a set of their slice images. The problem is being actively researched by both the academic community and commercial companies due to its high relevance and demand in various fields.

The direct approach is to treat each voxel of the image stack as a hexahedral mesh element. This approach is widely used in biomedical applications [8] and geophysics [9]. The obvious advantages of the method are the optimal quality of the resulting partition and the conformity to the internal boundaries. However, the application of the algorithm in practice is greatly hampered by the prohibitive size of the resulting mesh. Therefore, modifications to this method are usually made for its use in real-world applications. For example, the mesh is coarsened in the areas remote from the interfaces. In the media with a large number of internal boundaries (all native objects), such methods provide only a limited reduction of the mesh size. It is also noted in the literature that mesh models built based on the direct approach tend to overestimate the surface area of the media interface [10]. This has a significant impact on the numerical simulation results.

There is a wide class of two-step algorithms for reconstructing the internal structure of a heterogeneous medium. The first step of such methods is aimed at extracting the internal interfaces in the media as iso-surfaces (.stl-surfaces) where the color intensity function has a constant value. The second step is to construct conformal mesh partitions with respect to the selected iso-surfaces. Algorithms for constructing iso-surfaces from a set of CT images are now well developed. The most popular methods are marching cubes [11] or marching tetrahedrons [12]. These algorithms require the use of some primary mesh. The triangular interfaces are constructed by connecting the points where the approximation of the interface between different materials in the image intersects the edges of the primary mesh (and triangulating them, if needed) [13]. The obtained surfaces can be used directly to generate a mesh partition, e.g. by the advancing-front method, or processed to obtain a higher-order CAD model. These algorithms are quite stable and provide the construction of coherent models. Their main disadvantage is the two-step procedure. It takes considerable time, and the different steps may require different software products.

The geophysical problems involving core analysis are characterized by complex pore space connectivity, chaotic internal structure, and significant difference in the scale of inclusions. Multiscale, virtual, or heterogeneous finite-element methods proved to be efficient tools for numerical simulations in highly heterogeneous media [14–16]. The idea behind these methods involves some hierarchy of different scales and they are usually not restricted to standard geometric shape of the finite elements [16]. Therefore, the finite element mesh model intended for these methods should comply with these requirements. In our previous work [19], we considered an algorithm based on a marching simplex technique for constructing hierarchical mesh partition of the heterogeneous media suitable for the finite element methods of the multiscale family. In this paper, we propose a modification to this algorithm that combines the marching simplex approach with local refinement of the initial micro-level mesh partition to include the extracted interfaces between the components of the media into the fine-scale mesh model. We considered a two-level hierarchy, although the approach is straightforwardly generalized to more levels. Within the multiscale framework, we introduce the macro-scale partition and a primary tetrahedral meshing inside each macro-scale element, which is then rebuilt with respect to the interface boundaries. Information about interfaces is obtained from a set of slice images of the medium. This approach allows us both to capture the fine-scale internal structure of the medium more accurately and to avoid a catastrophic growth of the mesh partition size through independent work with each macro-scale element.

3   Problem Statement

CT and MRI scanning methods yield a set of slice grayscale images describing the internal structure of the medium. The number and resolution of the scans vary in a wide range and are highly dependent on the equipment. Usually, it is a significant amount of data, about 10^9 voxels.

In general, the algorithm for recovering the internal geometry of the area should include several preliminary steps:

1. Noise reduction and elimination of non-physical artifacts [17].

2. Increasing the image contrast.
3. Contour extraction.
4. Image segmentation into a finite number of subdomains with the same physical properties [18].

These procedures are of significant complexity and academic interest but are beyond the scope of this study and will not be discussed in this paper. Steps 1–3 are optional and depend on the quality of the original images.

We assume the input data to be a set of segmented images (Fig. 1a). We consider a domain with a certain internal structure (Fig. 1b), for which the volume fraction of the inclusions and the surface area of the inclusions are known a priori. To verify the proposed algorithm, we use the set of slice images generated for a synthetic computational domain as the input data. Two main subdomains are defined in the considered sample: the matrix (black in Fig. 1a) and the inclusions (white in Fig. 1a).

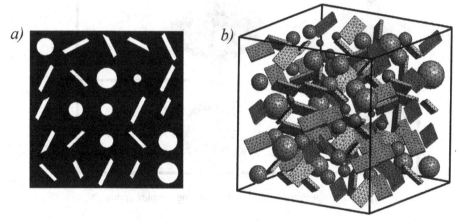

Fig. 1. Example of a scan from the generated stack (a) and internal structure (synthetic model) of a heterogeneous medium (b).

4   An Algorithm for Constructing a Discrete Geometric Model

Previously, the authors implemented two algorithms for solving this problem [19]:

1. An algorithm based on the primary internal meshing of the computational domain. The first step was to construct a conforming mesh for the entire computational domain. The second step was to establish a one-to-one correspondence between the mesh tetrahedra and the pixels of the segmented scan. This allowed us to determine which physical subdomain belongs to the finite element.
2. An algorithm based on the extraction of interface boundaries by the marching simplex method (Fig. 2).

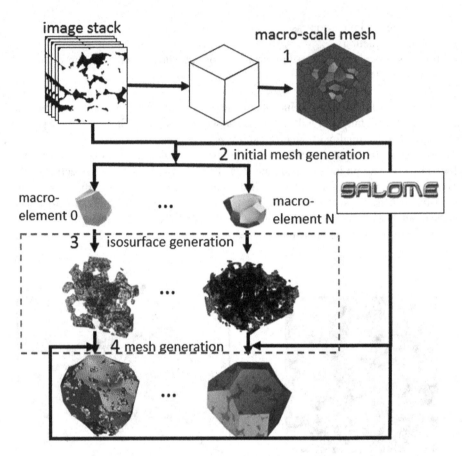

Fig. 2. An algorithm based on the marching simplex approach.

The algorithm shown in Fig. 2 is also based on a multiscale approach, with a two-level hierarchy of the domain partition. At the macroscopic level, the domain is meshed into a finite number of homogeneous elements – generally, polyhedrons. This step might be carried out after the image clustering procedure. In rock core samples, the pores or inclusions are often distributed uniformly in the matrix, hence, the clusterization yields partitions of Voronoi type. On the other hand, polyhedral elements may prove challenging in finite element applications, thus, in this work, we focus on hexahedral macro-scale elements. Inside each macro-scale element, the initial tetrahedral partition is built. This partition does not account for the interfaces inside the object, but one should use the image data to adjust the size of the element so that the marching simplex algorithm captures the essential microscopic features. In our work, we used an open-source SALOME software with the Netgen meshing kernel to construct the initial tetrahedral partitions.

The main drawbacks of the algorithms above were the high surface approximation error (up to 17% for smooth inclusions) and the low speed due to structure reconstruction procedure. In the hybrid algorithm proposed in this paper, we aimed at eliminating step

3 shown in Fig. 2, since it implies explicit construction of the stl-model representing the interfaces between the materials as triangulated iso-surfaces. This step requires the optimization of the triangulations, prescribing material tags to the volumes enclosed by the surfaces, and, finally, the construction of the new volume mesh with respect to these interfaces and, hence, the potential use of the meshing software once again. Instead, we skipped this step by locally refining the initial mesh to incorporate the interfaces into the model.

The hybrid algorithm includes the following basic steps:

1. Construct a regular partitioning of the entire computational domain, without considering its internal structure (macro-level mesh). Let us call these finite elements macro-scale elements

$$P = \left\{ \bigcup_{i=1}^{N} p_i \right\},$$

where N is a number of macro-scale elements;

2. For each macro-scale element $p_i, i = 1, N$:

   a. Construct the primary tetrahedral microlevel mesh

   $$T = \left\{ \bigcup_{j=1}^{M_i} t_j \right\}$$

   inside each macro-scale element p_i. Here M_i is a number of tetrahedrons in the macro-scale element p_i.

   b. For each tetrahedron tj, j = 1, Mi:

      (1) Use the values of the color intensity function $\rho(x, y, z)$ to determine the intersection points of the tetrahedron tj with the iso-surface S corresponding to the media interface. Incorporate them in the mesh as hanging nodes.
      (2) Make a local refinement of the primary mesh.
      (3) To each of the tetrahedrons obtained by refinement, prescribe the color value of the color intensity function $\rho(x, y, z)$ in the barycenter and vertices.
      (4) Identify the physical subdomain for the tetrahedron based on step (3).

   c. Optimize the mesh partition.

3. Check the quality of the discrete model.

Local refinement of the primary mesh (step 2.2.2) by including iso-surfaces corresponding to the media interface may result in ill-quality mesh elements. Consequently, optimization in step 2.3 means improving the quality of the mesh and rebuilding the tetrahedrons that do not satisfy the standard quality criterion for an individual tetrahedron, which consists in evaluating the ratio of the radii of the circumscribed and inscribed spheres [20]. Here, we note that the local refinement with respect to the interfaces may distort the initial mesh and worsen its overall quality. Optimization may be of limited help. For the resulting mesh to be of better quality, we can recommend that the initial mesh should mostly contain elements close to regular.

5 Validation of the Model

Figure 1b shows the cubic test sample of size 2 cm containing 125 inclusions: 42 spherical inclusions with varying radii, and 83 plate inclusions.

The inclusions volume fraction is 5.91 %. The integral surface area of the interfaces in the sample is $7.10 \cdot 10^{-3}$cm^2. The image sequence contains 600 slice images with a resolution of 600×600 pix (Fig. 1a). The macroscopic level mesh contains 216 hexahedral elements, each of them containing an average of 52 000 micro-level elements.

Figure 3 shows the result of the test medium internal structure reconstruction – a discrete three-dimensional mesh model obtained by the proposed algorithm. One layer of macro elements with reconstructed inclusions inside is shown in Fig. 3a. In Fig. 3c, the mesh partition in the macro-scale element in the cross-section passing through the center of the inclusion is zoomed in to demonstrate the refinement of the mesh to the "host medium-inclusion" interface.

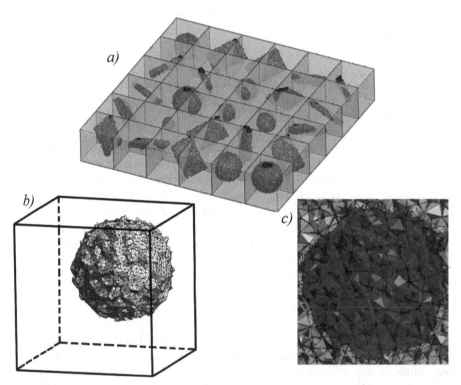

Fig. 3. The reconstructed medium internal structure: a) a layer of macro-scale elements; b) a certain macro-scale element; c) a cross-section passing through the inclusion (zoom-in).

To estimate the quality of the 3D mesh model built using a stack of slice images, one may rely on various criteria. These criteria concern the quality of approximation of the internal geometry of the object, including connectivity preservation, which is of great importance in fluid dynamics applications, as well as the mesh quality of the resulting mesh itself. For geometry approximation estimates, one may use the following criteria:

1. Pore space preservation:

    a. Volume fraction of different materials in the sample.
    b. Porosity of the sample (including isolated pores).
    c. Effective porosity (in terms of fluid flow).

2. Surface area approximation.
3. Effective physical properties.

The volume fraction of the material constituting the sample and the surface area of the interfaces can be, somewhat roughly, estimated based on slice images, without the use of any additional a priori information. Porosity and effective porosity and the effective physical properties of the reconstructed virtual object may be obtained by numerical simulations in this virtual object, and then compared to laboratory measurements, although the latter are not always available. In this work, we used criteria 1.1 and 2 to validate the algorithm.

The error between the volume concentration of the inclusions known a priori and an estimate obtained for the constructed 3D mesh model is 0.67%. The approximation error for the surface area is 11.54%. It is a known fact that algorithms of marching simplex type tend to overrate the surface area of the smooth interfaces. Additional smoothing may be required in this case. In the native media, the overshoot in the surface area is usually smaller.

A proposed algorithm is naturally parallel, since modern nonconforming finite element methods do not require mesh conformity through the macro-scale element interfaces, nor even the strict conformity of the objects intersecting these boundaries. Therefore, it may be effectively distributed between the threads in the shared memory, or between the processes working with distributed memory. In this paper, we use the OpenMP technology with a balanced load, as different macro-scale elements may contain different volume fractions of the inclusions. The processor characteristics are AMD Ryzen™ 7 5800 HS 4.4 GHz. The algorithm demonstrates good scalability (see Table 1).

Table 1. Parallel performance of the algorithms (OpenMP).

Threads	Time, sec	Speed-up
1	7044.72	–
2	3254.01	2.16
4	2014.98	3.49
8	1428.98	4.92

6 Conclusion

In this paper, we proposed an extension of the algorithm for constructing hierarchical mesh models of the heterogeneous media represented as a stack of slice 2D images obtained via digital imaging methods (CT, MRI, etc.). The current modification does not require explicit reconstruction of the geometrical model of the medium. Instead, it refines the initial mesh partition, so that the complex curvilinear interfaces between the materials, typical of the geological media, are incorporated into the mesh. The algorithm proposed preserves the volume fraction of the inclusions better than previous modifications do (0.67% vs 5%) [19]. It also improves the accuracy of the interface approximation (11.54% vs 17%), which is important in many applications, since the interface surface area considerably affects the estimations of the effective physical properties yielded by numerical simulation in the digital twins of the heterogeneous objects. The algorithm is naturally parallel and demonstrates good scalability, although it is not linear due to each thread disc reading. The scalability may be improved if distributed memory parallelism is employed. In this paper, we also did not address the question of the resulting discrete model mesh quality. Further research is required.

Acknowledgments. The work was performed within the framework of the FSI project FWZZ-2022–0030.

References

1. Chen, C., Amelon, R.E., Heiner, A., Saha, P.K.: Assessment of trabecular bone strength at in vivo CT imaging with space-variant hysteresis and finite element modelling. In: 2016 IEEE 13th International Symposium on Biomedical Imaging (ISBI), pp. 872–875 (2016)
2. Sencu, R.M., et al.: Generation of micro-scale finite element models from synchrotron X-ray CT images for multidirectional carbon fibre reinforced composites. Compos. Part A Appl. Sci. Manuf. **91**, 85–95 (2016)
3. Abdul-Aziz, A., Roth, D.J., Cotton, R., Studor, G.F., Christiansen, E., Young, P.C.: Material characterization and geometric segmentation of a composite structure using microfocus X-ray computed tomography image-based finite element modeling. J. Mater. Eval. **71**(2), 167–175 (2013)
4. Gelb, J., et al.: Multi-length scale X-ray microscopy: a unique solution for digital rock physics. In: EAGE/FESM Joint Regional Conference Petrophysics Meets Geoscience, vol. 2014, no. (1), pp. 15 (2014)

5. Algive, L., Bekri, S., Vizika, O.: Pore-Network modeling dedicated to the determination of the petrophysical-property changes in the pres-ence of reactive fluid. SPE J. **15**(3), 618–633 (2010)
6. Xong, Q., Baychev, T.G., Jivkov, A.P.: Review of pore network modelling of porous media: Experimental characterisations, network constructions and applications to reactive transport. J. Contam. Hydrol. **192**, 101–117 (2016)
7. Zhu, L.Q., et al.: Challenges and prospects of digital core reconstruction research. Geofluids **2019**, 1–29 (2019)
8. Wang, Z.L., et al.: Computational biomechanical modelling of the lumbar spine using marching-cubes surface smoothened finite element voxel meshing. Comput. Methods Programs Biomed. **80**(1), 25–35 (2005)
9. Koketsu, K., Fujiwara, H., Ikegami, Y.: Finite-element simulation of seismic ground motion with a voxel mesh. Pure Appl. Geophys. **161**(11), 2183–2198 (2004). https://doi.org/10.1007/s00024-004-2557-7
10. Berg, C.F., Lopez, O., Berland, H.: Industrial applications of digital rock technology. J. Petrol. Sci. Eng. **157**, 131–147 (2017)
11. Lorensen, W.E., Cline, H.E.: Marching cubes: a high resolution 3d surface construction algorithm. In: Computer Graphics (Proceedings of SIGGRAPH 87), vol. 21, no. 4, pp. 163–169 (1987)
12. Cong, A., Liu, Y., Kumar, D., Cong, W., Wang, G.: Geometrical modeling using multiregional marching tetrahedral for bioluminescence tomography. In: Medical Imaging 2005: Visualization, Image-Guided Procedures, and Display. International Society for Optics and Photonics, vol. 5744, pp. 756–764 (2005)
13. Kobbelt, L.P., Botsch, M., Schwanecke, U., Seidel, H.-P.: Feature-sensitive surface ex-traction from volume data. In: Proceedings of SIGGRAPH 2001, Computer Graphics Proceedings, Annual Conference Series, pp. 57–66 (2001)
14. Karimpouli, S., Faraji, A., Balcewicz, M., Saenger, E.H.: Computing heterogeneous core sample velocity using digital rock physics: a multiscale approach. Comput. Geosci. **135**, 104378 (2020)
15. Torres, J., Hitschfeld, N., Ruiz, R.O., Ortiz-Bernardin, A.: Convex polygon packing based meshing algorithm for modeling of rock and porous media. In: Krzhizhanovskaya, V.V., et al. (eds.) ICCS 2020. LNCS, vol. 12141, pp. 257–269. Springer, Cham (2020). https://doi.org/10.1007/978-3-030-50426-7_20
16. Brown, D., Efendiev, Y., Hoang, V.: An Efficient hierarchical multiscale finite element method for stokes equations in slowly varying media. Multiscale Model. Simul. **11**(1), 30–58 (2013)
17. Li, Q., Sone, S., Doi, K.: Selective enhancement filters for nodules, vessels, and airway walls in two-and three-dimensional CT scans. Med. Phys. **30**(8), 2040–2051 (2003)
18. Roth, H.R., Farag, A., Lu, L., Turkbey, E.B., Summers, R.M.: Deep convolutional net-works for pancreas segmentation in CT imaging. In: Medical Imaging 2015: Image Processing. International Society for Optics and Photonics **9413**, 94131G (2015)
19. Shurina, E.P., Dobrolubova, D.V., Shtanko, E.I.: Special techniques for objects with complex inner structure based on a CT image sequence. Cloud Sci. **5**(1), 40–58 (2018). (in Russian)
20. Sukov, S.A.: Methods for generating tetrahedral meshes and their software implementations. Prepr. Keldysh Inst. Appl. Math. **23**, 1–22 (2015). (in Russian)

Modeling a Potential Plant Habitat by Ensemble Machine Learning

Alexei V. Vaganov[1] , Vladimir F. Zaikov[1] , Olga S. Krotova[2] ,
Andrey I. Musokhranov[2] , Zakhar V. Pokalyakin[2] ,
and Lyubov A. Khvorova[2](✉)

[1] South-Siberian Botanical Garden, Altai State University, Lenin Avenue 61, 656049 Barnaul,
Russia
[2] Altai State University, Lenin Avenue 61, 656049 Barnaul, Russia
KhvorovaLA@gmail.com

Abstract. The article is devoted to modeling the potential distribution habitat of *Pulsatilla turczaninovii Kull. et Serg.* (Turchaninova prostrate). Modeling of ecological niches of plants is the process of building models using modern computer algorithms and bioclimatic data to predict the habitat distribution of plant species. The result of modeling is a model that can be used to map the area where species grow or live, predict the habitat, or analyze the impact of the environment on species. Building effective models for predicting plant ecological niches requires data on both the presence and absence of species in an area. Species absence points (or background points) are not recorded in databases but can be generated using different approaches. This article describes the implementation of three approaches to selecting pseudo-absence points in a given area: 1) randomly selecting from all points in a given area, excluding existing points of presence; 2) randomly selecting any point located at least one degree of latitude or longitude from any point of presence; 3) random selection of points from all points outside the suitable area estimated based on bioclimatic variables. The article presents the result of modeling the potential distribution range of the species *Pulsatilla turczaninovii Kull. et Serg.* using the random forest algorithm, the most popular method of constructing ensembles of decision trees. The software implementation of the model is carried out in the high-level programming language Python.

Keywords: Ecological niche · Biological species · Pseudo-absence points · Bioclimatic features · Python programming language · Machine learning models · Random forest

1 Introduction

One of the important tasks of botany is the estimation of the spatial distribution of plant objects [1–4]. The objectivity of such assessment is possible only with an integrated approach that combines various applied and fundamental areas of botany, mathematics, information technology, and GIS capabilities. Modern global geoinformation technologies, machine learning methods, and predictive modeling capabilities are increasingly

V. Jordan et al. (Eds.): HPCST 2022, CCIS 1733, pp. 206–216, 2022.
https://doi.org/10.1007/978-3-031-23744-7_16

being used in biological sciences to identify distribution patterns and calculate potential plant habitats [5, 6]. With a proper approach, these methods can be applied for monitoring and assessment of plant resources of economically valuable plants.

The article is devoted to modeling the potential habitat of plants by machine learning methods on the example of the species *Pulsatilla turczaninovii Kull. et Serg. Pulsatilla turczaninovii Kull. et Serg.* is used in folk medicine and is a rare plant. Recent studies have made it possible to detail the modern range of the species.

The process of modeling the potential distribution habitat of a plant species involves several steps:

1. Selection of pseudo-absence points of a plant species;
2. model building and optimization;
3. visualization and estimation of simulation results.

Data on the presence of a biological species in a given area is a set of geographic coordinates called species registration points in space. Species registration points are usually present on botanical labels of herbarium sheets. Observations in the natural environment with photographic fixation of plants and field diaries with records of coordinates can also be used as data.

Building effective ecological niche prediction models requires data on both the presence and absence of species in an area. Species absence points (or background points) are not recorded in databases but can be generated using different approaches. In this article three approaches to the sampling of pseudo-absence points of a species are considered. A random forest algorithm is trained on each of the generated datasets, resulting in three models. Point data on the distribution of *Pulsatilla turczaninovii Kull. et Serg.* species in the study area, which cover a long time period, were taken from the global GBIF – Global Biodiversity Information Facility [7].

2 Selecting Pseudo-absence Points of a Plant Species

A comprehensive comparative analysis given in the article [8] showed that the modeling results are sensitive to the method of selecting pseudo-absence points of the species and depend on the number of selected points. The data set of registration points of *P. turczaninovii* species includes 122 points. It was decided to use 1220 pseudo-absence points for the study.

In the process of this study, three approaches to selecting pseudo-absence points were implemented in the Python programming language [9–11]:

1. randomly selecting from all points in the study area, excluding existing points of presence;
2. randomly selecting any point located at least one degree of latitude or longitude from any point of presence;
3. random selection of points from all points outside the suitable area estimated based on bioclimatic variables.

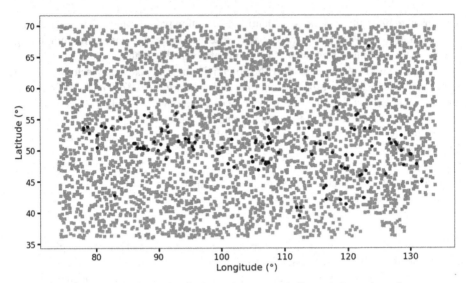

Fig. 1. Randomly selecting points in the study area, excluding existing points of presence.

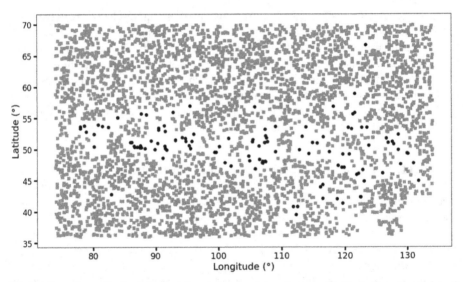

Fig. 2. A random selection of points located at least one degree of latitude from any point of presence

The first approach is the most popular, but has the serious drawback that pseudo-absence points can coincide with places where the species under study occur, and false-absence data can have a negative impact on species distribution models (Fig. 1).

The second approach, also called the buffer methodology, is to select the absence points beyond a certain radius around each presence point. A buffer zone equal to 1 in latitude was used to solve the problem (Fig. 2).

The third approach is to identify territorial areas where the values of bioclimatic indicators are close to those of the known habitats of the species. Any location with environmental conditions similar to those of the species is included in the potential habitat for that species. Points of pseudo-absence are selected outside such areas. The similarity of areas by bioclimatic conditions was determined as follows. If the value of a certain indicator for a point fall within the interquartile range (Q3–Q1) is calculated for the indicator for all the points of presence, then the point is defined as a potential point of species presence and is excluded from the set of possible points of species absence (Fig. 3).

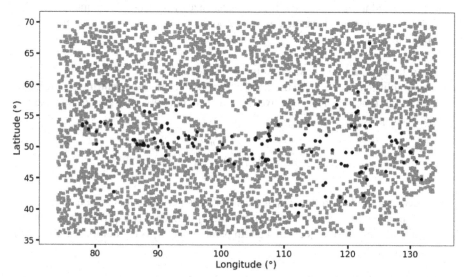

Fig. 3. Random selection of points from all the points outside the suitable area, estimated based on bioclimatic variables

To implement this approach and compare the results of the study with the results presented in the article [2], 19 bioclimatic characteristics from the WorldClim database [12] were used, such as the average annual air temperature, annual precipitation, the amount of precipitation of the wettest month, the amount of precipitation of the driest month, and others.

Thus, the authors obtained three datasets, each of which includes 122 points of presence and 1220 points of pseudo-absence of the species, selected using the three approaches described.

3   Algorithm of the Random Forest Method

The authors chose the random forest algorithm to build models on the obtained data sets. Random Forest is one of the most popular ways of combining decision trees into ensembles [13]. Ensemble learning is based on the idea of combining multiple machine learning models in order to obtain a more powerful model than each of the models individually. The basic idea of a random forest is that each tree can solve a given problem quite well, but it is very likely to be over-trained on parts of the data. If you build many trees that work well and overtrain to different degrees, it helps to reduce overtraining by averaging their results.

The random forest algorithm can be described by the following steps [14].

Step 1. You need to extract a bootstrap sample of size n from the original dataset. Using bootstrap, a random object is taken from the original sample of size l and written to the training sample. The next object is also taken randomly from the initial sample of size l. This is repeated n times, where n is the desired size of the training sample.

Step 2. Each decision tree is trained on one bootstrap sample. At the same time, in each node of the tree:

1) *s* features are randomly selected in a nonrepeatable way;
2) node splitting occurs, with the feature that provides the best splitting according to the target function. The objective function is to maximize the gain of information at each splitting:

$$IG(D_p, f) = I(D_p) - \sum_{j=1}^{m} \frac{N_j}{N_p} I(D_j) \rightarrow \max,$$

where f is the feature for which the splitting is performed, D_p is a dataset of the p-th parent node, D_J is a dataset of the child j-th node, I is a splitting criterion, N_p is a total number of objects in the p-th parent node, N_j is the number of objects in the child j-th node.

The following splitting criteria are usually used for binary decision trees:

- Entropy:

$$I_H(t) = -\sum_{i=1}^{c} p(i|t)\log_2 p(i|t),$$

where $p(i|t)$ is the ratio of objects, which belongs to the class i for a single node t. Entropy equals 0, if all objects in the node belong to the same class, and entropy is maximal if the classes are evenly distributed.

- Gini coefficient is a criterion that minimizes the probability of misclassification:

$$I_G(t) = \sum_{i=1}^{c} p(i|t)(1 - p(i|t)) = 1 - \sum_{i=1}^{c} p(i|t)^2.$$

- Misclassification:

$$I_E(t) = 1 - \max\{p(i|t)\}.$$

Step 3. Steps 1 and 2 are repeated k number of times, where k is the number of trees in the forest.

Step 4. To assign a class label to an object, the tree responses are aggregated on a majority vote basis.

You can build a random forest model for classification using the RandomForestClassifier() class of the sklearn.ensemble module of the Python programming language. The number of trees is specified by the parameter n_estimators, the splitting criterion and maximal depth of each tree can be specified by the parameters criterion and max_depth, the parameter max_features defines the number of randomly selected features considered for splitting [15].

4   Modeling Plant Ecological Niches

The task of ecological niche modeling is to discover the relationships between the location of species in nature and environmental factors [16, 17]. Thus, the input variables for such models are bioclimatic variables characterizing the area where *P. turczaninovii* species grows. It should be emphasized that we are only modeling the probability distribution of climatic conditions favorable for the growth of the species. Biological features, competitive ability of the species and other factors are not considered.

An important step in data preprocessing is the selection of features. Often, multicollinearity is the reason why the inclusion of certain features in the model can lead to unsatisfactory results. Multicollinearity is a problem where there is a strong correlation between features. In machine learning, multicollinearity leads to over-training of the model, excess coefficients increase the complexity of the model and its training time. It is reasonable to include variables in the model, the correlation coefficient between which does not exceed the value of 0.7 [18–20]. Thus, 5 variables were selected from 19 variables: mean annual temperature, daily temperature fluctuations, mean annual temperature fluctuation amplitude, mean annual precipitation. The results obtained differ significantly from the results obtained in the article [2].

In the next stage of the study, three models based on the random forest algorithm were built using selected features for each of the three datasets. Each model includes 100 trees. A Gini heterogeneity measure was used as the splitting criterion. The following classification quality metrics were chosen to estimate the modeling results: the error curve (ROC) and the AUC measure known as the area under the ROC curve.

An error curve or ROC-curve is a graphical method for evaluating the quality of a binary classifier and selecting a threshold for class separation. The ROC-curve describes the relationship between two quantities: model sensitivity and model specificity. The diagonal of the ROC curve can be interpreted as a random guess of the class label. Classification models that fall below the diagonal are considered worse than the random guess. A quantitative interpretation of the ROC is given by the AUC, the area on the ROC curve. AUC takes values from 0 to 1. It is assumed that the closer the AUC value is to 1, the better the quality of the model.

Figure 4 shows the ROC curves for the models built. The blue curve corresponds to the model trained on the dataset in which pseudo-absence points were selected using the buffer technique (AUC = 0.91). The orange curve corresponds to the model trained on the dataset in which pseudo-absence points were selected at random from all points in the study area, excluding the available presence points (AUC = 0.92). The green curve corresponds to a model trained on the dataset in which pseudo-absence points were selected outside a suitable habitat area for the species estimated from bioclimatic variables (AUC = 0.94).

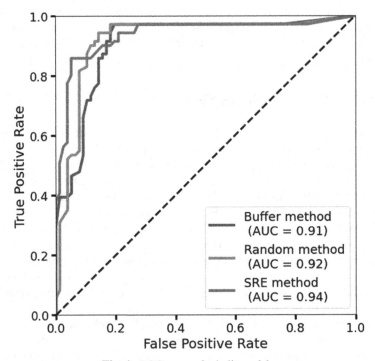

Fig. 4. ROC-curves for built models.

The results of the simulation are shown in Figs. 5, 6 and 7. The red color reflects the higher probability of species presence in the area. The lighter the color, the less likely the species is present.

Interpretation of the obtained results was carried out by experts in botany. Each of the constructed models reflects distribution of *Pulsatilla turczaninovii Kull. et Serg.* In the opinion of experts, the map shown in Fig. 5 most accurately reflects the distribution of the species.

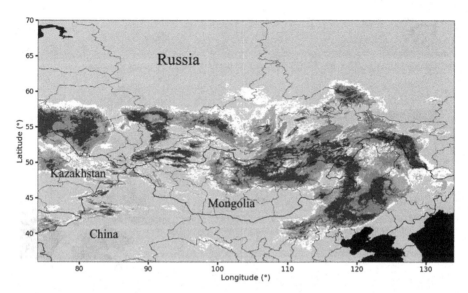

Fig. 5. The potential habitat of *P. turczaninovii* derived from model training on the dataset, in which pseudo-absence points of the species were selected based on bioclimatic variables.

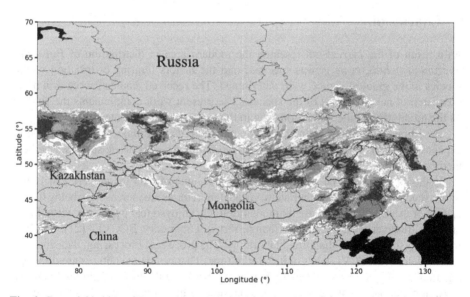

Fig. 6. Potential habitat of *P. turczaninovii*, derived from model training on a dataset in which the pseudo-absence points of the species were selected using a buffer methodology.

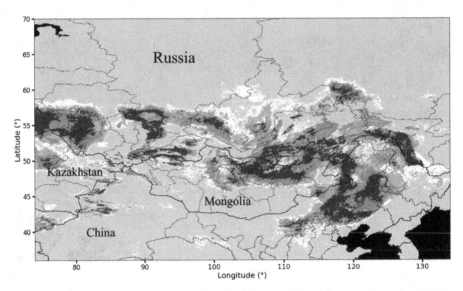

Fig. 7. Potential habitat of *P. turczaninovii*, derived from model training on a dataset in which the pseudo-absence points of the species were randomly selected.

5 Conclusion

As a result of the carried out research the modern area of distribution of *Pulsatilla turczaninovii Kull. et Serg.* was specified and the factors limiting distribution of the species to the greatest degree were determined. The received data can serve as a basis for search of new locations of the species. The present work complements the studies on the monitoring of the vegetation state in the territory of the Greater Altai.

Based on the results obtained, we can conclude that the modeling result directly depends on the method of selecting the pseudo-absence points of the species. The best method of generation based on bioclimatic data.

It should be noted that with the increasing volume of data for modeling with the limited capacity of personal computers, virtual laboratories are increasingly being developed. For example, for bioclimatic modeling today enjoys great popularity virtual laboratory "Biodiversity and Climate Change Virtual Laboratory, BCCVL" [21]. BCCVL is the English version of the virtual laboratory. The process of processing, data analysis and modeling in BCCVL takes quite a lot of time. That is why the authors are developing a unique IT product – a virtual laboratory for solving the tasks of digital inventory of Altai biota, bioclimatic modeling, and research of global biodiversity of the Greater Altai regions. The virtual laboratory developed by the Altai State University team and partners is based on advanced digital and intellectual tools and includes the development of algorithms and programs, databases, computer methods and models for processing, analyzing and visualizing biological data for more effective work with them.

In addition, once a predicted estimate of the potential habitat of a species has been obtained, it is important to estimate its area. In the case of DIVA-GIS and BCCVL, the results have only the model consistency and a map with the probability of occurrence of the species. For the resource assessment, it is important to obtain information on the quantitative value of the potential range area of an economically valuable taxon. To be able to integrate the whole chain of actions for resource assessment, the authors plan to develop within the national virtual laboratory an algorithm and an interface solution for estimating the area of the real and predicted range of the studied taxon with a set of additional services.

Acknowledgments. This work was supported in the framework of the "Priority-2030" Program by Altai State University.

References

1. Guisan, A., Thuiller, W.: Predicting species distribution: offering more than simple habitat models. Ecol. Lett. **8**(9), 993–1009 (2005)
2. Zaikov, V.F., Vaganov, A.V., Shmakov, A.I.: Climate modeling of the potential range of *Pulsatilla turczaninovii* Kryl. et Serg. (Ranunculaceae) on the territory of Eurasia. Theor. Appl. Ecol. **1**, 140–144 (2022). (in Russian)
3. Vaganov, A.V., Pokalyakin, Z.V., Khvorova, L.A.: Integrated solution of problems of plant resources assessment using GIS and climate modeling. Probl. Bot. South. Siberia Mongolia **20**(1), 87–91 (2021). (in Russian)
4. Makunina, N.I., Egorova, A.V., Pisarenko, O.: Construction of potential habitats of plant communities for the purpose of botanical and geographical zoning (on the example of Tuva forests). Siberian Ecol. J. **4**, 517–524 (2020). https://doi.org/10.1134/s1995425520040095. (in Russian)
5. Dudov, S.V.: Modeling of species distribution using topography and remote sensing data, with vascular plants of the Tukhuringra Range low mountain belt (Zeya Nature Reserve, Amur Region) as a case study. J. Gen. Biol. **7**(1), 16–28 (2016). (in Russian)
6. Elith, J., Leathwick, J.R.: Species distribution models: ecological explanation and prediction across space and time. Annu. Rev. Ecol. Evol. Syst. **40**, 677–697 (2009). https://doi.org/10.1146/annurev.ecolsys.110308.120159
7. Global Biodiversity Information Facility (GBIF) Occurrence Download [Internet resource]. Accessed 20 Dec 2019. https://doi.org/10.15468/dl.4khq61
8. Barbet-Massin, M., Jiguet, F., Thuiller, W.: Selecting pseudo-absences for species distribution models: how, where and how many? https://doi.org/10.1111/j.2041-210X.2011.00172.x
9. Downey, A.B.: Exploring Complex Systems with Python. DMK Press, Moscow (2019). (in Russian)
10. Vander, P.: Python for Complex Problems: Data Science and Machine Learning. Piter, St. Petersburg (2018). (in Russian)
11. Raska, S.: Python and Machine Learning: A Tutorial. DMK Press, Moscow (2017). (in Russian)
12. WorldClim. – Access mode. Head. from the screen. https://www.worldclim.com/node/1. Accessed 30 Apr 2022
13. Thuiller, W., Lafourcade, B., Engler, R., Araújo, M.B.: BIOMOD – a platform for ensemble forecasting of species distributions. Ecography **32**(3), 369–373 (2009)
14. Poletaeva, N.G.: Classification of machine learning systems. Bull. Baltic Fed. Univ. **1**, 5–22 (2020). (in Russian)
15. Geron, O.: Applied Machine Learning with Scikit-Learn and TensorFlow: Concepts, Tools and Techniques for Building Intelligent Systems. St. Petersburg (2018). (in Russian)

16. Anderson, R.P., Lew, D., Peterson, A.T.: Evaluating predictive models of species' distributions: criteria for selecting models. Ecol. Model. **162**, 211–232 (2003). https://doi.org/10.1016/s0304-3800(02)00349-6

17. Barthlott, W., Biedinger, N., Braun, G., Feig, F., Kier, G., Mutke, J.: Terminological and methodological aspects of the mapping and analysis of the global biodiversity. Acta Bot. Fennica **162**, 103–110 (1999)

18. Austin, M.: Species distribution models and ecological theory: a critical assessment and some possible new approaches. Ecol. Model. **200**, 1–19 (2007). https://doi.org/10.1016/j.ecolmodel.2006.07.005

19. Brown, J.L.: SDMtoolbox: a Python-based GIS toolkit for landscape genetic, biogeographic and species distribution model analyses. Methods Ecol. Evol. **5**(7), 694–700 (2014). https://doi.org/10.1111/2041-210X.12200

20. Korznikov, K.A.: Climate envelope models of Kalopanax septemlobus and Phellodendron amurense var. sachalinense in the insular part of the Russian Far East. Izvestiya RAN. Seriya biologicheskaya **6**, 648–657 (2019). https://doi.org/10.1134/S1062359019040083 (in Russian)

21. Hallgren, W., et al.: The biodiversity and climate change virtual laboratory: where ecology meets big data. Environ. Model. Softw. **76**, 182–186 (2016). https://doi.org/10.1016/j.envsoft.2015.10.025

Information and Computing Technologies in Automation and Control Science

Adaptive Traffic Signal Control Based on a Macroscopic Model of the Transport Network

Sergey V. Matrosov[1]([✉]) [iD] and Nikolay B. Filimonov[1,2] [iD]

[1] Lomonosov Moscow State University, Leninskie Gory 1/2, 119991 Moscow, Russia
matrosik14@gmail.com
[2] V.A. Trapeznikov Institute of Control Sciences of Russian Academy of Sciences,
Profsoyuznaya str. 65, 117997 Moscow, Russia

Abstract. This paper presents a new adaptive traffic signal control algorithm within the model predictive control framework. It aims to increase throughput of the traffic network under high traffic congestion. In order to do that, a detailed traffic flow model and specially designed target metric are used. The predictive model of the transport network is based on the second-order macroscopic traffic model. It can predict the formation of waves and other nonlinear effects occurring in the traffic. Also, it is possible to fine-tune the model with historical data to improve the quality of predictions. The paper describes both the model and the numerical scheme for its computation. Proposed optimization metric is based on the fundamental diagram of traffic flows and considers the characteristics of traffic dynamics. This metric minimization leads to a more uniform distribution of vehicles in the transport network. To solve the discrete optimization problem that emerges during the search for the optimal control, a genetic algorithm based on the local search heuristic was used. We carried out computer approbation of the proposed traffic signal control algorithm in the traffic simulation package SUMO. The accuracy of the transport network model predictions and performance of the control system were tested in several synthetic scenarios.

Keywords: Transportation network · Transport modeling · Macroscopic traffic model · Control system · Traffic signal control · Model predictive control · Genetic algorithm · Local search

1 Introduction

Modern urban traffic infrastructure allows us to get real-time information about the situation in the network and regulate it. We can control traffic lights, collect detectors data, camera records, GPS tracks of service vehicles, operational information about accidents, and much more. With this infrastructure, it is possible to create an automatic traffic signal control system maximizing the capacity of the traffic network.

One of the most common traffic signal control methods is fixed-time control. For this approach, we choose a program with a repeated sequence of signals for each controlled

V. Jordan et al. (Eds.): HPCST 2022, CCIS 1733, pp. 219–229, 2022.
https://doi.org/10.1007/978-3-031-23744-7_17

intersection. This method is practical in its simplicity, but it does not allow adjusting the behavior of traffic lights according to the current situation on the road. In this regard, adaptive traffic signal control algorithms are of great interest.

The world community has accumulated extensive practical experience using traffic signal control systems [1–3]. Based on this experience, we can assume that adaptive control algorithms significantly increase the transport network's capacity compared to fixed-time control.

But many popular traffic control algorithms use simple traffic models [4–10] and do not take into account nonlinear effects, which occurs in dense traffic [11–13]. We assume that consideration of the effects like hysteresis or capacity drop can improve the performance of the control system.

We would also like to note the importance of long-term planning when considering nonlinear phenomena in traffic flows. Traffic wave propagation velocity is finite, so the impact of our decision can manifest itself after some time. Thus, locally optimal control can lead to a general drop in network capacity.

This paper proposes a new algorithm for adaptive traffic signal control based on model predictive control methods and using a macroscopic hydrodynamic model of traffic flows as a predictive model. The work draws on the results of the articles [14–16].

2 Transport Network Model

2.1 Traffic Controllers

The traffic controller manages the traffic lights. It has several phases between which it can switch. Each phase corresponds to one or more permitted directions of movement.

The traffic controller will operate according to the installed program. As with fixed-time control, it sets the sequence and duration of phases within the traffic light cycle. We will perform traffic signal control by setting new programs for controllers. We ensure that current program terminates only when its current cycle ends, and immediately after that new program starts.

We will assume that the phase sequence for each intersection is fixed. This assumption is made to reduce the search space and increase the performance of the control algorithm. Let us denote the program for controller ν as $u^\nu = (\tau_k^\nu)_{1,\dots,K^\nu}$, where τ_k^ν is the duration of phase k, and K^ν is the number of phases. We denote the set of programs for all controllers as $u = (u^1, \dots, u^\nu)$. The set of all allowed program sets for the controlled section will be denoted as U.

2.2 Macroscopic Traffic Flow Model

To model the behavior of traffic flows, we will use the macroscopic approach, in which the state of the road with length L at a time t is described as the distribution of density $\rho(x, t)$, average speed $v(x, t)$ and intensity $Q(x, t)$ on the segment $x \in [0, L]$. Traffic dynamics are specified by the system of partial differential equations.

The transport network is defined as a directed graph, where the vertices correspond to roads and the edges to directions of movement at intersections. Each edge of the graph is assigned a weight α_{ij} that specifies the proportion of vehicles turning from the road i exit to the road j entrance. The controllers operate by allowing movement on edges.

In this paper, a generalized ARZ model (GARZ) [17] is used to describe the traffic dynamics. The distinctive feature of this model is the ability to fine-tune its parameters based on historical data. As independent macroscopic variables, the traffic density ρ and the parameter ω, characterizing the maximum speed of the driver in the flow, are chosen:

$$\begin{cases} \rho_t + (\rho v)_x = 0; \\ (\rho \omega)_t + (\rho \omega v)_x = 0; \\ v = V(\rho, \omega). \end{cases}$$

The first equation in the system is analogous to the law of conservation of mass (in this case the number of cars on the road). The second and third equations describe the change in speed of the flow. The average speed of the traffic flow $V(\rho, \omega)$ is based on the collected historical data. It is related to the intensity $Q(\rho, \omega)$ by the expression $Q(\rho, \omega) = \rho V(\rho, \omega)$. The function $Q(\rho, \omega)$ is sometimes called the fundamental diagram [18, 19]. The fitting procedure for $V(\rho, \omega)$ is shown below (Fig. 1).

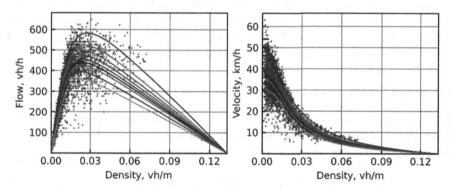

Fig. 1. Data-fitted fundamental diagram for the GARZ model

The curve of the fundamental diagram is given by the expression:

$$Q_{\alpha,\lambda,p}(\rho) := \alpha\left(a + (b - a)\frac{\rho}{\rho_{max}} - \sqrt{1 - y^2}\right);$$
$$a = \sqrt{1 + \lambda^2 p^2}; \quad b = \sqrt{1 + \lambda^2(1 - p)^2}; \quad y = \lambda(\rho/\rho_{max} - p).$$

The historical data will be referred to as

$$(\rho_{data}, Q_{data}) = \{(\rho_i, Q_i) : i = 1, \ldots, N_{data}\}.$$

To determine the parameters α, λ, p, we solve the following optimization problem:

$$argmin_{\alpha,\lambda,p}\left(\beta\left\|Q_{\alpha,\lambda,p}(\rho_{data}) - Q_{data}\right\|_+^2 + (1-\beta)\left\|Q_{\alpha,\lambda,p}(\rho_{data}) - Q_{data}\right\|_+^2\right)$$

$$\left\|Q_{\alpha,\lambda,p}(\rho_{data}) - Q_{data}\right\|_+^2 = \sum_{i=1}^{N_{data}} \max\{Q_{\alpha,\lambda,p}(\rho_i) - Q_i, 0\}$$

$$\left\|Q_{\alpha,\lambda,p}(\rho_{data}) - Q_{data}\right\|_-^2 = \sum_{i=1}^{N_{data}} \max\{Q_i - Q_{\alpha,\lambda,p}(\rho_i), 0\}$$

Parameter β defines the proportion of historical points under the curve $Q_{\alpha,\lambda,p}(\rho)$. Solving the above optimization problem for different $\beta_i \in (0,1)$, we obtain a set of points $\alpha(\beta_i)$, $\lambda(\beta_i)$, $p(\beta_i)$ and a family of functions $Q(\rho, \beta_i)$.

We consider that ω is the maximum speed of the driver, i.e., $\omega = V(0, \omega)$. Then we can construct a mapping from β to ω:

$$V(\rho, \beta) = \begin{cases} Q_\beta(\rho)/\rho, \rho > 0 \\ Q'_\beta(\rho), \rho = 0 \end{cases} ; \quad \omega = V(0, \beta) = \frac{\alpha_\beta}{\rho_{max}}\left(b_\beta - a_\beta + \frac{\lambda_\beta^2 p_\beta}{a_\beta}\right).$$

Now we can apply the mapping $\omega(\beta)$, fit polynomial regression on the resulting points and obtain functions (ω), $\lambda(\omega)$, $p(\omega)$ for $\omega \in [\omega_{min}, \omega_{max}]$. Thus, we obtain required functions $Q(\rho, \omega)$ and $V(\rho, \omega)$ in domain $[0, \rho_{max}] \times [\omega_{min}, \omega_{max}]$. We can extend this domain to $[0, \rho_{max}] \times R_+$ projecting ω on the segment $[\omega_{min}, \omega_{max}]$.

2.3 Numerical Scheme

To find an approximate solution, Godunov's method is used - the computational domain is divided into volumes and averaged traffic parameters are assigned to each cell. The finite-difference scheme for the above system is written as follows:

$$\rho_i^{n+1} = \rho_i^n - \frac{\Delta t}{\Delta x}\left(Q_{i+1/2}^n - Q_{i-1/2}^n\right);$$
$$y_i^{n+1} = y_i^n - \frac{\Delta t}{\Delta x}\left(\omega_{i-1}^n Q_{i+1/2}^n - \omega_i^n Q_{i-1/2}^n\right);$$
$$\omega_i^n = y_i^n/\rho_i^n.$$

Flows between cells $Q_{i\pm1/2}^n$ are found as solutions of the Riemann problem. For numerical calculations, the formalism of sending and receiving functions [20] is used.

For a pair of cells $U_L = (\rho_L, \omega_L)$ and $U_R = (\rho_R, \omega_R)$, the sending function $S(U_L, U_R)$ specifies the maximum flow that can leave the cell U_L. The receiving function $R(U_L, U_R)$ specifies the maximum flow that cell U_R can receive. Thus, the flow between cells is given by the expression $Q = \min\{S, R\}$.

To correctly solve the Riemann problem in the case of the GARZ model it is necessary to introduce an intermediate state $U_M = (\rho_M, \omega_M)$ between the cells. In the case of $\omega_L < v_R = V(\rho_R, \omega_R)$ the intermediate state is $U_M = (0, \omega_L)$. Otherwise, it is calculated from the equations:

$$\rho_M : u_R = V(\rho_M, \omega_L); \omega_M = \omega_L.$$

To calculate the sending and receiving functions, we need the following expressions:

$$\rho_c(\omega) = argmax_\rho Q(\rho, \omega); \quad \rho_c(\omega) = max_\rho Q(\rho, \omega).$$

Then the sending and receiving functions for the GARZ model are specified as:

$$S(\rho_L, \omega_L) = \begin{cases} \rho_L \omega_L, & \rho_L \leq \rho_c(\omega_L) \\ Q^{max}(\omega_L) & \rho_L > \rho_c(\omega_L) \end{cases}; \quad R(\rho_M, \omega_L) = \begin{cases} Q^{max}(\omega_L) & \rho_M \leq \rho_c(\omega_L) \\ \rho_M \omega_L & \rho_M > \rho_c(\omega_L) \end{cases}.$$

2.4 Intersection Model

To model the traffic dynamics at an intersection, it is necessary to describe how traffic flows are separated and merged. To do this, let us generalize the solution of the Riemann problem described above. Let us define that the sending function for the direction from the road i to the road j is equal to $S_{ij} = \alpha_{ij} \delta_{ij} S_i$, where α_{ij} is the percentage of cars traveling in this direction, δ_{ij} is the indicator that movement in this direction is allowed, and S_i is the sending function for the last road cell i.

In the case of traffic separation, the generalization is trivial. As a sending function we use S_{ij}, and we calculate the receiving function for the first road cell j using the standard procedure.

The process of merging several traffic flows at the road entrance is described as an optimization problem with constraints:

$$max_{f_{ij}} \left(\sum_{i=1}^{M} f_{ij} \right);$$
$$0 \leq f_{ij} \leq S_{ij}, \forall i;$$
$$\sum_{i=1}^{M} f_{ij} \leq R(\rho_{j-}, \omega_{j-}).$$

In this case we consider M incoming flows. f_{ij} is the flow from the i-th road to the j-th road. To calculate receiving function R for road j we need to find the intermediate state of the flow at the entrance $U_{j-} = (\rho_{j-}, \omega_{j-})$. It is given by the equations:

$$\omega_{j-} = \left(\sum_{i=1}^{M} \omega_i f_{ij} \right) / \left(\sum_{i=1}^{M} f_{ij} \right);$$
$$v_{j-} = \min\{\omega_{j-}, v_j\};$$
$$\rho_{j-} : v_{j-} = V(\rho_{j-}, \omega_{j-}).$$

Considering the non-cooperative behavior of drivers at the intersection, let us assume that the share β_{ij} of each exit in the total flow is proportional to the ratio of the maximum flow S_{ij} through the exit to the possible maximum flow. Let us denote the final flow through the entrance as $F_j = \sum_{i=1}^{M} f_{ij}$ and rewrite the flows f_{ij} with our assumption in mind:

$$f_{ij} = \beta_{ij} F_j; \quad \beta_{ij} = S_{ij} / \left(\sum_{i=1}^{M} S_{ij} \right).$$

It is easy to see that the obtained optimization problem could be solved analytically after substituting the expression for f_{ij} into the original optimization problem.

3   Model Predictive Control

We will control the system at discrete time moments $t_i = i\Delta T$. It is assumed that at the moment t_i we know the current state of the transport network s_i and the input flows d_i. Using the transport network model F, we can predict the behavior of the system within some prediction horizon T for the chosen control u_i. Based on this prediction we can estimate the value of the target metric $J_F(s_i, d_i, u_i)$. Then the choice of the optimal control can be formulated as an optimization problem $argmin_{u_i \in U} J_F(s_i, d_i, u_i)$.

3.1   Target Metric

The target metric should correctly describe the parameters of system behavior we are interested in, and it should also be easily computable. In traffic signal control tasks, target functions such as average travel time, average intensity/speed of traffic flows, average queue length at intersections, or total flow through a section of the transport network are often used.

In this paper, we are primarily interested in traffic signal control in a congested traffic network. It was already noted earlier that dense traffic is less predictable, and non-linear effects can spontaneously arise in it, reducing the capacity of the road. A reasonable strategy seems to be to "smear" the density across the transportation network. In this way, you can reduce the likelihood of traffic waves forming and prevent (or at least delay) the concentration of vehicles on one section of the network.

To implement the strategy described, we propose the following target function:

$$J_F(s_i, d_i, u_i) = \sum_{t=0}^{T} \sum_{l \in L} w(l) \cdot \left(\frac{\overline{\rho}_l^t}{\rho_c(\omega_{max})} \right)^2; \ \overline{\rho}_l^t = \frac{1}{M_l} \sum_{m=1}^{M_l} \rho_m^t.$$

It reflects the process of accumulating a penalty for excessive density in the transportation network. We get an estimate of congestion by comparing the average density $\overline{\rho}_l^t$ on road $l \in L$ at time t with the critical density ρ_c, and then squaring the resulting ratio. Thus, we will penalize more when the average density is greater than the critical density. The function $w(l)$ sets the weight for each road in the graph. It can be used to increase the priority of certain sections of the transport network. By default, the weight is equal to the length of the road, because we assume that long sections are more likely to generate waves of traffic.

3.2   Genetic Search Algorithm

The search for a suitable control for a transport network section is reduced to a discrete optimization problem $argmin_{u_i \in U} J_F(s_i, d_i, u_i)$. In this case, the search space U can be large even for relatively compact sections of the transport network. Because of this, brute force is not applicable in most cases. Therefore, it is necessary to use various heuristics to find the optimal control.

We will use the local search technique as the foundation for the optimization algorithm. For the set of programs $u \in U$ we will set the δ-neighborhood by the following expression:

$$B_\delta(u) = \left\{ \tau_k^v \in \left[\tau_k^v - \delta, \tau_k^v + \delta \right] \cap \left[\tau_{k,min}^v, \tau_{k,max}^v \right] \right\} \cap U$$

Iteration of the local search consists of several steps - we sample a new control at random $\widehat{u} \in B_\delta(u)$, compute its target function $J_F\left(s_i, d_i, \widehat{u}\right)$ and compare it with the old control. If the new control turns out to be better, then at the new iteration we use it as the initial control.

This approach makes it possible to gradually improve the solution without disrupting the coordination between intersections, if it already existed. Moreover, the algorithm can be improved by modifying the shape of the neighborhood or by specifying on it a random distribution other than uniform. However, it tends to stop at local optima.

To reduce the probability of entering the local optimum, we propose to use a genetic algorithm that searches in parallel for several sets of programs. At each iteration, a new trial control is generated for the next population member, and if it turns out to be better than the current one, it replaces it. New controls are generated according to the following rule:

- with probability p_{ls} a local search iteration is performed for a population member u.
- with probability p_{cross} the new control $\hat{u}$ is generated by mixing programs for u and u_{other}.
- with the probability p_{new} a new control $\hat{u}$ is randomly generated.

The cycle is repeated until we run out of time to find the control. It should be noted that the resulting answer may be suboptimal, but we can guarantee that with each iteration its quality does not degrade.

4 Numerical Experiments

To test the proposed traffic signal control algorithm, we used software package of microscopic traffic simulation SUMO [21]. Real-world detectors data was substituted by the synthetic data generated by SUMO simulator tools.

The netedit utility was used to build several road networks, and the traci utility was used to collect road detector data (inductive loops) and to simulate the control process of the road controllers. The interaction of SUMO-simulation with the software module of the proposed algorithm was carried out in the Python programming language.

Three scenarios were used to test the correctness of the traffic network model predictions: a straight road with two traffic lights (Fig. 2.a), a straight road with two regulated T-junctions (Fig. 2.b), and a rectangular 2×2 network, i.e., with four regulated intersections at road junctions (Fig. 2.c). When streams are split, they are divided evenly among the directions. The length of each lane is 200 m.

First, a SUMO simulation was run, and detectors data was collected. Three traffic detectors (at the beginning, in the middle, and at the end) were installed on each lane in the traffic network. Then, we run proposed macroscopic simulation and recorded traffic parameters. For each detector, the RMS error of the model prediction at the detector location was calculated. The estimate of the total error of the transport network model is determined by averaging the error over all lanes. The percentage of error to the maximum possible value provided by the fundamental diagram is given in brackets. The error estimates for this model are given in Table 1.

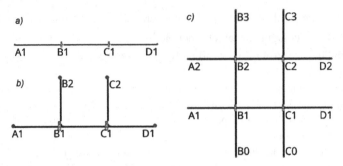

Fig. 2. Test sections of the transport network

Table 1. Model error (RMSE) compared to SUMO simulation

	Intensity	Velocity	Density
Straight road	0.026 vh/s (5.9%)	2.608 m/s (13.4%)	0.014 vh/m (7.0%)
T-junctions	0.0314 vh/s (7.1%)	2.856 m/s (14.6%)	0.007 vh/m (3.5%)
Grid	0.038 vh/s (8.6%)	2.889 m/s (14.9%)	0.021 vh/m (10.5%)

Tests for control algorithm were performed only for a straight road and grid network. Control step = 1 min, forecast horizon = 300 s. The same duration limits were set for all phases - from 10 to 100 s.

For the straight road scenario, the static plan was not chosen optimally. The durations of red and green traffic lights were equal to 30 s, and the phase shifts of the traffic lights were not coordinated with each other. Table 2 shows the results of comparing the performance of the proposed control algorithm with the static plan.

Table 2. Performance of the control methods for straight road scenario.

	Static plan	Predictive control
Total output	880 vh	1011 vh (14.89%)
Travel time	184 s	90 s (−50.9%)
Wait time	36 s	2 s (−94.5%)

During the work of proposed control algorithm, the durations of red signals of traffic lights decreased to the minimum allowable values, and the shifts were coordinated.

When testing the grid network, the traffic flows were set as follows: in the beginning there was low-intensity background traffic in each of the traffic directions, and then there was a 10 min intensity spike in each direction in turn. The comparative results of the proposed algorithm are presented in Table 3.

Table 3. Performance of the control methods for grid scenario

	Static plan	Predictive control
Total output	1770 vh	1957 vh (10.56%)
Travel time	229 s	169 s (-25.9%)
Wait time	144 s	91 s (-37.2%)

Figure 3 shows the data of detector located in front of the B1 intersection at the end of lane A1B1 (the lower west entrance to the zone, see Fig. 2.c).

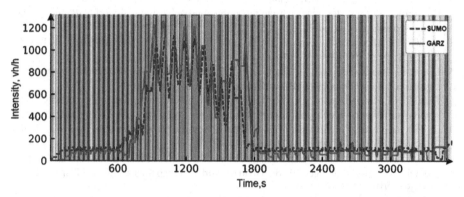

Fig. 3. Intensity measured by the detector located at the end of lane A1B1 (Color figure online)

In the figure, the different colors highlight the operating times of the different phases of the lane exit controller. The blue dashed line shows the intensity of flows collected from the SUMO detectors; the red line is the result of the model prediction.

The proposed algorithm puts the longest time for the phase that passes the most traffic.

5   Conclusion

The proposed traffic signal control algorithm showed its consistency in synthetic tests. The transport network model predicts the traffic behavior with reasonable accuracy. The optimization algorithm finds the correct control in an acceptable time. The algorithm provides an increase in throughput even when the transport network is heavily congested.

Further research should focus on studying the behavior of the algorithm in more complex scenarios, as well as directly comparing its performance with other traffic light control algorithms.

Another area of work is the search for additional heuristics for local search. Of particular interest are heuristics that allow one to determine the direction of local search more quickly, as well as algorithms for constructing initial sets of programs.

We also note that the proposed algorithm is easily modifiable, since the transport network model is used as a black box in the optimization problem. Thus, we can freely change the macroscopic model of the transport network. This can be another direction for algorithm development.

References

1. Bretherton, R.D.: SCOOT urban traffic control system—philosophy and evaluation. IFAC Proc. **23**(2), 237–239 (1990)
2. Samadi, S., et al.: Performance evaluation of intelligent adaptive traffic control systems: a case study. J. Transp. Technol. **2**(3), 248 (2012)
3. Khattak, Z.H., Magalotti, M.J., Fontaine, M.D.: Operational performance evaluation of adaptive traffic control systems: a Bayesian modeling approach using real-world GPS and private sector PROBE data. J. Intell. Transp. Syst. **24**(2), 156–170 (2020)
4. Varaiya, P.: Max pressure control of a network of signalized intersections. Transp. Res. C: Emerg. Technol. **36**, 177–195 (2013)
5. Lioris, J., Kurzhanskiy, A.B., Varaiya, P.: Adaptive max pressure control of network of signalized intersections. IFAC-PapersOnLine **49**(22), 19–24 (2016)
6. Diakaki, C., Papageorgiou, M., Aboudolas, K.: A multivariable regulator approach to traffic-responsive network-wide signal control. Control Eng. Pract. **10**(2), 183–195 (2002)
7. de Oliveira, L.B., Camponogara, E.: Predictive control for urban traffic networks: initial evaluation. IFAC Proc. **40**(20), 424–429 (2007)
8. Aboudolas, K., et al.: A rolling-horizon quadratic-programming approach to the signal control problem in large-scale congested urban road networks. Transp. Res. C: Emerg. Technol. **18**(5), 680–694 (2010)
9. Lin, S., et al.: Fast model predictive control for urban road networks via MILP. IEEE Trans. Intell. Transp. Syst. **12**(3), 846–856 (2011)
10. Lin, S., et al.: Efficient network-wide model-based predictive control for urban traffic networks. Transp. Res. C: Emerg. Technol. **24**, 122–140 (2012)
11. Edie, L.C.: Car-following and steady-state theory for non-congested traffic. Oper. Res. **9**(1), 66–76 (1961)
12. Newell, G.F.: Instability in dense highway traffic, a review. In: The 2nd International Symposium on the Theory of Traffic Flow (1963)
13. Treiterer, J., Myers, J.: The hysteresis phenomenon in traffic flow. Transp. Traffic Theory **6**, 13–38 (1974)
14. Matrosov, S.V.: The concept of predictive traffic signal control with trainable model of transport network. J. Adv. Res. Tech. Sci. **25**, 56–62 (2021). (in Russian)
15. Matrosov, S.V.: Algorithm prognostic control of system of crossroads based on macroscopic models of traffic flows. J. Adv. Res. Tech. Sci. **27**, 80–87 (2021). (in Russian)
16. Matrosov, S.V., Filimonov, N.B.: Intersection system traffic signal control with macroscopic model prediction of a transport network. High Perform. Comput. Syst. Technol. **6**(1), 166–171 (2022). (in Russian)
17. Fan, S., Herty, M., Seibold, B.: Comparative model accuracy of a data-fitted generalized Aw-Rascle-Zhang model. arXiv preprint arXiv:1310.8219 (2013)
18. Greenshields, B.D., Thompson, J.T., Dickinson, H.C., Swinton, R.S.: The photographic method of studying traffic behavior. In: Highway Research Board Proceedings, vol. 13 (1934)
19. Greenshields, B.D., Bibbins, J.R., Channing, W.S., Miller, H.H.: A study of traffic capacity. In: Highway Research Board Proceedings, vol. 1935. National Research Council (USA), Highway Research Board (1935)

20. Lebacque, J.-P., Haj-Salem, H., Mammar, S.: Second order traffic flow modeling: supply-demand analysis of the inhomogeneous Riemann problem and of boundary conditions. In: Proceedings of the 10th Euro Working Group on Transportation (EWGT) **3**(3) (2005)
21. Lopez, P.A., et al.: Microscopic traffic simulation using SUMO. In: 2018 21st International Conference on Intelligent Transportation Systems (ITSC), pp. 2575–2582 (2018)

Optimization and Robustization of Angular Motion Control of an Automatic Maneuverable Aerial Vehicle

Phong Quoc Pham[1] (iD) and Nikolay B. Filimonov[1,2](✉) (iD)

[1] Bauman University, 2nd Baumanskaya str., 5/1, 105005 Moscow, Russia
nbfilimonov@mail.ru
[2] Lomonosov Moscow State University, Leninskie Gory, 1/2, 119991 Moscow, Russia

Abstract. The article is devoted to the questions of the optimization and robustization of the control of surface-to-air and air-to-air of an automatic maneuverable aerial vehicle (AMAV). AMAV is a unique and rapidly developing type of the aviation and rocket technology designed for autonomous withdrawal and rapprochement with aerial targets having on-board homing systems. In the first part of the article, the problem of parametric synthesis of the AMAV optimal homing system is considered according to the accuracy and energy consumption criteria of the guidance. The two-criterion optimization of the homing system is carried out in the Matlab environment using a genetic algorithm in the conditions of possible three typical maneuvers of an aerial target: a slide, a dive and a dead loop. The second part of the article analyzes the problem of synthesizing robust control algorithms of the AMAV attack angle. Two algorithms for AMAV robust tracking for programmed change in the attack angle are developed: the first implements the control method in sliding modes, and the second deals with the adaptive backstepping control method. The developed algorithms for controlling the AMAV attack angle in the Matlab environment was approbated using a computer.

Keywords: Automatic maneuverable aerial vehicle · Homing system · Parametric optimization · Genetic algorithm · Adaptive robust control · Controlling angle · Sliding mode method · Backstopping

1 Introduction

Currently, an actual and rapidly developing type of the aviation and rocket technology designed for autonomous withdrawal and rapprochement with air targets of a conditional enemy is an automatic maneuverable aerial vehicle (AMAV) [1, 2]. A key role in ensuring a stable and high-precision guidance of AMAV to the target, eliminating the movement's deviations from the program caused by disturbances, belongs to the on-board homing system (HS) [3–5].

The AMAV functional effectiveness to a large degree is determined by the accuracy and quality of guiding at a maneuvering target for intercepting or destroying it. In this regard, one of the promising ways to improve the efficiency of using this aerial vehicle

V. Jordan et al. (Eds.): HPCST 2022, CCIS 1733, pp. 230–249, 2022.
https://doi.org/10.1007/978-3-031-23744-7_18

(AV) is to improve flight control algorithms based on using achievements of the modern automatic control theory.

In the modern automatic control theory, the problems of optimization and robustization of control processes are brought to the fore [6]. They have penetrated into all applied developments, including the aviation and rocket technology. By optimizing the design parameters of the HS AMAV, one may ensure its high-precision guidance at maneuvering air targets. And robustizing the flight control algorithm allows providing the HS AMAV operation in conditions of its parametric uncertainty.

The present article, developing the results of the authors [7–9], is devoted to the optimization and robustization of the AMAV class "air-to-air" control. In the first part of the article, the parametric synthesis problem of the optimal HS AMAV is considered according to the accuracy and energy consumption criteria of the guidance process. The two-criterion optimization of the HS is performed in the Matlab environment using the NSGA-II genetic algorithm under the conditions of three possible typical maneuvers of an aerial target: a slide, a dive and a dead loop. In the second part of the article, the problem of synthesizing algorithms for controlling the AMAV attack angle is analyzed. Two algorithms of AMAV robust tracking for a programmatic change of the attack angle are developed: the first implements the control method in sliding modes, and the second realizes the adaptive backstepping control method. The developed algorithms for controlling the AMAV attack angle in the Matlab environment was tested using a computer.

2 Optimization of the AMAV Homing Process

2.1 Functional Scheme and Mathematical Model of the HS AMAV

The HS AMAV represents the process of the AV autonomous deduction to a zone of contact with the air target based on the energy acceptance, radiated or imaged by the target (radio, light, infrared or sound waves). The HS AMAV is the automatic system intended for realizing the method of the AV guidance towards the target. This method defines the law of the specific rapprochement with the target. The minimum distance between the target and AMAV during the guidance process is called an error.

Without losing generality, we suppose that the AMAV homing guidance is effected by the proportional guidance method, such that during the flight the angular velocity touching the AV movement trajectory is proportional to the angular velocity of the sight target line (the "AMAV – target" line).

The HS AMAV generalized functional scheme, realizing the proportional guidance, is represented in Fig. 1. The homing head (HH), the shaper of guidance law (SGL), the kinematic link (KL) and two-contour stabilization system (SS), providing control by the AV corner movement according to the orders of the guidance, are the HS part. The steering drive (SD), the shaper of the stabilization law (SSL) and sensitive elements (angular velocity (SAV) and linear acceleration (SLA) sensors) are the SS part. Figure 1 presents the following notations: ϕ is the sight angle line; u_g is the guidance signal; u_s is the stabilization signal; u_{sp} and u_{ac} are SAV and SLA signals; u_c is the SD input; δ is the rudder's rotation angle; ϑ is the pitch angle, $w_z = \dot{\vartheta}$; a_n is normal acceleration, x, y are coordinates; V is the AV speed; x_T, y_T are coordinates, V_T is the target speed.

Homing System

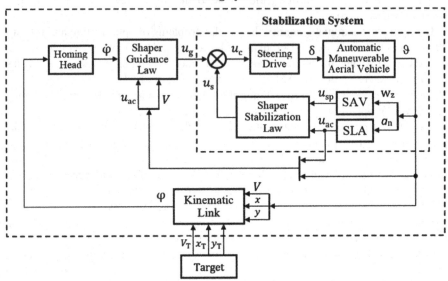

Fig. 1. HS AMAV generalized functional scheme.

It is obvious that the mathematical HS AMAV model is essential nonlinear. But to research the effectiveness of the guidance methods, we restrict our consideration to the linearized HS model (see, for example [7, 8, 10–13]). Let us reduce the linear mathematical models of HS dynamic elements (AMAV, SD, SAV, SLA, HH) to the form of their transfer functions [3, 10]:

$$W_{AMAV}(s) = \frac{\vartheta(s)}{\delta(s)} = \frac{a_{13}s + a_{13}a_{42}}{s[s^2 + (a_{11} + a_{42})s + a_{12} + a_{11}a_{42}]};$$

$$W_{SD}(s) = \frac{\delta(s)}{u_c(s)} = \frac{k_{dr}}{T_{dr}^2 s^2 + 2\xi_{dr}T_{dr}s + 1};$$

$$W_{SAV}(s) = \frac{u_{sp}(s)}{w_z(s)} = \frac{k_{av}}{T_{av}^2 s^2 + 2\xi_{av}T_{av}s + 1};$$

$$W_{SLA}(s) = \frac{u_{ac}(s)}{a_n(s)} = \frac{k_{la}}{T_{la}^2 s^2 + 2\xi_{la}T_{la}s + 1};$$

$$W_{HH}(s) = \frac{\dot{\varphi}(s)}{\varphi(s)} = \frac{k_h s}{s + k_h}.$$

Let us also reduce the linear mathematical models of static elements HS (KL, SSL, SGL) to the form of algebraic equations [3, 10]:

$$\varphi = \frac{57.3 \arcsin(y_T - y)^2}{\sqrt{(x_T - x)^2 + (y_T - y)^2}};$$

$$u_s = k_s u_{sp} + k_a u_{ac}; \quad u_g = k(k_w \dot{\varphi} - V u_{ac}),$$

where a_{11} is the damping coefficient; a_{12} is the weather vane coefficient; a_{13} is the rudder's effectiveness coefficient; a_{42} is the normal force coefficient; $k_{dr}, k_{av}, k_{la}, k_h$ are transform coefficients; k_s, k_a, k, k_w are SSL and SGL amplification coefficients; T_{dr}, T_{av}, T_{la} are time constants; $\xi_{dr}, \xi_{av}, \xi_{la}$ are damping coefficients.

2.2 Parametric Optimization Problem of the HS AMAV

Let us consider a vector of free HS AMAV parameters found in the process of its parametric synthesis:

$$\mathbf{K} = \text{col } (k_s, k_a, k, k_w),$$

where k_s, k_a are formation law coefficients of stabilization signal u_s; k, k_w are formation law coefficients of guidance signal u_g.

We suppose that the main exponents of the AMAV guidance process are the HS *precision* and the AV *overload*.

The first exponent characterizes the error value and the second one characterizes its maneuver possibilities. Let us introduce the following two optimality criteria, being the functions of the unknown vector of parameters K:

– the first is a final error value in the guidance process $J_1(\mathbf{K})$:

$$J_1(\mathbf{K}) = \sqrt{[x(t_f) - x_T(t_f)]^2 + [y(t_f) - y_T(t_f)]^2};$$

– the second is a power input value of the guidance process $J_2(\mathbf{K})$:

$$J_2(\mathbf{K}) = \int_0^{t_f} |n_y(t)| dt,$$

where $x(t_f), y(t_f), x_T(t_f), y_T(t_f)$ are the AV and target coordinates at the instant time $t = t_f$ of the guidance process; n_y is the AV available overload.

Let us normalize criteria $J_1(\mathbf{K})$ and $J_2(\mathbf{K})$:

$$J_i^{n}(\mathbf{K}) = \frac{J_i(\mathbf{K}) - J_i^{\min}(\mathbf{K})}{J_i^{\max}(\mathbf{K}) - J_i^{\min}(\mathbf{K})}, i = 1, 2.$$

It is obvious that:

$$0 \leq J_i^{n}(\mathbf{K}) \leq 1, i = 1, 2.$$

Let us pose the problem of the HS AMAV parametric synthesis, that is, the problem of detecting the vector with parameters $\mathbf{K}$ as the following *two-criterion problem of the parametric optimization with the limitation*:

$$J_i^{n}(\mathbf{K}) \to \min_{\mathbf{K} \in \Omega}, i = 1, 2, \ \Omega = \{\mathbf{K} | \mathbf{K}_{\min} \leq \mathbf{K} \leq \mathbf{K}_{\max}\}, \tag{1}$$

where K_{min} and K_{max} are minimum and maximum admissible values of the vector K, respectively.

The problem of HS AMAV parametric optimization by the criterion of the final miss of the guidance process is noted to be considered in [14]. The solution of this problem is given in the Matlab environment by the direct scanning method of the desired parameters and integrating the differential equations of the HS mathematical model.

The main problem in the multi-criteria optimization is known to be ambiguity of the "optimal solution" [15, 16]. One of the most effective principles in solving multi-criteria optimization problems is the Pareto principle. Based on this principle, the optimal solution according to the so-called Pareto optimal solutions, that is, the solution set, which composes the area of compromise, is chosen. In the Pareto set, all the solutions are equivalent: the value of every solution from criteria cannot be improve without deterioration of the other criteria. In doing so, the compromise curve, named Pareto front and being the form of the Pareto set in the space of criteria, correspond to the Pareto set.

To detect the Pareto set of the two-criterion optimization problem (1), it is advisable to *scalar* it, that is, reduce it to a scalar one-criterion problem. The scalar method, based on the structure of the auxiliary linear criterion $J^n(K)$, is the most effective:

$$J^n(\mathbf{K}) = h_1 J_1^n(\mathbf{K}) + h_2 J_2^n(\mathbf{K}),$$
$$h_i = \text{const} > 0, \ h_1 + h_2 = 1, \ i = 1, 2; \tag{2}$$

including the auxiliary Chebyshev criterion $J^n(K)$:

$$J^n(\mathbf{K}) = \max_i h_i |J_i^n(\mathbf{K}) - z_i|, \ h_i = \text{const} > 0,$$
$$z_i < \min_{K \in \Omega} J_i^n(\mathbf{K}), \ i = 1, 2. \tag{3}$$

Changing the coefficients h_i, $i = 1, 2$, we may obtain the Pareto front.

To solve one-criterion optimization problems (2) and (3), the *genetic algorithms* are very effective (see, for example [17]). They exploit key notions of Darvin's evolution: population, mutation, crossing, survival of the strongest. Here every space point of the synthesized parameters is the member of some population P of order N. The initial population is chosen accidentally. Using the adaptation function, we may estimate how well a certain number of population is customized to the "medium condition". The subset $P_{parents}$ of the most adjusted parents is chosen from the population P. Then, by means of the mutation and crossing operations, it is possible to obtain the new population P from them. This process is repeated until the stop criterion is realized.

To solve problems (2) and (3), one of the most popular genetic algorithms NSGA-2 was used (non-dominated GA sorting). It was developed by K. Deb [18] to solve the problems of multi-criteria optimization. In this case, the base genetic algorithm with the initial population of 50 individuals was used with the following operators: panmixia (the choice of the parents from the population) with the normal distribution law and the uniform crossingover (crossing); the absolute mutation (stochastic change) with the probability of 10%.

2.3 Researching Computer Effectiveness of the Optimal HS AMAV

To study the effectiveness of the synthesized optimal HS AMAV, the numerical simulation of the process of the proportional AV guidance in a vertical plane to the maneuver air
target in the Matlab environment is considered. In doing so, there are cases of shooting
AMAV at the target towards the back and front hemispheres so that the target can perform some type of maneuvers. They are the *gain in height* at a $+60°$ angle, the *reduction*
at a $-60°$ angle, the *hill* in the form of the abrupt ascent by 500 m at a $+60°$ angle, the
nosedive in the form of the abrupt reduction by 500 m at a $-60°$ angle and the *dead loop*
with the overload $n_T = 5$.

The target begins to maneuver at a distance to AMAV of less than 3000 m. The initial
coordinates of AV and the target are equal to $x_0 = 0$, $y_0 = 0$, $x_{T0} = 5000$ m, $y_{T0} =$
3000 m respectively. The AV and target speeds are equal to $V = 2M$ and $V_T = 1M$
respectively. The available AV overload is $|n_y| \le 32$.

The parametric values of the mathematical HS model correspond to hypothetical
AMAV like a cruise missile taken from the paper [13]:

$a_{11} = 1.2\,s^{-1}$, $a_{12} = 20\,s^{-2}$, $a_{13} = 30\,s^{-2}$, $a_{42} = 1.5\,s^{-2}$, $k_{dr} = 1$, $\xi_{dr} = 0.6$, $T_{dr} = 0.01$;

$|\delta| \le 20$, $k_{av} = 1$, $\xi_{av} = 0.6$; $T_{av} = 0.01$; $k_{la} = 1$, $\xi_{la} = 0.6$, $T_{la} = 0.01$; $k_h = 50$;

$0.06 \le k_s \le 0.4$; $0.001 \le k_a \le 0.01$; $1 \le k \le 20$; $3 \le k_w \le 5$.

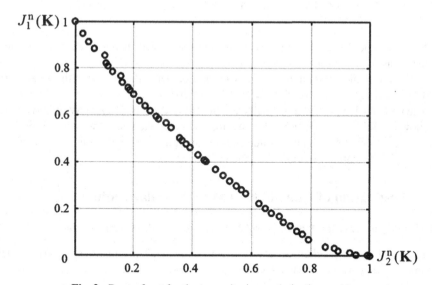

Fig. 2. Pareto front for the two-criterion optimization problem

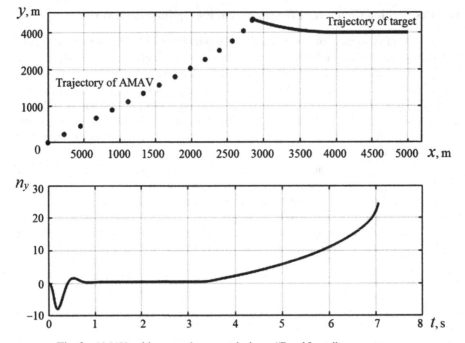

Fig. 3. AMAV guidance to the target during a "Dead Loop"-type maneuver

Figures 2 and 3 demonstrate the case of shooting AMAV towards the target in such maneuver as a "dead loop".

If we choose the point on the front as the solution being the nearest to the ideal point $(0, 0)$, we will obtain $k_s = 0.29$, $k_a = 0.001$, $k = 70.08$, $k_w = 4.28$.

The parametric optimization results and the analysis of the HS AMAV effectiveness are shown in Table 1, where the following notations are taken. t_{tt} is the transient time; σ is overregulation; ΔA is the amplitude stability margin; e_f is the guidance error; n_{ymax} is the amplitude stability margin; t_g is the guidance time.

3   Robustization of Control by AMAV Angular Motion

3.1   Setting a Tracking Control Problem by the AMAV Attack Angle

Let us consider the hypothetical AMAV reported in [19]. The equations of longitudinal dynamics of this AV as a controlled rigid body are of the form:

$$\dot{V} = -\frac{X_a}{m} - g \sin \theta + \frac{\cos \alpha}{m} P, \tag{4}$$

$$\dot{\alpha} = -\frac{Y_a}{mV} + \frac{\cos \theta}{V} g - \frac{\sin \alpha}{mV} P + w_z, \tag{5}$$

$$\dot{w}_z = \frac{M_{z1}}{J_z} + \frac{M_{z2}}{J_z} w_z + \frac{M_{z3}}{J_z} \delta, \tag{6}$$

Table 1. Effectiveness results of the AMAV guidance process.

	Shooting at the rear hemisphere of the target			
	k_s	k_a	k	k_w
	0.37	0.01	53.43	3.3
	t_{tt}, s	σ, %		ΔA, dB
	1.01	0		7.8
	e_f, m	$n_{y\max}$		t_g, s
	0.009	6.77		17.23
Lack of target maneuver	Shooting at the front hemisphere of the target			
	k_s	k_a	k	k_w
	0.25	0.01	44.06	5.00
	e_f, m	$n_{y\max}$		t_g, s
	0.428	-7.927		6.640
	Shooting at the rear hemisphere of the target			
	k_s	k_a	k	k_w
	0.25	0.01	81.19	3.31
	t_{tt}, s	σ, %		ΔA, dB
	0.668	0.288		6.7
	e_f, m	$n_{y\max}$		t_g, s
	0.003	16.99		16.985
Maneuver of the target "Hill"	Shooting at the front hemisphere of the target			
	k_s	k_a	k	k_w
	0.26	0.01	33.30	3.76
	t_{tt}, s	σ, %		ΔA, dB
	0.70	0.118		6.79
	e_f, m	$n_{y\max}$		t_g, s
	0.5	12.812		7.240

(*continued*)

Table 1. (*continued*)

	Shooting at the rear hemisphere of the target			
	k_s	k_a	k	k_w
	0.19	0.01	68.34	3.33
	t_{tt} , s	σ , %	ΔA , dB	
	0.927	3.466	6.154	
	e_f , m	$n_{y\max}$	t_g , s	
	0.002	9.766	16.02	
Maneuver of the target "Nosedive"	Shooting at the front hemisphere of the target			
	k_s	k_a	k	k_w
	0.27	0.01	53.40	4.69
	t_{tt} , s	σ , %	ΔA , dB	
	0.732	0.03	6.884	
	e_f , m	$n_{y\max}$	t_g , s	
	0.115	-32.000	6.425	
	Shooting at the rear hemisphere of the target			
	k_s	k_a	k	k_w
	0.31	0.01	44.24	3.57
	t_{tt} , s	σ , %	ΔA , dB	
	0.86	0	7.249	
	e_f , m	$n_{y\max}$	t_g , s	
	0.025	13.233	17.85	
Maneuver of the target "Dead Loop"	Shooting at the front hemisphere of the target			
	k_s	k_a	k	k_w
	0.29	0.01	70.08	4.28
	t_{tt} , s	σ , %	ΔA , dB	
	0.798	0	7.066	
	e_f , m	$n_{y\max}$	t_g , s	
	0.275	24.494	7.050	

where the aerodynamic forces and the moments acting on the AV are determined by the equalities:

$$X_a = X_a^* + \Delta X_a,\ Y_a = Y_a^* + \Delta Y_a,\ M_{z_1} = M_{z_1}^* + \Delta M_{z_1},$$

$$X_a^* = QS\left\{C_{xa_0}^* - \sin\alpha\left[a_n^*\alpha^3 + b_n^*\alpha|\alpha| + c_n^*(2 - \frac{M}{3})\alpha\right]\right\},$$

$$\Delta X_a = QS\left\{\Delta C_{xa_0} - \sin\alpha\left[\Delta a_n\alpha^3 + \Delta b_n\alpha|\alpha| + \Delta c_n(2 - \frac{M}{3})\alpha\right]\right\},$$

$$Y_a^* = -QS\cos\alpha\left[a_n^*\alpha^3 + b_n^*\alpha|\alpha| + c_n^*(2 - \frac{M}{3})\alpha\right],$$

$$\Delta Y_a = -QS\cos\alpha\left[\Delta a_n\alpha^3 + \Delta b_n\alpha|\alpha| + \Delta c_n(2 - \frac{M}{3})\alpha\right],$$

$$M_{z_1}^* = QSd\left[a_m^*\alpha^3 + b_m^*\alpha|\alpha| + c_m^*(-7 + \frac{8M}{3})\alpha\right],$$

$$\Delta M_{z_1} = QSd\left[\Delta a_m\alpha^3 + \Delta b_m\alpha|\alpha| + \Delta c_m(-7 + \frac{8M}{3})\alpha\right],$$

$$M_{z_2} = M_{z_2}^* + \Delta M_{z_2} = QSde_m^* + QSd\,\Delta e_m,$$

$$M_{z_3} = M_{z_3}^* + \Delta M_{z_3} = QSdd_m^* + QSd\,\Delta d_m.$$

The following notations are taken for the designation. V is the AV speed; θ is the trajectory inclination angle; w_z is the pitch angular velocity; α is the attack angle; δ is the deviation angle of the rudder's height; P is the thrust force; J_z is the AV inertia moment; g is the free incidence acceleration. $Q = (1/2)\rho V^2$ is the dynamic pressure; ρ is the atmosphere density; $M = (V/a)$ is Mach number; a is the sound velocity; m is the AV mass; S is the calculated space; d is the calculated distance.

In the Eqs. (4–6) of the AMAV dynamics, the nominal value of some parameter π is denoted by π^*, and its interval indefiniteness is denoted by $\Delta\pi$:

$$-\gamma\pi^* \leq \Delta\pi \leq \gamma\pi^*,\ \gamma = \text{const} > 0.$$

Such parameters are constants $C_{xa_0}, a_n, a_m, b_n, b_m, c_n, c_m, e_m, d_m$.

Let us consider the attack angle α as the controlled AMAV variable and tail stabilizers δ as the controlling action.

The following *task of tracking control* is to be analyzed. The AMAV control with the dynamic Eqs. (4–6) requires synthesizing the controlling action δ, provided during the flight reproduced by HS of a given program $\alpha_p(t)$ with the change in the AV attack angle α:

$$\lim_{t\to\infty}|e(t)| = 0,\ e(t) = \alpha(t) - \alpha_p(t).$$

In this case we suppose that the AMAV control problem is considered in the conditions of a priori indefiniteness of its model parameters with the coefficient γ: $0 \leq \gamma \leq 0.3$. That is, the model's parameters of the AV dynamics are universally denoted by π being in the range of $[-30\%; +30\%]$.

In this way, the *problem of robust tracking control by the AMAV attack angle* in the conditions of the parametric indefiniteness of the mathematical model of its dynamics is considered. In doing so, the current control of the AV δ is formed as a function of deviations $e(t)$ of its actual trajectory of movement $\alpha(t)$ from the nominal program trajectory of movement $\alpha_p(t)$. The AMAV velocity and, consequently, Mach number M are believed to be the constant:

$$\dot{V} = 0 \Rightarrow P = \frac{m}{\cos \alpha}\left(\frac{X_a}{m} + g \sin \theta\right),$$

and introducing notations:

$$F_1^* = F_1^*(\alpha) = \frac{Y_a^*}{mV} + \frac{\tan \alpha}{mV}X_a^*, \quad F_1 = F_1(\alpha) = F_1^* + \Delta F_1 = \frac{Y_a}{mV} + \frac{\tan \alpha}{mV}X_a,$$

$$G = \frac{\cos \theta}{V}g - \frac{\tan \alpha \sin \theta}{V}g, \quad F_2^* = F_2^*(\alpha) = \frac{M_{z1}^*}{J_z} + \frac{M_{z2}^*}{J_z}w_z,$$

$$F_2 = F_2(\alpha) = F_2^* + \Delta F_2 = \frac{M_{z1}}{J_z} + \frac{M_{z2}}{J_z}w_z,$$

let us transform the initial model (4–6) of the AMAV hypothetical movement into the following form:

$$\dot{e} = -F_1 + G - \dot{\alpha}_p + w_z, \tag{7}$$

$$\dot{w}_z = F_2 + \frac{M_{z3}}{J_z}\delta. \tag{8}$$

As far as the dynamics model (7), (8) of the AMAV hypothetical movement is essentially a nonlinear one, solving the set problem of the tracking control is very difficult. In this connection, being the dynamics model of the similar objects, they often take the linearized model and attract the methods of the linear automatic control theory (see, for example [20, 21]) to solve the control problem. To solve the set problem, widely known formalized methods of the recent nonlinear control theory are used in the present paper. They are the control method of sliding modes and the method of adaptive backstepping control.

3.2   Algorithm of AMAV Robust Tracking Control by the Sliding Mode

The *sliding modes control* (SMC) *method* is based on synthesizing the system with the variable structure having the premeditated organization of sliding modes [22, 23] on the "attracting" hyper-plane, being the system desired phase trajectory.

This method changes the system's dynamics by means of applying the relay controlling signal. This signal makes the system "slide" over the hyper-plane to the desired target point. In the sliding mode, the controlling device switching theoretically happens with an infinitely high frequency. Besides, the system in motion over the sliding line has the property of robust control for parametric perturbations.

It is well known that the sliding mode condition in the general case has the form:

$$\frac{1}{2}\frac{d}{dt}s^2 = s\dot{s} \le -\eta|s|, \quad \eta = \text{const} > 0.$$

For the AMAV subsystem (4), let us choose the sliding hyper-plane:

$$s = e - e_d = e \text{ because } e_d = 0,$$

where the point $e = 0$ is the attracting one. The sliding mode condition takes the form:

$$\frac{1}{2}\frac{d}{dt}s^2 = s\left[-F_1 + G - \dot{\alpha}_p + w_z\right].$$

Considering w_z by the controlling action for the subsystem (4), let us find for it the following desired signal:

$$w_{zd} = F_1^* - G + \dot{\alpha}_p - k_1\text{sign}(\alpha - \alpha_p) - k_2(\alpha - \alpha_p),$$

where:

$$k_1 = k_1(\alpha) = \gamma\left|F_1^*(\alpha)\right| + \eta_1, \quad \eta_1 > 0; \quad k_2 = \text{const} > 0.$$

At this point, the coefficient k_1 guarantees the achievement of sliding mode $s = 0$ and determines the system's robustness degree in conditions of the parametric indeterminacy. And the coefficient k_2 provides the system's speed increase and determines the exponential decrease degree of the transients.

Now let us choose the following sliding hyper-plane for the AMAV subsystem (8):

$$s = w_z - w_{zd}.$$

In this case, the sliding mode condition has the form:

$$\frac{1}{2}\frac{d}{dt}s^2 = s\left[F_2 + \frac{M_{z3}}{J_z}\delta - \dot{w}_{zd}\right],$$

and the expression for the AMAV sliding control algorithm δ takes the following view:

$$\delta = \frac{J_z}{\tilde{M}_{z3}}\left[\dot{w}_{zd} - F_2^* - k_3\text{sign}(w_z - w_{zd}) - k_4(w_z - w_{zd})\right], \tag{9}$$

where:

$$k_3 = k_3(\alpha) = \beta\gamma\left|F_2^*(\alpha)\right| + \beta(\beta - 1)\left|F_2^*(\alpha) - \dot{w}_{zd}\right| + \beta\eta_2, \quad \eta_2 > 0; \quad k_4 = \text{const} > 0;$$

$$\tilde{M}_{z3} = \sqrt{M_{z3\,\text{max}}M_{z3\,\text{min}}}; \quad \beta = \sqrt{\frac{M_{z3\,\text{max}}}{M_{z3\,\text{min}}}} = \sqrt{\frac{1+\gamma}{1-\gamma}}.$$

The coefficients k_3 and k_4 are similar to the coefficients k_1, k_2 determining the system robustness degree and the exponential decrease degree of its transients respectively.

The computer approbation results of the found control algorithm δ in the form (9) in the Matlab environment are represented in Fig. 4. This algorithm provides the reproduction (4–6) by the AMAV attack angle α defined by the guidance system of the following programmed changes of $\alpha_p(t)$:

$$\alpha(t) = \begin{cases} 5°, & 0 \le t \le 3\,\text{s}, \\ -5°, & 3 \le t \le 6\,\text{s}. \end{cases}$$

The numerical values (7), (8) of the parameters of the AMAV mathematical model are taken from [19]:

$J_z = 247.43662\ \text{kg m}^2,\ d = 0.2286\ \text{m},\ S = 0.0409\ \text{m}^2,\ m = 204.108\ \text{kg},$

$V = 2M,\ g = 9.81\ \text{ms}^{-2},\ a = 315.89472\ \text{ms}^{-1},\ a_n = 19.373,\ b_n = -31.023,$

$c_n = -9.717,\ d_n = -1.948,\ C_{xa_0} = 0.3,\ a_m = 40.440,\ b_m = -64.015,$

$c_m = 2.922,\ d_m = -11.803,\ e_m = -1.719,\ \gamma = 0.3,$

$k_1 = 0.05,\ k_2 = 2,\ k_3 = 0.05,\ k_4 = 2.$

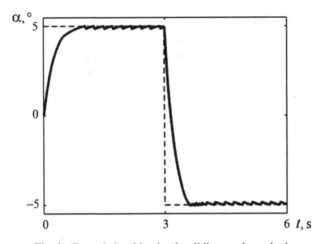

Fig. 4. Control algorithm by the sliding mode method.

It is not difficult to see that the AMAV synthesized control algorithm has the infinite number of the switching points at the final time interval; that is, there is a "mode of quickened switching" in the system. The given appearance is well known as the "Fuller phenomenon" (see, for example [24]) and the corresponding control mode in the foreign literature was named as "chattering". To avoid the given undesirable mode of the system functioning, in connection with smoothing off the explosive control in the boundary ε layer of the switching surface, there is:

$$|s| \le \varepsilon.$$

By changing the control algorithm (9) from the function sign(s) to the function sat(s/ε):

$$\text{sign}(s) = \begin{cases} 1, & s > 0, \\ 0, & s = 0, \\ -1, & s < 0. \end{cases} \qquad \text{sat}\left(\frac{s}{\varepsilon}\right) = \begin{cases} \left(\frac{s}{\varepsilon}\right), & |s| < \varepsilon, \\ \text{sign}\left(\frac{s}{\varepsilon}\right), & |s| < \varepsilon. \end{cases}$$

As a result, the attracting point $e = 0$ is replaced by the attracting circle of the radius ε, surrounding the given point. Being in the given area, the phase trajectories of the system in the sliding process do not leave it. In this case, the condition of the sliding mode takes the form:

$$\begin{cases} s \geq \varepsilon \Rightarrow \dfrac{d}{dt}(s - \varepsilon) \leq -\eta, \\ s < -\varepsilon \Rightarrow \dfrac{d}{dt}(s + \varepsilon) \geq \eta, \end{cases}$$

or

$$|s| \geq \varepsilon \Rightarrow \frac{1}{2}\frac{d}{dt}s^2 \leq (\dot{\varepsilon} - \eta)|s|.$$

For the subsystem (7), let us choose the ε_1 circle, inside of which we have:

$$\dot{s} = -\Delta F_1 - k_1(\alpha)\frac{s}{\varepsilon_1}.$$

As far as in the area $\left|\alpha - \alpha_p\right| \leq \varepsilon_1$ the following equality is correct,

$$\dot{s} = -k_1(\alpha_p)\frac{s}{\varepsilon_1} - \Delta F_1(\alpha_p) + O(\varepsilon_1),$$

it is evident that s is the output of the 1st order filter with the input signal:

$$-\Delta F_1(\alpha_p) + O(\varepsilon_1)$$

and with the strip of the transmitter λ_1:

$$\lambda_1 = \frac{k_1(\alpha_p)}{\varepsilon_1}.$$

In this case:

$$k_1(\alpha_p) = \gamma\left|F_1^*(\alpha_p)\right| + \eta_1 - \dot{\varepsilon}_1 = \lambda_1\varepsilon_1,$$

or

$$\dot{\varepsilon}_1 = -\lambda_1\varepsilon_1 + \gamma\left|F_1^*(\alpha_p)\right| + \eta_1.$$

In a similar way, choosing the ε_2 circle for the subsystem (8), the following expression for the AMAV sliding control algorithm δ is true:

$$\delta = \frac{J_z}{\tilde{M}_{z3}}\left[\dot{w}_z - F_2^* - k_3\text{sat}\,(w_z - w_{zd}) - k_4(w_z - w_{zp})\right], \tag{10}$$

where

$$k_3 = \beta\gamma\left|\hat{F}_2(\alpha_p)\right| + \beta(\beta - 1)\left|\hat{F}_2(\alpha_p) - \dot{w}_{zd}\right| + \beta\eta_2, \quad k_4 = \text{const} > 0.$$

And the following conditions are correct:

$$k_3 \geq \frac{\lambda_2\varepsilon_2}{\beta} \Rightarrow \dot{\varepsilon}_2 = -\lambda_2\varepsilon_2 + \beta k_3,$$

$$k_3 < \frac{\lambda_2\varepsilon_2}{\beta} \Rightarrow \dot{\varepsilon}_2 = -\lambda_2\beta^{-2}\varepsilon_2 + \frac{k_3}{\beta}.$$

One may show that here s is the output of the 1st order filter with the input signal:

$$\Delta F_2(\alpha_p) + \left[\frac{M_{z3}(\alpha_p)}{\tilde{M}_{z3}} - 1\right]\left(-F_2^*(\alpha_p) + \dot{w}_{zd}\right) + O(\varepsilon_1, \varepsilon_2),$$

and with the strip of the transmitter λ_2:

$$\lambda_2 = \beta\frac{k_3(\alpha_p)}{\varepsilon_2}.$$

It is not difficult to see that the sliding exactness (mistakes $\varepsilon_1, \varepsilon_2$) and strips of the filter passing (λ_1, λ_2) are inversely proportional.

The results of the computer simulations of the control process in AMAV with parameters:

$$k_1 = 0.05; \quad k_2 = 2; \quad k_3 = 0.05; \quad k_4 = 2;$$
$$\lambda_1 = 30; \quad \varepsilon_1(0) = 0.005; \quad \lambda_2 = 30; \quad \varepsilon_2(0) = 0.005$$

are represented in Fig. 5. They show that the application of the sliding control algorithm δ in the form (10) provides for the Eqs. (4–6) of the effective suppression in the system of the undesirable chattering mode.

3.3 Algorithm of AMAV Robust Tracking Control Using the Adaptive Backstepping Method

The *backstepping method,* known as the integrator circuit method, is the interactive procedure of synthesis where the problems of determining the Lyapunov functions and the corresponding control law [25, 26] are combined. The essence of the given method consists in decomposing the initial control problem by the system into a sequence of the corresponding sub-problems of a smaller order by means of the system's representation in the form of the embedded subsystems chain. The auxiliary controlling signals are formed for each of them, and the Lyapunov functions of dependence on these signals are composed. As a result, the stability is imposed on every integrator of the object by means of adding the corresponding feedback. This fact provides high quality of transients in the system without increasing the control amplitude.

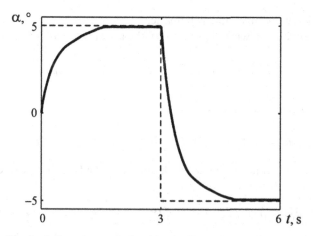

Fig. 5. Adjusted control algorithm by the sliding mode method.

Let us transform the system of the Eqs. (4–6) into the following form:

$$\dot{e} = Y_1^T \theta_1 + G - \dot{\alpha}_p + w_z, \tag{11}$$

$$\dot{w}_z = Y_2^T \theta_2 + Y_3^T \theta_3 \delta, \tag{12}$$

where:

$$Y_1 = \frac{QS}{mV} \begin{bmatrix} -\tan(\alpha) \\ [\tan(\alpha)\sin(\alpha) + \cos(\alpha)]\,\alpha^3 \\ [\tan(\alpha)\sin(\alpha) + \cos(\alpha)]\,\alpha|\alpha| \\ [\tan(\alpha)\sin(\alpha) + \cos(\alpha)](2 - \frac{M}{3})\,\alpha \end{bmatrix}, \quad \theta_1 = \begin{bmatrix} C_{xa0} \\ a_n \\ b_n \\ c_n \end{bmatrix},$$

$$Y_2 = \frac{QSd}{J_z} \begin{bmatrix} \alpha^3 \\ \alpha|\alpha| \\ (-7 + \frac{8M}{3})\alpha \\ w_z \end{bmatrix}, \quad \theta_2 = \begin{bmatrix} a_m \\ b_m \\ c_m \\ e_m \end{bmatrix}, \quad Y_3 = \frac{QSd}{J_z}, \quad \theta_3 = d_m.$$

In the formula transformation below for the estimates $\hat{\theta}_i$ of parameters θ_i, the notations of errors $\tilde{\theta}_i = \theta_i - \hat{\theta}_i$, $i = 1, 2, 3$ are used.

Choosing for the subsystem (11) the controlling action w_z in the form of:

$$w_z = -Y_1^T \hat{\theta}_1 - G + \dot{\alpha}_p - k_5 e + v, \ k_5 > 0,$$

let us obtain the following system:

$$\dot{e} = Y_1^T \tilde{\theta}_1 - k_5 e + v,$$

$$\dot{v} = Y_2^T \tilde{\theta}_2 + Y_2^T \hat{\theta}_2 + Y_3^T \tilde{\theta}_3 \delta + Y_3^T \hat{\theta}_3 \delta + \dot{Y}_1^T \hat{\theta}_1 + Y_1^T \dot{\hat{\theta}}_1 + \dot{G} - \ddot{\alpha}_p + k_5 \dot{e}.$$

Let us form the following Lyapunov function:

$$V = \frac{1}{2}e^2 + \frac{1}{2}v^2 + \frac{1}{2}\tilde{\theta}_1^T \Gamma_1^{-1} \tilde{\theta}_1 + \frac{1}{2}\tilde{\theta}_2^T \Gamma_2^{-1} \tilde{\theta}_2 + \frac{1}{2}\tilde{\theta}_3^T \Gamma_3^{-1} \tilde{\theta}_3,$$

where Γ_1, Γ_2, Γ_3 are positive diagonal matrices.

Let us choose the following algorithm δ of the tracking control of AMAV:

$$\delta = \frac{1}{\hat{\theta}_3}\left[-e - Y_2^T\hat{\theta}_2 - \dot{Y}_1^T\hat{\theta}_1 - Y_1^T\dot{\hat{\theta}}_1 - \dot{G} + \ddot{\alpha}_p - k_5\dot{e} - k_6v\right], \quad k_6 > 0, \qquad (13)$$

with the following adaptation laws:

$$\dot{\hat{\theta}}_1 = (e + k_5v)\Gamma_1 Y_1, \quad \dot{\hat{\theta}}_2 = v\Gamma_2 Y_2, \quad \dot{\hat{\theta}}_3 = \delta v\Gamma_3 Y_3. \qquad (14)$$

Then the expression for the complete derivative of the Lyapunov's function has the form:

$$\dot{V} = e\dot{e} + v\dot{v} - \tilde{\theta}_1^T\Gamma_1^{-1}\dot{\hat{\theta}}_1 - \tilde{\theta}_2^T\Gamma_2^{-1}\dot{\hat{\theta}}_2 - \tilde{\theta}_3^T\Gamma_3^{-1}\dot{\hat{\theta}}_3 = -k_5e^2 - k_6v^2 \leq 0, \quad \forall k_5, k_6 > 0.$$

Using the lemma of I Barbalat allows showing the correctness of the following conditions of the asymptotic stability in the reproduction process of a certain program $\alpha_p(t)$ change in the AMAV attack angle α:

$$\lim_{t\to\infty} e = 0 \text{ and } \lim_{t\to\infty} v = 0 \text{ or } \lim_{t\to\infty} \alpha = \alpha_p(t).$$

The computer simulation results of the AMAV flight dynamics (4–6) with the adaptive algorithm of the backstepping control (13, 14) and parameters:

$$k_1 = 20; \quad k_2 = 20; \quad \Gamma_1 = \Gamma_2 = \Gamma_3 = \begin{bmatrix} 5 & 0 & 0 & 0 \\ 0 & 5 & 0 & 0 \\ 0 & 0 & 5 & 0 \\ 0 & 0 & 0 & 5 \end{bmatrix},$$

are shown in Fig. 6. The graph shows that the control system by the attack angle α quickly and exactly reproduces its programmed change $\alpha_c(t)$.

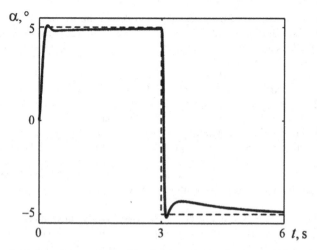

Fig. 6. Adjusted control algorithm by the adaptive backstepping method.

4 Conclusion

The work analyzes the optimization and robustization issues of the AMAV surface-to-air and air-to-air control.

In the first part of the research, the problem of the optimal HS AMAV parametric synthesis according to the accuracy and energy consumption criteria of the guidance process is considered. The two-criterion HS optimization is carried out in the Matlab environment using the NSGA-II genetic algorithm in conditions of three possible typical maneuvers of an aerial target: a hill, a nosedive and a dead loop.

In the second part of the study, the problem of synthesizing algorithms for robust control of the AMAV attack angle is examined. Two algorithms for the AMAV robust tracking of a programmatic change in the attack angle are described. The first one implements the control method in sliding modes, and the second algorithm deals with the method of adaptive backstepping control. The computer approbation of the developed algorithms for controlling the AMAV attack angle in the Matlab environment was performed.

The optimization methodology of AMAV control, presented in the work, can be widely used for the tasks of controlling mobile objects in the aviation, rocket and space technology.

References

1. Dynamic design of control systems of automatic maneuverable aircraft. In: Fedosov, E.A. (ed.) Mechanical Engineering, Moscow, p. 336 (1997). (in Russian)
2. Veremeenko, K.K., Veremeenko, K.K., Golovinsky, A.N., et al.: Management and guidance of unmanned maneuverable aircraft based on modern information technologies. In: Krasilshchikov, M.N., Serebryakov, G.G. (eds.) Fizmatlit, Moscow, p. 280 (2005). (in Russian)

3. High-precision homing systems: calculation and design. In: Pupkov, K.A., Egupov, N.D. (eds.) Computational Experiment. Fizmatlit, Moscow, p. 512 (2011). (in Russian)
4. Siouris, G.M.: Missile Guidance and Control Systems, p. 666. Springer, New York (2004). https://doi.org/10.1007/b97614
5. Zarhan, P.: Tactical Strategic Missile Guidance (Progress in Astronautics and Aeronautics), vol. 239, p. 1026. AIAA (American Institute of Aeronautics & Ast. (2012)
6. Polyak, B.T., Khlebnikov, M.V., Rapoport, L.B.: Mathematical Theory of Automatic Control: Textbook, p. 500. LENAND, Moscow (2019). (in Russian)
7. Pham, P., Filimonov, N.B.: Computer analysis of the effectiveness of cruise missile homing methods on maneuvering aerial target. Mechatron. Autom. Robot. **9**, 17–22 (2022). (in Russian)
8. Pham, P., Filimonov, N.B.: Pareto optimization of a cruise missile homing system for maneuvering air targets by a genetic algorithm. J. Adv. Res. Tech. Sci. **30**, 46–53 (2022). (in Russian)
9. Pham, P., Filimonov, N.B.: Nonlinear robust control of the angle attack of an automatic maneuverable aircraft. High Perform. Comput. Syst. Technol. **6**(1), 172–179 (2022). (in Russian)
10. Tolpegin, O.A., Kashin, V.M., Novikov, V.G.: Mathematical Models of Missile Guidance Systems: Textbook, p. 154. Baltic State Technical University, St. Petersburg (2016). (in Russian)
11. Tokar, A.D.: Comparative analysis of dynamic errors of the proportional guidance method and the modified guidance algorithm for aerial targets. Bull. Ryazan State Radio Eng. Univ. **20**, 44–47 (2007). (in Russian)
12. Rozhkov, I.V.: Synthesis of invariant circuit angular stabilization of unmanned aircraft at the pitch angle. Syst. Anal. Appl. Inf. Sci. **3**, 14–20 (2020). (in Russian)
13. Thong, D.Q.: Synthesis of high-precision missile homing system using proportional guidance method. Mekhatronika, Avtomatizatsiya, Upravlenie. **20**(4), 242–248 (2020)
14. Thong, D.Q.: Improving the accuracy of the missile homing system in the near zone and when shooting highly maneuverable targets. In: Scientific Forum "Technical and Physical and Mathematical Sciences": collection of articles based on the materials of the LI International Scientific and Practical Conference. No. 1(51), pp. 9–24. Publishing House "MCNO", Moscow (2022). (in Russian)
15. Lotov, A.V., Pospelova, I.I.: Lecture Notes on the Theory and Methods of Multicriteria Optimization, p. 127. MAKS Press, Moscow (2005). (in Russian)
16. Ehrgott, M.: Multicriteria Optimization, p. 324. Springer, Heidelberg (2005). https://doi.org/10.1007/978-3-662-22199-0
17. Filimonov, N.B., Belousov, I.V.: Application of genetic algorithms in optimization problems of terminal control of dynamic objects. Rep. Russ. Acad. Sci. Volga Interregional Branch **3**, 68–80 (2002). (in Russian)
18. Deb, K., Pratap, A., Agarwal, S., Meyarivan, T.: A fast and elitist multiobjective genetic algorithm: NSGA-II: evolutionary computation. IEEE Trans. Evol. Comput. **6**(2), 182–197 (2002)
19. Golubev, A.E.: Tracking a process of scheduled change in the angle of attack for longitudinal dynamics of an air-to-air missile with the use of an integrator back-stepping method. Sci. Educ. **11**, 401–414 (2013). (in Russian)
20. Isidori, A.: Lectures in feedback design for multivariable systems. In: Advanced Textbook in Control and Signal Processing, p. 414. Springer, London (2016). https://doi.org/10.1007/978-3-319-42031-8
21. Khalil, H.K.: Nonlinear Control, p. 304. Pearson, London (2015)
22. Degtyarev, G.L., Meshchanov, A.S.: Control methods on sliding modes: monograph, p. 104 (2014). (in Russian)

23. Recent trends in sliding mode control. In: Fridman, L., Barbot, J.-P., Plestan, F. (eds.) Stevenage, p. 439. The Institute of Engineering and Technology, UK (2016)
24. Suxinin, B.V., Surkov, V.V., Filimonov, N.B.: Fuller phenomenon in problems of analytical design of optimal regulators. Mechatron. Autom. Control $22(3)$, 339–348 (2021). (in Russian)
25. Krstic, M., Kanellakopoulos, I., Kokotović, P.V.: Nonlinear and Adaptive Control Design, p. 563. Wiley, New York (1995)
26. Miroshnik, I.V., Nikiforov, V.O., Fradkov, A.L.: Nonlinear and Adaptive Control of Complex Dynamic Systems, p. 549. Nauka Publisher, Saint Petersburg (2000). (in Russian)

Synthesis of the Rational Analyzing Function for Feature Extraction of Signals from the Electrostatic Location System

Yurii Skryabin$^{(\boxtimes)}$ and Dmitry Potekhin

MIREA – Russian Technological University, Vernadsky Avenue 78, 119454 Moscow, Russia
skryabin@mirea.ru

Abstract. The article deals with an algorithm for extracting features of an electrostatic signal from the electrostatic location system. The algorithm is based on converting a signal into a time-frequency distribution by means of convolution with analyzing function. In this article a software algorithm for the synthesis of a special function and a method for analyzing the time-frequency distribution are considered. The precision of the method for extracting features of the electrostatic signal with different signal-to-noise ratios is estimated. The precision of the method is evaluated with respect to the precision of the standard method. The standard method is based on the Morlet wavelet function as an analyzing one.

Keywords: Electrostatic monitoring technology · Electrostatic location · Electrostatic signal · Signal processing · Time-frequency analysis

1 Introduction

The problem of unmanned aerial vehicle detection has become a major aspect. This problem is related to the rapid number growth of unmanned aerial vehicles (UAVs) and relatively liberal laws on the UAVs' regulation and flight management. Thus, it is vitally important to ensure safety of critical facilities and highly crowded places from possible threats associated with the UAV application. It is also necessary to monitor the air space any time and under any weather conditions. It is proposed to employ electrostatic monitoring technology for solving the problem.

2 Problem Statement

It is worth mentioning that conventional detection methods have problems with the small low-altitude UAVs recognition. The disadvantages of radiolocation are low probability for detecting small low-flying targets and the need to comply with environmental regulations on the level of electromagnetic radiation. The electrostatic location system does not generate electromagnetic radiation and has a high probability of detecting low-flying targets. Combining the method of radar and electrostatic location makes it possible to eliminate the shortcomings of each of the systems.

© The Author(s), under exclusive license to Springer Nature Switzerland AG 2022
V. Jordan et al. (Eds.): HPCST 2022, CCIS 1733, pp. 250–261, 2022.
https://doi.org/10.1007/978-3-031-23744-7_19

The electrostatic location system is consisted of a sensor array and a computing device. A sensors signal is proportional to the rate of the electric field changes. The computing device calculates a UAV trajectory due to the processed sensors data. Electrostatic monitoring technology lies in the fact that this technology is capable of providing information on low-altitude targets passively by monitoring some electrostatic field changes in several points on the Earth surface. Besides, combination of electrostatic monitoring technology with other existing detecting techniques may increase the probability of targets detecting.

Currently, the operation of similar systems based on sensor arrays is thoroughly investigated in the field of meteorological and seismic activity research [1, 2] and as the basis of electrostatic monitoring technology [3]. However, a full cycle of scientific works in the field of electrostatic location of objects has not been found. In the Almaz-Antey paper [4] influence of the electrode shape on the detection sensitivity of an aircraft was researched. There are some other studies in this field [5]. However, there is no a full cycle of works in the field of electrostatic location. Successful development in the field of increasing the sensitivity of probes can make [6] the usage of electrostatic location systems more real.

A very large number of studies has been conducted in the world in the field of using electrostatic probes in the systems of electrostatic monitoring. In these system the intensity of charged particles movement in a liquid or gas stream is estimated to monitor the technical condition of bearings, aircraft engines and other objects. Here it is worth highlighting the works of Chinese scientists from the British University of Kent under the leadership of Yong Yan and their latest review article [7].

In the field of electrostatic monitoring the feature extraction of an electrostatic signal was carried out by transforming a signal into time-frequency distribution. At the early stage of research a time-frequency distribution was obtained vie the convolution with the Morlet wavelet function. The feature extraction of signals was carried out through finding the maximum in the amplitude-frequency response [8].

Recently, specially synthesized functions have been used to obtain a time-frequency distribution. Convolution of the measuring signal affords to obtain a time-frequency distribution with special properties vie such functions [9]. These properties contribute to increasing the precision and reliability of digital processing.

The same strategy is also used in this research. The first stage of the work is the development of the synthesis algorithm and the analysis function synthesis. The second stage of the work is the development of an algorithm of the time-frequency analysis.

The function of an ideal measuring signal was used as a measuring signal. That function was obtained within the framework of a mathematical model of a UAV flight over a flat plain. The UAV appeared to be a point electric charge. The Earth's surface was represented by an infinite conducting plane. The graph of the measuring signal function within the framework of these assumptions is shown in Fig. 1. The equation of the function is as follows [10]:

$$\mathrm{Ism}(t) = -\mathrm{Amp}\frac{t + \mathrm{tp}}{((t + \mathrm{tp})^2 + \mathrm{ht}^2)^{\frac{5}{2}}}, \tag{1}$$

where ht and tp are the features of the electrostatic signal (see Fig. 1).

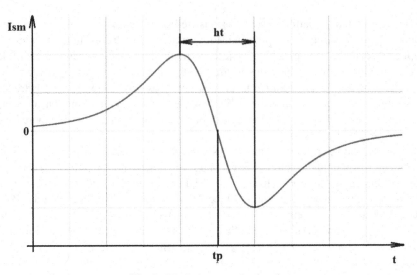

Fig. 1. Ideal electrostatic signal.

The purpose of this work ito develop an algorithm for digital signal processing to obtain ht and pt signs. The efficiency of the algorithm is compared with the standard method based on the Morlet wavelet function.

3   Algorithm for Synthesis of the Analyzing Function

The synthesized analyzing function must have the properties of the Morlet wavelet function. These functions have proven themselves in the field of spectral analysis [9]. For this purpose, the analyzing function is selected as a complex one:

$$Fc(t) = Fcre(t) + i \cdot Fcim(t). \tag{2}$$

The calculation of the measuring signal amplitude and the amplitude phase should be carried out according to the following formulas [6]:

$$CR(\tau) = \int_{-\infty}^{+\infty} Ism(t - \frac{\tau}{2}) \cdot Fcre(t + \frac{\tau}{2})dt, \tag{3}$$

$$CI(\tau) = \int_{-\infty}^{+\infty} Ism(t - \frac{\tau}{2}) \cdot Fcim(t + \frac{\tau}{2})dt, \tag{4}$$

$$Amp = \sqrt{CR(\tau)^2 + CI(\tau)^2}, \tag{5}$$

$$Ph = Arctg(CI(\tau)/CR(\tau)), \tag{6}$$

where τ is the convolution window position, $Ism(t)$ is the measuring signal.

The function should be indifferent to the constant component of the measuring signal if the average integral value of the imaginary and real parts is zero [9]. Thus, the following conditions must be met:

$$\int_{-\infty}^{+\infty} Fcre(t)dt = 0; \tag{7}$$

$$ZFim = \int_{-\infty}^{+\infty} Fcim(t)dt = 0; \tag{8}$$

To fulfill Eqs. (4) and (5), it is necessary that the real and imaginary parts would change as single-frequency harmonic functions offset from each other by $\pi/2$. The fulfillment of these equations will definitely be possible only in a limited area. The difference between the function $CR(\tau)$ and the cosine function should be minimized in the interval$[-\pi/2:\pi/2]$. The difference between the function $CI(\tau)$ and the sine function should be minimized in the same interval.

Achieving minimization in the algorithm is carried out vie the gradient search method. In order to limit the number of arguments of the objective function, it is necessary to parameterize the analyzing function. The analyzing function should be similar to the measuring signal. It was decided that the function should be written as a rational fraction. This solution is due to the fact that the measuring signal is represented with a function (1) close to rational. In addition, this type of functions contributes to a programmatic increase in precision and synthesis acceleration. This type of function affords to represent convolutions (2) and (3) through the sum of residuals at singular points in the upper complex half-plane.

The type of the analyzing function is presented in the form:

$$Ismp(t) = \frac{S_2 \cdot t^5 + S_1 \cdot t^3 + S_0 \cdot t}{(t^2 + A^2) \cdot (t^2 + B^2)^2 \cdot (t^2 + C^2)^2}, \tag{9}$$

$$Fcre(t) = \frac{K_3 \cdot t^7 + K_2 \cdot t^5 + K_1 \cdot t^3 + K_0 \cdot t}{(t^2 + A^2) \cdot (t^2 + B^2)^2 \cdot (t^2 + C^2)^2}, \tag{10}$$

$$Fcim(t) = \frac{P_3 \cdot t^6 + P_2 \cdot t^4 + P_1 \cdot t^2 + P_0}{(t^2 + A^2) \cdot (t^2 + B^2)^2 \cdot (t^2 + C^2)^2}, \tag{11}$$

where K_j, P_j, S_j, A, B, C are parameters.

The synthesis algorithm of the analyzing function for an electrostatic probe needs an efficient calculation of convolutions (2), (3) and (8) and their derivatives. The block diagram of the algorithm is shown in Fig. 2. The algorithm is based on the methods of the symbolic core of the Maple computer algebra system.

The first stage of the algorithm shown in Fig. 2 is the simplification of expressions (2), (3) and (8), represented as residues at singular points:

$$t_1 = \tau/2 + A \cdot i, t_2 = -\tau/2 + A \cdot i, t_3 = \tau/2 + B \cdot i,$$
$$t_4 = -\tau/2 + B \cdot i, t_5 = \tau/2 + C \cdot i, t_6 = -\tau/2 + C \cdot i.$$

For each of these points, a residue is calculated and the brackets in the numerator are expanded via the normal command. Then each pair of deductions (1st with 2nd, 3rd with

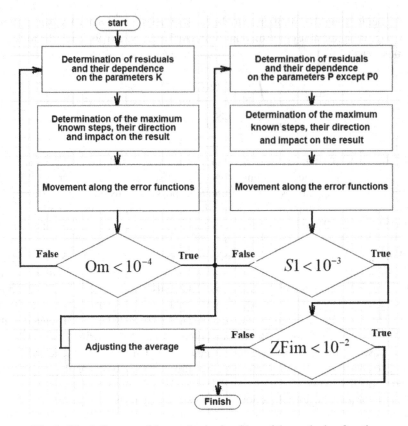

Fig. 2. Block diagram of the synthesis algorithm of the analyzing function.

4th, 5th with 6th) is brought to a common denominator, after that the brackets in the numerator are revealed and similar summands are reduced. Three obtained expressions are also combined to a common denominator and simplified. After that, the brackets in the denominator are expanded by applying the expand command separately to it. This procedure is the most effective in Maple 2021 and allows you to simplify expressions in less than a minute on a personal computer.

The intermediate step of the algorithm is the parametrization of the measuring signal, i.e. finding the coefficients A, B, C and the coefficients S_j. It is easily accomplished by the classical gradient descent over the objective function:

$$\text{Om} = \frac{1}{N} \sum_{sx=1}^{N} \left| -25 \cdot \sqrt{5} \cdot \frac{10 \cdot sx/N}{((sx/N)^2 + 4)^{2.5}} - \text{Ismp}(10 \cdot sx/N) \right|. \qquad (12)$$

The objective function (12) is the sum of the residuals between the parameterized function (8) and the canonized (tp $= 0$ and ht $= 2$) ideal signal (1). The search for this function is not difficult.

Gradient search for K_j parameters should be performed using the following objective function:

$$S1 = \frac{1}{N} \cdot \sum_{sx=0}^{N} OsR_{sx} = \frac{1}{N} \cdot \sum_{sx=0}^{N} \left| CR(2 \cdot sx/N, K_j) - \cos(2 \cdot sx \cdot \pi/N) \right|. \qquad (13)$$

The partial derivative functions (13) with respect to the parameters K do not depend on the parameters themselves. Thus, the algorithm easily calculates the optimal step of coordinate-wise gradient search by means of K parameters. It is necessary to calculate the values of partial derivatives of each summand (13) for the coordinate gradient descent in the program (see Fig. 2).

Then, the partial derivatives are summed, the sign of the residual for each K_j is taken into account:

$$dS1_j = \frac{1}{N} \cdot \sum_{sx=1}^{N} dOsR_{sx} = \frac{1}{N} \cdot \sum_{sx=1}^{N} \left[\frac{\partial CR(2 \cdot sx/N, K_j)}{\partial K_j} \cdot \text{sign}(OsR_{sx}) \right]. \qquad (14)$$

The sign of the sum (14) determines the necessary movement direction along the coordinate to minimize the sum (13). It should be understood that the sum (14) can determine the reduction of the objective function (13) until one of its members changes the sign. To determine this event the ratio of OsR to dOsR is calculated for each sx. This ratio is a descent step to achieve a zero discrepancy for a given sx. Obviously, these steps are multidirectional. Only steps reducing the total amount are taken into account, i.e. the steps with coinciding signs: $\text{sign}(dS1) = \text{sign}(OsR_{sx}/dOsR_{sx})$. The minimum for each K_j is selected from them. The gradient search will perform a step along the K_j that corresponds to the maximum reduction of the objective function (13), i.e. the product of dS_1 by the value of a certain step. Then the procedure is repeated again until S_1 becomes less than the threshold value (see Fig. 2).

In this algorithm, it is also necessary to exclude residuals that are almost close to zero. After calculating $dS1_j$, we determine steps, which are close to zero and meet the conduction $\text{sign}(dS1) = \text{sign}(OsR_{sx}/dOsR_{sx})$. If these steps have been found, then the value of the sum is reduced by twice the value of the corresponding residual. The doubled value is used, because it is still necessary to subtract the element from amount taken into account earlier. If the $dS1_j$ value changes sign, movement along this coordinate will make no sense at this iteration of the gradient search.

Gradient search for the search parameters P occurs similarly. The difference from the previous stage is that it is necessary to ensure the average Fcim value equal to zero. The average value is also calculated through residues.

The synthesized functions Icmp(t), Cr(t) and Fcim(t) are shown in Fig. 3.

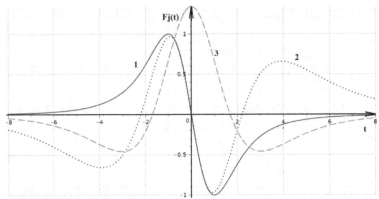

1   $Ism(t)$ *npu* $A = 1,603; B = 3,276; C = 5,183; S_2 = -1033; S_1 = -7695; S_0 = -372900$

2   $Fcre(t)$ *npu* $K_3 = 0; K_2 = 17782; K_1 = -11391; K_0 = -372900$

3   $Fcim(t)$ *npu* $P_3 = 50; P_2 = -37655; P_1 = -53337; P_0 = 151748$

Fig. 3. Measuring signal 1 and analyzing function real 2 and imaginary 3 part.

4   Time-Frequency Distribution

Digital processing of an electrostatic signal begins with the signal conversion into a time-frequency distribution, i.e. obtaining a function of two arguments associated with the necessary features. The first argument is the time, which is associated with the center position of the signal in time. The second argument is the half-period. The using of a half-period is connected with the fact that the concept of frequency or period for an electrostatic signal can only be abstract.

The next stage of digital signal processing is the analysis of its time-frequency distribution. In fact, the function obtained during the transformation is a vector that can be decomposed into an amplitude and a phase. The standard practice of time-frequency analysis is the analysis of the amplitude distribution. It is related to the fact that the coordinates of the maximum on this distribution coincide with the determined signal parameters up to a constant factor. This maximum can be found due to one of the methods of analyzing the function of two arguments, for example, by the method of ridge analysis [9]. For a successful analysis, the severity of the peak is necessary and its uniqueness is desirable.

It is possible to analyze the phase distribution of the signal. In this case, it is taken into account that the product of the phase by the signal period should physically mean the time of discrepancy between the signals in phase. If the close-standing nodes of the phase function of the same half-period have phases of different sign, then the product of the phase by the signal period will show the distance between the center of the convolution window and the center tp of the electrostatic signal. In fact, the last statement will only make sense for a special analyzing function synthesized for a specific signal. Such a function is a rational fraction, the synthesis of which is described in the previous chapter.

To check the applicability of the analyzing function, a calculation program using graphical interfaces provided by the Visual Studio environment was compiled. To simulate the UAV detection process, method of mirror charges has been used. The intensity of the electrostatic field is calculated with this method, as a vector addition of the intensity created by the UAV and the intensity created by mirror charges located under the Earth's surface [11]. Mirror charges create a field identical to the field of induced charges on the earth's surface. This solution is fully consistent with the formula (1) for the case of a perfectly conducting flat surface. The UAV flight is modeled as a set of solutions of electrostatic tasks for each point where the UAV is located on the flight path. The rate of change of the field or the electrostatic probe signal was found as the central finite difference of the 1st order. The program simulated the process of digital signal processing.

In digital signal processing, the mother analyzing function was first generated separately by the imaginary and separately by the real component. According to the recommendations in [9], it is necessary to calculate vectors consisting of an odd number of integer elements. Their sum should be zero. The calculation of values, for example, the imaginary component of the rational analyzing function, occurs according to the following formula:

$$V\_Fcim_j = \text{cell}(\text{Ampd} \cdot \text{Fcim}(\frac{j - N/2}{N\_dis}) + \text{corr}), \tag{15}$$

where N_dis is the number of steps per half–period of the analyzing function ht, corr ϵ $[-0.3..0.3]$ is the corrective coefficient.

In Eq. (2), the index j varies from 0 to N equal to the product of the number of steps by the number of half-periods of the analyzing function ht. Thus, for odd functions, the sum of all elements will be zero, in the case of even functions, strict equality to zero will be provided by the coefficient corr.

After generating the analyzing functions digital signal processing was carried out, i.e. convolutions (3) and (4) were performed, in which the indefinite integral was replaced by a sum. For example, for the imaginary component, the point of the nodal function was considered as follows:

$$\text{FCI}(\tau, T/2) = \sum_{j=0}^{N} \left[V\_Fcim_j \cdot Ism(\tau + (j - N/2) \cdot \frac{T/2}{N\_dis}) \right]. \tag{16}$$

The signal function (16) was calculated for an array of nodal points randomly shifted relative to the true values of tp and ht of the signal. To check the noise immunity, white noise was added to the signal.

The amplitude diagram is more informative in the case of using the wavelet function, since in this case, with acceptable signal-to-noise ratios, only one pronounced maximum is observed. This is evident from the diagram shown in Fig. 4. The amplitude diagram obtained through the rational analyzing function does not possess this property.

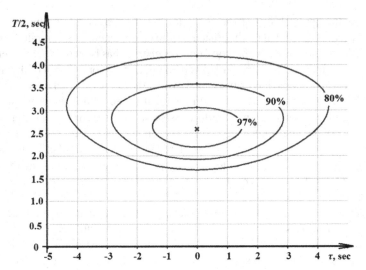

Fig. 4. Amplitude diagram under the wavelet transform.

To analyze the time-frequency distribution obtained vie convolution transformation of the measuring signal of a rational function, it is proposed to use the phase diagram shown in Fig. 5.

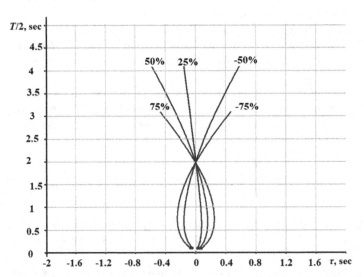

Fig. 5. The phase diagram obtained from the transformation of a rational function.

As can be seen from Fig. 5, the expectation of the position of the center of the tp signal in the time-frequency representation obtained vie the rational function is true only for the correct $ht = 2$. This makes it possible to construct an algorithm for finding the parameters of an electrostatic signal based on a phase diagram.

To extract the ht and tp signs of an electrostatic signal, its convolution with analyzing functions of various periods is carried out in real time. The periods of the analyzing functions are separated from each other with a certain step Th. The analyzing function is recorded in memory as a series of its values. The different periods of the analyzing function are set by assigning these values to the argument t with a different step between neighboring values.

For each individual period of the analyzing function each new convolution begins after some part of its period, for example, after $T/4$. That is, the centers of the convolution windows are located from each other at a given distance. The result of each convolution will be the amplitude and phase. As a result of spectral analysis, we obtain a grid two-dimensional function of the amplitude As and phase Phs, the nodal values of which have a different time step for each period. These functions can be mapped to the corresponding nodal functions of the period T and the time of the center of the convolution window τ. The flowchart algorithm for isolating tp(T) isolines from these grid functions (see Fig. 5) is shown in Fig. 6.

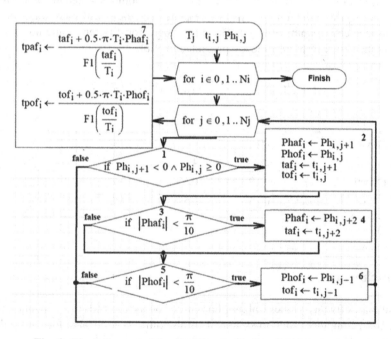

Fig. 6. Block diagram of the algorithm for finding signal parameters.

The essence of the algorithm for determining the tp(T) functions is as follows. It is necessary to select the phases of neighboring Phs windows in each period, which are located on different sides from the center of the electrostatic signal (condition 1, Fig. 6). If the phase is too small, then its neighboring phase is taken in the same side from the center of the signal for increasing its accuracy (condition 3 and 5, Fig. 6). Ideally, the tpof and tpaf functions should correspond to the functions shown in Fig. 5. These functions are symmetric with respect to the center of the signal. The intersection point of these functions will give the correct parameters of the electrostatic signal $tp = tpaf(T/2) = tpof(T/2)$; $ht = T/2$. The algorithm for the phase diagram processing is described in more detail in [12].

The methods were compared through multiple numerical experiments - 10 for selected values of σ (from 0 to 2) white noise with the same ideal measuring signals ($ht = 2$, $tp = 0$, $Ism(-1) = 1$ – signal amplitude). With the wavelet analysis detailed frequency-time distribution of the amplitude was studied, the technique of restoring the ridge was not reproduced (the error of the technique was not taken into account). The electrostatic signal features were determined through the peak coordinates on the frequency response (see Fig. 4). With the analysis of the rational function, detailed phase lines were studied (see Fig. 5) and the algorithm for finding the signal parameters was reproduced (see Fig. 6). The error in feature extracting of the electrostatic signal was performed using the Student's criterion for the reliability of the confidence interval 0.9. The relative error of ε_{tp} and ε_{ht} was calculated with respect to the actual ht.

When white noise was added, the error in feature extraction of the electrostatic signal using a rational function manifested itself at $\sigma > 0.3$. It should be noted that this affected the fact that while maintaining a qualitative picture of the phase lines (Fig. 5), their common intersection point changed. The error values for different white noise values inverse to the signal-to-noise ratio are presented in Table 1.

Table 1. Relative errors in determining ht and tp.

		SNR = 3	SNR = 2	SNR = 1,5	SNR = 1
Morlet wavelet	ε_{ht}	0%	0%	0%	0%
	ε_{tp}	2%	4%	>5%	>5%
Rational function	ε_{ht}	0.1%	0.5%	3%	8%
	ε_{tp}	0.1%	0.5%	3%	8%

5   Conclusions

The resulting software synthesis algorithm is based on the methods of the symbolic core of the Maple computer program system. This algorithm is applicable for the synthesis of rational analyzing functions. Given type of functions is suitable for the signals, which can be represented as a rational function with a limited number of parameters. To optimize the algorithm, methods of analytical representation of indefinite integrals and a combination of numerical and symbolic calculations have been applied. These methods made it possible to quickly and accurately calculate the values of function convolutions and average integral values in an analytical form. It allowed to quickly and accurately calculate the necessary features for the analyzing function. The algorithm on the basis of the synthesized analyzing function has been tested for noise immunity in a numerical experiment. The algorithm based on this function turned out to be more efficient than the algorithm based on the Morlet wavelet function at signal-to-noise ratios of more than 1.5.

Acknowledgements. The reported study was funded by RFBR, project number 20-37-90028.

References

1. Efimov, E., Polushin, P., Grunskaya, L.: Measurement of the electrostatic component of electrostatic fields. Electrostatic field meters: Monograph, Berlin (2008)
2. Chubb, J.: The measurement of atmospheric electric fields using pole mounted electrostatic fieldmeters. J. Electrostat. **72**, 295–300 (2014)
3. Wen, Zh., Hoa, J., Atkin, J.: A review of electrostatic monitoring technology: the state of the art and future research directions. Prog. Aerosp. Sci. **2**(1), 1–11 (2017). ISSN 2250-2459
4. Lastovetskii, A.E., Klepka, S.P., Ryabokon', M.S.: Remote measurement of electric charges of air objects. Vestnik Koncerna VKO Almaz-Antej **3**, 59–69 (2015)
5. McFarland, M.B., Zachery, R.A., Taylor, B.K.: Motion planning for reduced observability of autonomous aerial vehicles. In: IEEE International Conference on Control Applications, 22–27 August 1999, Hawaii, vol. MP 5(3), pp. 231–235 (1999)
6. Wang, W., et al.: High-performance graphene-based electrostatic field sensor. IEEE Electron Device Lett. **38**, 1136–1138 (2017)
7. Yan, Y., Hu, Y., Wang, L., et al.: Electrostatic sensors—their principles and applications. Measurement **169** (2021). https://doi.org/10.1016/j.measurement.2020.108506
8. Iatsenko, D., McClintock, P., Stefanovska, A.: Linear and synchrosqueezed time—frequency representations revisited: overview, standards of use, resolution, reconstruction, concentration, and algorithms. Digit. Sig. Process. **42**, 1–26 (2015). https://doi.org/10.1016/j.dsp.2015.03.004
9. Addabbo, T., Fort, A., Mugnaini, M., Panzardi, E., Vignoli, V.: A combined polynomial chirplet transform and synchroextracting technique for analyzing nonstationary signals of rotating machinery. IEEE Trans. Instrum. Meas. **69**(4), 1505–1518 (2020)
10. Skryabin, Yu.M., Potekhin, D.S.: Determination of the horizontal flight path of an unmanned aerial vehicle through a line of electrostatic sensors. Trudy MAI 106 (2019). http://trudymai.ru/published.php?ID=105747
11. Jackson, J.D.: Classical Electrodynamics, 3rd edn. Wiley, New York (1999)
12. Skryabin, Yu.M., Potekhin, D.S.: Rational analyzing function for precise feature extraction from an electrostatic signal. Trudy MAI 119 (2021). https://trudymai.ru/published.php?ID=159792

Application of Computer Numerical Control in Eddy-Current Study Methods

Vladimir Malikov[1]([✉]) [iD], Nickolay Tihonskii[1] [iD], Victoria Kozlova[1] [iD], and Alexey Ishkov[2] [iD]

[1] ASU – Altai State University, Lenin Avenue 61, 656057 Barnaul, Russia
osys11@gmail.com

[2] ASAU – Altai State Agricultural University, Krasnoarmeiskii Avenue 98, 656049 Barnaul, Russia

Abstract. This study aims to develop a software and hardware system that implements modern hardware and digital methods of working with devices and signals. The software and hardware system is based on sensors using the eddy-current principles and focused on scanning of metallic materials. The scanning unit described in this study includes a compact transformer-type eddy-current transducer and allows making local measurements of non-ferromagnetic materials based on local conductivity recording. The unit combines hardware and software parts. The article describes in detail the methods and approaches that make it possible to design the device. It shows the development of digital signal generation and amplification system based on Arduino microcontrollers, automated methods for sensor positioning with simultaneous signal input, methods for signal reception during the operation of the positioning system. And it describes the software that makes it possible to control the measuring system and provide high-quality scanning results in the form of visual images of objects under control and their operation through machine vision using modern microcircuits that allow processing signals with a frequency of up to 1.8 GHz.

Keywords: Software and hardware system · Arduino microcontrollers · Positioning system · Digital signal

1 Introduction

In the rolling industry, there are various methods of metal product control, such as ultrasonic, radiation, eddy currents, etc. The eddy-current method of metal inspection is widely used due to its high sensitivity, high detection rate and contactless nature. When an eddy-current transducer coil carrying alternating current is close to a metal item, eddy current is generated in the metal, which affects the initial magnetic field so that the impedance and induced voltage of the coil changes. Information about the material under test is produced by analyzing the impedance or induced voltage. The development trends of eddy current (EC) testing technology mainly lie in several directions. A theoretical model of eddy-current detection is built using numerical calculation and simulation by the

V. Jordan et al. (Eds.): HPCST 2022, CCIS 1733, pp. 262–276, 2022.
https://doi.org/10.1007/978-3-031-23744-7_20

finite element method [1–3]. At this stage, it is extremely important to obtain information about the theoretical depth of penetration of the electromagnetic field into the object under study and its surface and depth distribution. Vertical software (Elcut, Comsol Multiphysics, Ansys Maxwell, etc.) is used for modeling purposes. These mathematical packages allow one to make efficient and fast calculation of electromagnetic fields from an eddy-current transducer. Next, the design and development of eddy-current sensors is carried out taking into account the earlier performed computer modeling, depending on the required tasks. An important part of the operation of the eddy-current measuring system allowing scanning metal objects is a system for signal processing and extraction of useful information about the object of control. Such system should provide signal processing in the time domain, signal processing in the frequency domain, and frequency-time analysis in relation to information received from the eddy-current sensors. At that, the software that controls the operation of the sensor should eliminate the gap effect on the measurements obtained and support the ability to classify and recognize multiple defects simultaneously. In papers [4–6], a three-dimensional imaging of corrosion defects was performed using computer vision systems.

Material defect data may be extracted by processing a sequence of defectograms. In order to extract defect information, many image-processing algorithms are applied [7–9]. The Principal Component Analysis (PCA) [10] was applied to extract the eddy-current sensor signal distribution patterns [11] and to determine the depth of defects in composite structures. However, as the physical meaning of PCA is not fully defined, further analysis of this information is difficult. Besides, PCA is also not a criterion for judging the presence of defects [12]. To solve these problems, the Independent Component Analysis (ICA) is used to process a sequence of defectograms in order to isolate anomalous features of surface defects [13]. The ICA considers that the original received information is the result of interaction of several independent components (ICs). The goal of ICA is to compute all ICs that contained the base attributes of the original data. ICA may be used to reduce the volume of data and increase the efficiency of further data processing. When ICA deals with large volumes of defectograms, the efficiency of their processing can be reduced [13]. Therefore, studies on the creation of new algorithms for defectogram images processing are relevant to improve the efficiency of such processing.

Automated analysis [14] of large data sets (defectograms) coming from the relevant equipment is still a relevant task. The analysis refers to the process of determining the presence of defective areas along with the identification of structural elements of metal products on defectograms. At that, fast and efficient algorithms of data analysis are of the greatest interest under conditions of significant volumes of incoming information.

Automation of the scanning process is also of interest. In paper [15], the systems of automation of eddy-current sensor motion during scanning of railway rails were described. However, the task of developing a system that makes it possible to scan metal objects with the possibility of sensor motion along the three axes and plotting the image of the scanned object is still extremely relevant.

This study is intended for identification of the optimal technical and software solutions for the development of a software and hardware system that would allow one to detect defects in parts made of different alloys using an eddy-current transducer, as well as to determine the most effective core shape to be used in it.

2 Materials and Methods

The developed software and hardware system is focused on the control of a given parameter of metal product in small areas. Such parameter is the electrical conductivity of a substance and its surface and depth distribution. The set includes a transducer connected to an Arduino board, and a special Python-based software intended for installation PCs under Windows OS. Interaction between the Arduino board and the PC is done via a virtual COM port.

The role of the software is to flexibly control the voltage supply to the excitation circuit of the transducer, control of the positioning system of the transducer, recording the output voltage values in conditional units, calculate the conductivity values on the basis of the obtained data taking into account the pre-calibration, and construct the part images based on the computer vision technology.

Based on the tasks set, the hardware was designed to be as portable as possible and PC-controlled.

The developed circuit of the measuring system is shown in Fig. 1.

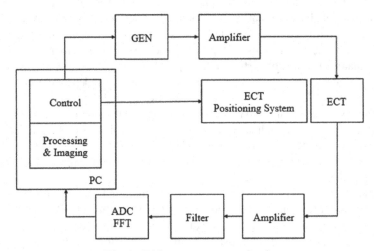

Fig. 1. Eddy-current system circuit.

To determine the absolute and relative spatial coordinates of the ECT measuring units on the control object surface, a 3D positioning system was designed and developed, which allows one to move the sensor along three axes with an accuracy of 0.05 mm in automatic mode, taking off data from the eddy-current transducer. The software is provided with a corresponding software module for a strap ECT positioning, the analysis of signal amplitude from the ECT measuring channels is carried out, the results of which are further used in the graphical representation of the measured information.

The positioning system is based on Cartesian kinematics. Based on the Cartesian coordinate system, this technology works on the basis of three axes, namely X, Y, and Z (Fig. 2).

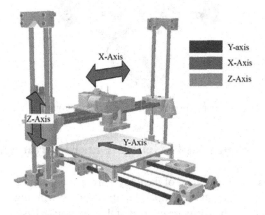

Fig. 2. Cartesian kinematics diagram.

The platform for securing the test object moves along the Y-axis, the sensor holder moves along the X-axis and in the heightwise direction.

Each direction has its own motor, the Y and X axes are belt-driven, and the Z axis is driven by a screw system (Figs. 3 and 4).

Fig. 3. Stepper motor and a screw for moving along the Z axis.

The maximum sensing area is 22 × 22 cm, the maximum travel speed is 180 mm/s, and the travel accuracy is 100 μm.

The stepper motors are controlled by a motherboard with a 32-bit processor. The motherboard is fitted with a USB connector via which it is connected to a PC to receive control commands. The motherboard used has Marlin firmware and is controlled using G-code commands.

The positioning system is equipped with a sensor, which is an eddy-current transducer that interacts with the object under control using the generated electromagnetic field.

Fig. 4. Stepper motor and a belt for moving along the Y axis.

3 Digital Signal Synthesis to Generate an Electromagnetic Field

To generate an electromagnetic field of the eddy-current transducer, a system was developed for generating signals to be transmitted to the exciting winding of the eddy-current transducer. To implement the software control of the generation system being developed, a generator made as an AD 9850-based module was used in the software and hardware system (Fig. 5). It is a highly integrated chip that uses a combination of advanced Direct Digital Synthesis (DDS), a high-quality digital-to-analog converter and comparator that provides digital software-controlled frequency synthesis and clock-signal generation functions. When operated from an accurate reference clock-signal source, AD9850 provides a stable analog sine-wave output signal with programmable frequency and phase.

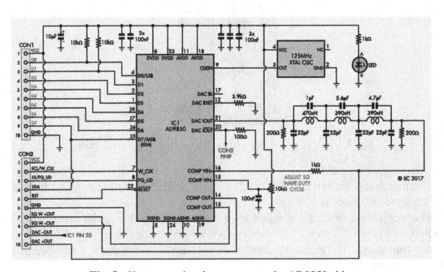

Fig. 5. Sine-wave signal generator on the AD9850 chip.

The developed module has the following characteristics: at a clock-signal generator frequency of 125 MHz, a sinusoid of 0 to 40 MHz can be obtained at the output of the module with a high stability, low noise, low supply voltage (3.3–5 V), and compactness.

The disadvantages include the lack of the output signal amplitude adjustment, and low power. Both disadvantages are solved by using an amplifier with an adjustable gain factor. The chosen AD 9850 module requires a control element. The Arduino hardware computing platform was chosen as a computer–generator interface. The platform consists of two basic components: an input-output board and a Processing/Wiring development environment (Fig. 6).

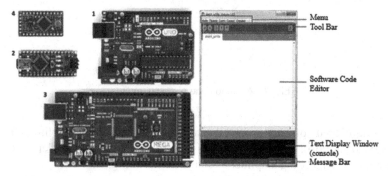

Fig. 6. Arduino Platform (1 – Arduino Uno, 2 – Arduino Nano, 3- Arduino MEGA, 4 – Arduino PRO mini, 5 – Arduino Development Environment Window).

The developed device used the Nano platform based on ATmega 328 microcontroller. ATmega 328 microcontroller has 32 kB ATmega328 for the software code storing, and 2 kB RAM. On the Nano platform, there are 8 analog inputs, with a resolution of 10 bits each (i.e. can take 1024 different values). Normally, the outputs have a measurement range of up to 5 V relative to ground. Figure 7 shows the board pinout.

Arduino board has pins for information both output and input. The availability of input pins makes it easy to use the Arduino as an element controlling the generator in the system. Arduino Nano has a 10-bit ADC (1024 discrete numbers) with a low sampling rate of 10 kHz. In the course of the research, it was found that these indicators are sufficient for the tasks set.

The eddy-current transducer measuring winding voltage values introduced by eddy currents are relatively small, so it became necessary to amplify the output signal of the ECT. Also, the generator module needed to amplify the output signal, as the generator winding of the ECT had a high resistance, and the generator was not able to adjust the amplitude of the generated signal.

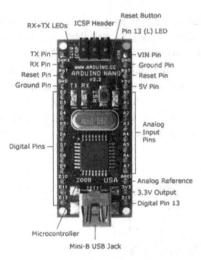

Fig. 7. Arduino nano pin assignments.

In the developed software and hardware system, an inverting amplifier based on an AD8051 operational amplifier (OA) was used for solving the problems of signal amplification. The inverting amplifier circuit is shown in Fig. 8.

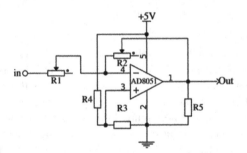

Fig. 8. Inverting amplifier based on AD8051 OA.

This amplifier has the following specifications:

wide bandwidth: 300 MHz;
ramp up rate: 650 V/μs;
low current consumption;
low supply voltage: from 2.7 V to 8 V.

4 Data Acquisition, Digitization and Pre-processing Unit

RTL2832U chip was used to digitize the output signal of the ECT. The RTL2832U chip contains an ADC, a digital processor, a USB interface, and filters. In addition, the

RTL2832U chip is fitted with a block that makes it possible to perform a fast Fourier transform. Characteristics of the RTL2832U chip are as follows:

Receiving Frequency Range: 24–1750 MHz.
Variable Filter Width
Sensitivity: 0.22 mKv (at 438 MHz in NFM mode)
Input gain control, Automatic Gain Control (AGC) on/off

The general scheme of operation of the RTL2832U-based device is shown in Fig. 9.

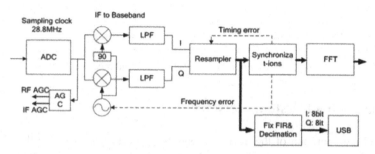

Fig. 9. RTL2832UC-based device operational diagram.

Based on the RTL2832U microcircuit, an ADC module was developed (Fig. 10), which allows digitizing the signal of the eddy-current transducer. The module is fitted with two connectors: SMA for input signal and USB connector for transmitting the output signal to a PC. The module is enclosed in an aluminum housing for protection. The housing also plays the role of a heat sink, because the receiver board has a connection with the housing through a thermally conductive silicone gasket, which acts as a shock absorber in addition to the heat sink.

Fig. 10. ADC module developed on the basis of the RTL2832U microcircuit.

The circuit diagram of the Module is shown in Fig. 11:
At the input of the chip, a three-section LC filter and a low-noise wideband preamplifier based on BGA2711 chip is placed. A low-noise voltage regulator based on AP2114 is used to power the receiver chips.

Python-based software was developed to receive and process the signal coming from the module.

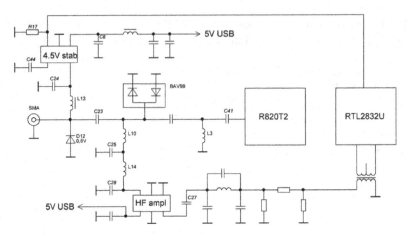

Fig. 11. Circuit diagram of the developed module.

5   Measurement System Software

To control the operation of the developed system, process and visualize the obtained values, software was written in Python 3.9 and the graphical shell was created in the Qt Designer software.

To control the generator, a software was written using Arduino IDE editor.

When writing the software, the following libraries were used:

NumPy provides support of multidimensional arrays; support of high-level mathematical functions designed to work with multidimensional arrays;

Serial is a library for working with Serial port;

Sys provides access to some variables and functions that interact with the Python interpreter. An interpreter is a software that converts instructions written in Python into bytecode and executes them.

PyQt5 is a set of Python libraries for creating a graphical interface based on the Qt5 platform from Digia.

QThread is a library for organizing threaded operation of a measuring system. Threaded operation is the parallel execution of several tasks, such as managing the sensor positioning system and updating the graphics window.

Time is a module for working with time in Python, used to get the system time; when the file is saved, the current date and time are automatically indicated in the file name.

RtlSdr is a library for controlling the module for analog-to-digital conversion of the measured signal.

Interpolation is used if there is extra data in a read buffer or the buffer is not full.

Plotly is a library for graphing.

Figure 12 shows the graphical interface of the primary window of the system software.

The window is divided into blocks; each of them is designed to interact with a separate module of the measuring system.

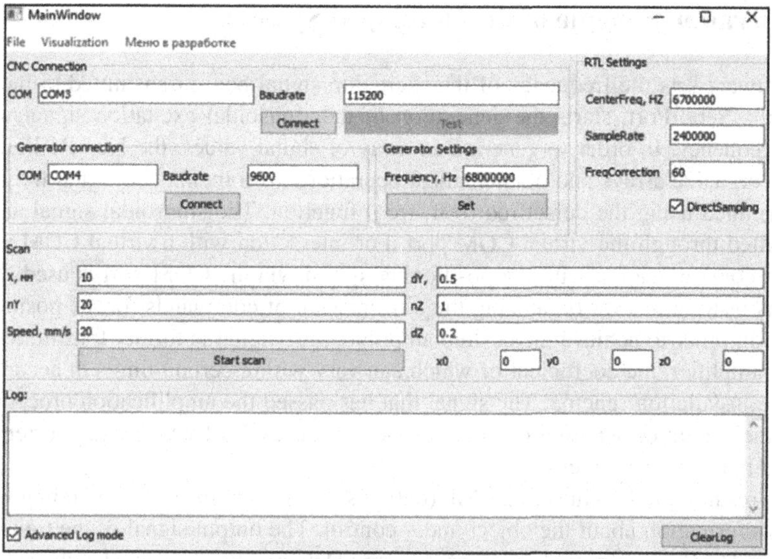

Fig. 12. Software interface.

CNC Connection is designed to configure the connection to the ECT positioning system. It is necessary to specify the name of a port to which the positioning system is connected (COM) and the data exchange rate (Baudrate). When pressing the Connect button, data is exchanged with the positioning system, and the correctness of the entered parameters is checked.

Generator connection is the connection setting block with the excitation signal generation module for ECT.

Generator Settings set the parameters of the excitation signal.

RTL Settings set the operation parameters of the analog-to-digital conversion module.

Scan sets the parameters for moving the sensor over the object under control. Along the X axis, scanning is continuous, X, mm is the width of the scanning zone, after each passage along the X axis, one-step skipping along the Y axis takes place. nY is the number of steps along the Y axis, and dY is the step width along the Y axis. The software allows performing layer-by-layer scanning with a change in the distance between the sensor and the test object, nZ is the number of layers, and dz is the value of one step along the Z axis.

Speed, mm/s sets the speed of sensor moving over the object under control.

X0, Y0, and Z0 set the initial position of the sensor in relative coordinates.

The log window is intended for displaying the error messages, as well as current scanning information (sensor position, read buffer length, percentage of completion, etc.).

6   Operation Scheme of the Developed System

The software sets the frequency of the excitation signal being transmitted to the generator via SerialPort, starts the generation of the sinusoidal excitation signal with a given frequency. In order to generate an array of signal values, the NumPy library is used to organize arrays and mathematical operations with them. The signal frequency is determined using the defsetfreq (self, freq) function. The sinusoidal signal array is transmitted through the virtual COM port. For interaction with a virtual COM port, a class of communication with the positioning system via the COM port is used, which contains instructions for organizing the transmission of commands for the positioning system. In order to achieve the desired amplitude, the signal is further transmitted to a special amplifier, the coefficient of which can vary within certain limits in accordance with the installation scheme. The signal that has passed the amplification procedure is fed to the excitation winding of the transducer, as a result of which eddy currents are induced in the object under test.

The resulting field induces an EMF (output signal) in the measuring winding, which carries information about the object under control. The output signal of the transducer (registered EMF) is amplified and filtered using a Deliann filter combined with a selective signal amplifier. After amplification, the signal is fed to the collecting, digitizing and primary data processing unit, where the voltage measurement and the analog-to-digital conversion of the result takes place.

The resulting voltage values are transmitted to a PC via SerialPort using the developed RtlSdr ADC module, where the signal is digitized and transmitted to a PC for visualization in a form convenient for processing. To do this, the class RTL (QThread) is used for reading data from the ADC module and primary data processing. In the process of using the class, the ADC is switched to discrete reading mode using the *self.sdr.set_direct_sampling (direct_sampling)* command, the sampling rate is set by the *self.sdr.sample_rate = SampleRate* command. The received data is transferred to the array by the self.data = data command. Further, using the *fft = np.fft.rfft (samples)* command, a fast direct discrete Fourier transform of the array is performed, and the result is written to the fft array having a complex form. The calculation of the signal component amplitudes $((Re^2 + Im^2)^{1/2}$ is performed using the *abs* function, the average value of the amplitude is also calculated, and the resulting value is written in the array.

The PC software receives and presents the measured information from the sensors of the measuring system in the form of an image of the conductivity distribution over the workpiece surface on the screen in real time mode.

The dimensions of the image correspond to the dimensions of the surface area of the object under test and are provided with coordinate axes. The color of the sections of the object under test corresponds to the amplitude of the informative signal coming from the sensors and reflects the defective areas of the object.

This allows detecting the places of the greatest damage to the metal visually.

The software of the measuring system allows analyzing the identified objects on the surface of the object under test quickly, for example, to identify zones corresponding to certain defects (weld, crack), carry out a detailed analysis of the surface area of the object under test with the ability to determine the size of the defect by their coordinates.

7 Experimental Results

In order to perform measurements, sequential execution of the operations of sensor movements and reading the EMF of the measuring circuit was used. Thus, the surface of the sample under study is divided into m and n discrete points, which the sensor passes in turn, where dx and dy are the displacement distances along the x and y axes, respectively (Fig. 13). The result is entered into a two-dimensional array for further processing, and, at the end of the operation, the result with the final image is displayed on the screen. The numpy and matplotlib libraries were used for data processing and visualization.

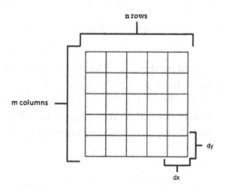

Fig. 13. Discrete surface model description.

As a result of algorithm execution, a significant drawback of its operation was noted: as the number of points for reading in the surface plane increases, there is a significant increase in the time required for execution. This is caused by the linear algorithm execution, where and the sensor was stationary for the duration of the reading before each shift. This shortcoming was eliminated by using the pyThreading module, taking the advantages of the multi-threaded code execution.

At that, the research algorithm was changed. After specifying the boundaries of the research area, a command is sent to make shift along the X-axis to the extreme right border of the research area, while two threads are executed in parallel in order to interact with the sensor. The final algorithm is shown in Fig. 14.

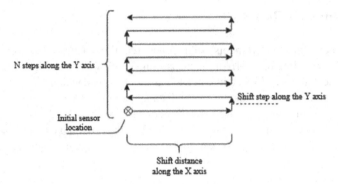

Fig. 14. Algorithm schematic view.

To check the performance of the developed hardware and software system, the model objects, represented by a metal disk and a ball with the radii of 1 and 1.5 cm, respectively, were used (Fig. 15.a). The results of the experiment are presented in Fig. 15.b. The graphically visualized scanning data allow restoring with sufficient accuracy the shape and dimensions of the objects being controlled by changes in the signal amplitude.

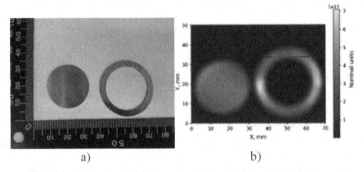

Fig. 15. Experimental results on metal disk and a ball. a) photo of the objects under control, b) visualization of scanning data of the objects under control.

Figure 16 represents the results of a thin metal film scanning (the film thickness was 100 nm). The defect in the form of a crack is clearly visible in the image obtained using the developed measuring system. The crack width was 0.05 mm.

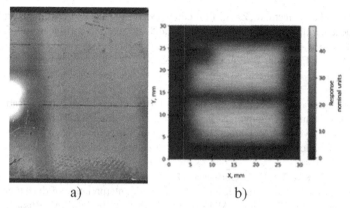

Fig. 16. Experimental results on thin metal film. a) photo of the objects under control, b) visualization of scanning data of the objects under control.

8 Conclusion

It can therefore be concluded that it is fundamentally possible to use automated control systems with eddy-current sensors for studying the conductive materials and alloys. The developed control data visualization system is also promising, as making it possible to see possible inhomogeneities and defects in materials clearly in a graphical mode.

References

1. Fan, M., Cao, B., Sunny, A.I., et al.: Pulsed eddy current thickness measurement using phase features immune to liftoff effect. NDT E Int. **86**, 123–131 (2017)
2. Yoshioka, S., Gotoh, Y.: Proposal of thickness measurement method of steel plate with high lift-off using pulsed magnetic field. IEEE Trans. Mag. **53**, 1–4 (2017)
3. Cheng, W.: Thickness measurement of metal plates using sweptfrequency eddy current testing and impedance normalization. IEEE Sens. J. **17**(14), 4558–4569 (2017)
4. Zheng, G., Xia, R., Chen, Y., et al.: Research on double coil pulse eddy current thickness measurement. In: Proceedings of the International Conference on Intelligent Computation Technology and Automation, pp. 406–409 (2017)
5. Azaman, K.N., Sophian, A., Nafiah, F.: Effects of coil diameter in thickness measurement using pulsed eddy current non-destructive testing. Mater. Sci. Eng. **260**(1), 012001 (2017)
6. Fu, Y., Lei, M., Li, Z., et al.: Lift-off effect reduction based on the dynamic trajectories of the received-signal fast Fourier transform in pulsed eddy current testing. NDT E Int. **87**, 85–92 (2017)
7. Yin, A., Gao,B., Tian, G.Y., Woo, W.L., Li, K.J.: Physical interpretation and separation of eddy current pulsed thermography. J. Appl. Phys. **113** (2013)
8. Xu, S., Jiang, X., Huang, J., Yang, S., Wang, X.: Bayesian wavelet PCA methodology for turbomachinery damage diagnosis under uncertainty. Mech. Syst. Sign. Process **80**(1), 1–18 (2016)
9. Lopes, A.M., Tenreiro, M.: Analysis of temperature time-series: embedding dynamics into the MDS method. Commun. Nonlinear Sci. Numer. Simul. **19**, 851–871 (2014)
10. Han, L., Li, C., Guo, S., Su, X.: Feature extraction method of bearing AE signal based on improved FAST-ICA and wavelet packet energy. Mech. Syst. Sign. Process **62**, 91–99 (2015)

11. Omar, M.A., Parvataneni, R., Zhou, Y.: A combined approach of selfreferencing and principle component thermography for transient, steady, and selective heating scenarios. Infrared Phys. Technol. **53**, 358–362 (2010)
12. Cheng, Y., Yin, C., Chen, Y., Bai, L., Huang, X.: ICA fusion approach based on fuzzy using in eddy current pulsed thermography. Int. J. Appl. Electromagn. Mech. **52**(1–2), 443–451 (2016)
13. Eisencraft, M., Fanganiello, R.D., Grzybowski, J.M.V.: Chaos-based communication systems in non-ideal channels. Commun. Nonlinear Sci. Numer. Simul. **17**, 4707–4718 (2012)
14. Tarabrin, V.F., Zverev, A.V., Gorbunov, O.E., Kuzmin, E.V.: About data filtration of the defectogram automatic interpretation by hardware and software complex ASTRA. NDT World **64**(2), 5–9 (2014)
15. Tarabrin, V.F., Kuzmin, E.V., Gorbunov, O.E., Zverev, A.V.: On the determination of the dynamic threshold of the signal level in the automatic interpretation of defectograms of the hardware and software complex ASTRA. In: Abstracts of the XXth All-Russian Scientific-Technical Conference on Non-Destructive Testing and Technical Diagnostics, Moscow, pp. 145–147 (2014)

Computing Technologies in Information Security Applications

Open-Source Solution for Identification and Blocking of Anomalous BGP-4 Routing Information

Natalya Minakova$^{(\boxtimes)}$ and Alexander Mansurov

AltSU – Altai State University, Lenin Avenue 61, 656049 Barnaul, Russia
minakova@asu.ru

Abstract. This study proposes a solution that performs the analysis of the propagated BGP routing information and detects the anomalous route prefixes using the public information from the routing databases about routing policies and connectivity of all ASes on the Internet. The proposed solution design and its main operation steps are presented and discussed. The analysis algorithm provides the in-depth checking and validation of the BGP path attribute and route prefix origin against the routing databases maintained by the Internet Routing Registries. It acts as an external tool that has no requirements for modifications or upgrades of existing BGP routing infrastructure of any telecom operator. The proposed solution can be adapted to any types of BGP routers and configuration file styles and uses the regular means of controlling of received and advertised BGP route prefixes. Test results demonstrate the effectiveness of the solution and its capability to identify and eliminate the common BGP route leaks and hijacking incidents.

Keywords: Border gateway protocol · BGP-4, · Route leaks · BGP hijacking · Route analysis · Network routing · Telecom operator

1 Introduction

The worldwide Internet is an interconnection of numerous telecommunication operators. In terms of internetworking, it can be described as maintaining connectivity and exchanging routing information among independent autonomous systems (ASes) using the Border Gateway Protocol ver. 4 (BGP-4) to ensure network reachability [1]. In terms of BGP-4, each AS has its unique number to identify a telecommunication operator with its connected neighbors (other operators) together with defined internal and external routing policy. BGP-4 is a path vector routing protocol used to exchange information about originated or transit route prefixes (routes) along with the AS_PATH path attribute that includes the origin AS number of each network prefix and all transit AS numbers that the route prefix traverses. The AS_PATH length is the main default criterion for selection of the optimal route, so the route prefix with the shortest AS_PATH is preferred and installed as the active route into routing tables of BGP routers inside an AS.

However, BGP-4 has some problems with its security. Its vulnerability concerns the BGP-4 decision-making process to select the most preferable route prefix among the

V. Jordan et al. (Eds.): HPCST 2022, CCIS 1733, pp. 279–290, 2022.
https://doi.org/10.1007/978-3-031-23744-7_21

received same ones due to the lack of validating mechanisms that can be applied to verify the originated and propagated through transit ASes route prefixes [2–4]. Thus, there are many route leaks and hijacking incidents observed on the Internet. Here, the BGP Route Leak is "the propagation of routing announcement(s) beyond their intended scope" (RFC7908) [5]. The BGP Route Hijacking is the kind of attack that allows an attacker to impersonate a network, using a legitimate network prefix as their own (full prefixes or more specific routes) [3, 4]. According to the Qrator Labs [6] and MANRS (Mutually Agreed Norms for Routing Security global initiative) [7] reports, there are thousands of vulnerable ASes and hundreds of thousands of route leaks and hijacking incidents registered every month. Therefore, a certain solution capable to identify such incidents and prevent their spreading across the Internet becomes of high importance.

There are two main approaches to solve the identified problems. The first one is to perform continuous monitoring of propagated BGP-4 routing information updates to analyze and reveal the anomalous routes [8–12]. The second one is to incorporate a cryptographic signature (or hash signature) into propagated BGP-4 routing information updates to ensure the authenticity of the originated network prefixes and legitimacy of their propagation paths [13–15]. The latter approach assumes the deployment of the BGPSec extension [13] over all the BGP routing infrastructure of a telecom operator and its external BGP peers. The additional BGPSec_PATH attribute of each route prefix contains digital signatures based on the Resource Public Key Infrastructure (RPKI) [16] ensures the authenticity of the propagated routes and validity of their distribution over the right transit path. Also, BGPSec helps maintain trusted relations between BGP peers to provide the secure path for the received and advertised routes. However, it can be seen as a new version of the BGP protocol that requires updating the routing software on all BGP routers inside the AS and to be supported by all external BGP peers of the AS. In this case, the former approach that uses the analysis of BGP routing updates certainly has its own bright sides as well.

Analysis based solutions mentioned above identify anomalous routing information by using their algorithms to conduct detailed analysis of received BGP routes. Data from external sources play a crucial role in the most of the employed analysis algorithms. Such data are typically obtained from the Internet Routing Registry (IRR) databases (DBs) [17] that contain routing policy and connectivity information for each active AS on the Internet. The analysis algorithms follow their peculiar decision-making logic to incorporate the external data and produce the final verdict. For example, Kruegel et al. [9] propose a solution based on geographical proximity of external BGP peers. The work [10] suggests the usage of calculated statistical characteristics of several datasets of propagated route prefixes between neighboring ASes. Also, there is successful implementation of analysis and decision-making processes based on reachability of BGP neighbors and compliance with routing policies [11] or following the ontological graphs of ASes connectivity and actual propagation paths of routes [12], etc. These solutions operate "externally" and independently of the routing software of BGP routers and provide alerts or instructions to adjust configurations of affected BGP peers and exclude anomalous routes using available means, like Access Control Lists (ACL), prefix lists, AS path lists, etc. The performance and effectiveness of the analysis based solutions can be further improved without affecting the actual operations of BGP routers within ASes.

There are many prospects for development of analysis based solutions since they have no strict limitations and requirements for standardization to ensure compatibility among many vendors of routing equipment. The authors started developing their solution to analyze routing updates and identify anomalous routes suitable for small and mid-scale telecom operators [18]. The proposed solution produced good results, but the analysis was limited to validating origins and checking AS_PATH attribute of short length (only direct BGP peers with strict distinction of uplinks, private peers, and customers among them). In this paper, further development of the analysis based solution proposed in [18] is considered. The analysis algorithm of the solution becomes more complex and includes improvements for processing the full AS_PATH length validation with no distinction among BGP peers. The analysis time is reduced due to optimization. It results in more effective and faster reaction of the controlled AS to eliminate and stop spreading the anomalous routing information, thus, preventing the incidents and Internet connectivity disruption.

2   Proposed Solution Design and Implementation

The developed and proposed solution acts as an "external" tool and interacts with the controlled AS as the internal BGP speaker. It operates under the same AS number as the number of the controlled AS (the condition is to have no confederations inside the controlled AS) and analyzes all BGP routing information updates received by the controlled AS from its external BGP peers. It has the full up-to-date data on BGP routers within the controlled AS and their BGP configuration details, as well as the full data on the connectivity and routing policy of the controlled AS along with the restrictive measures that can be applied to the routes by the routing policy (like controlling the advertisements using BGP-4 communities, etc.). It is assumed that the controlled AS is not a stub multi-homed AS, but acts as the transit AS with its connectivity and routing policies. The controlled AS receives and maintains the full-view BGP routing table that includes all available route prefixes. The proposed solution does not require any upgrades of the routing software or migration to different BGP protocol versions.

The overall design of the proposed solution is shown in Fig. 1. It includes several algorithmic modules responsible for receiving BGP routing updates, conducting analysis, and performing specific actions if any anomalous route is detected. Operations of the algorithmic modules are desynchronized thorough queries to ensure flexibility of the solution. Thus, the algorithmic modules can be treated as independent units of the solution.

The "BGP Routing Information Updates Pre-Processing Module" receives and works only with the BGP UPDATE (type = 2) [1] messages and performs their initial processing. All other BGP messages are discarded. Information about withdrawn routes in UPDATE messages is ignored. The following data are extracted from the UPDATE messages:

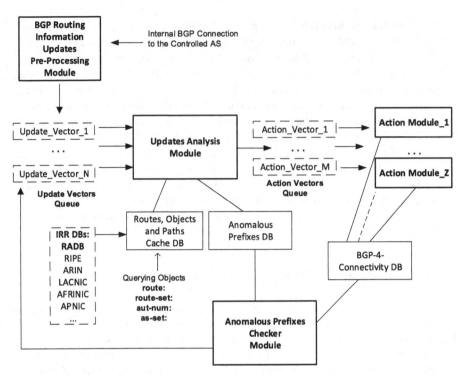

Fig. 1. The proposed solution design and its structure elements.

- the set of AS_PATH path elements $ASP[i]$, $i = 1 \ldots P$, where P is the length of the route prefix(es) path. All artificial AS_PATH lengthening using AS_PATH prepend [1] technique is reduced to the single AS number used for prepending. The order of AS numbers in the $ASP[]$ set is reversed for further convenience.
- the *NEXT_HOP* parameter with the IP address of the next hop router (external BGP neighbour IP address).
- the set of BGP-4 communities *CMM[]* provided in the UPDATE message.
- the set of the received route prefixes *NLRI[]* (Network Layer Reachability Information).

The extracted data produce the 'Update Vector' which is placed into the "Update Vectors Queue" for further processing. The queue is polled every 15 s.

The "Updates Analysis Module" processes each 'Update Vector' in the "Update Vectors Queue" to analyze and validate the received route prefixes. The analysis and decision making process is supported by external information extracted from the Routing Assets Database (RADB) [19] and other IRR DBs. The extracted information describes the connectivity and routing policies of ASes on the Internet using the Routing Policy Specification Language (RPSL, RFC2622) [20]. The design and structure elements of the "Updates Analysis Module" are shown in Fig. 2.

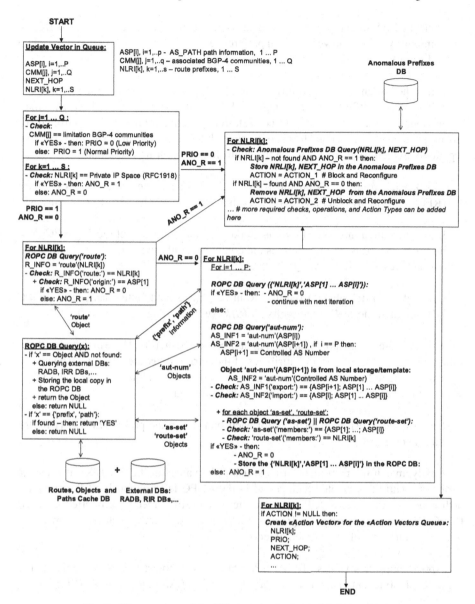

Fig. 2. The structure of the "Update Analysis Module" algorithm.

The analysis algorithm and the decision-making process operate according to the following steps:

1. Check iteratively the set of BGP-4 communities *CMM[]* associated with the route prefixes in the update message to match any restrictive communities like the 'no-advertise' community or other communities defined in the routing policy of the

controlled AS to apply any kinds of restrictive actions (like, a pre-defined BGP-4 community value to lower the local preference, to install the route prefix as a backup and so on).

If matched, then reset the 'priority' flag ($PRIO = 0$), otherwise set the flag ($PRIO = 1$). When reset ($PRIO = 0$), this flag should indicate that the received route prefixes in this update are highly unlikely to be preferred as active routes in the controlled AS and propagated further. Thus, they should be excluded from the analysis and processed separately to identify their impact on the controlled AS routing operations.

2. Check each route prefix in the $NLRI[]$ set to match the private address space (RFC1918) or any other address spaces that should never be advertised by any AS.

If matched, then this route prefix is marked as 'anomalous' (set the flag $ANO\_R = 1$) and further analysis for this prefix is not conducted. Otherwise, mark the route prefix as 'normal' (reset the flag $ANO\_R = 0$).

3. Query the **"route"** object from the "Routes, Objects, and Paths Cache DB" (ROPC DB) for each 'normal' route prefix $NLRI[k]$ (with the flag ANO_R $= 0$).

The route prefix and its originating AS number stored in ASP [1] are matched with the values of the 'route' and 'origin' attributes of the retrieved **"route"** object, respectively. If the matching fails, then this route prefix is marked as 'anomalous' (set the flag $ANO\_R = 1$) and further analysis for this prefix is not conducted.

4. Perform the transit path analysis for each 'normal' route prefix $NLRI[k]$ (with the flag $ANO\_R = 0$).

Firstly, match the path reconstructed using the values of the $ASP[]$ set is matched iteratively with the validated path stored in the ROPC DB previously. If the validated path is not found in the ROCP DB, then the **"aut-num"** objects are queried from the ROPC DB for neighboring $ASP[i]$ and $ASP[i + 1]$ AS number values. Compare the following attributes of the queried objects:

– 'export:' attributes values of the $ASP[i]$**"aut-num"** object to match the defined in the RPSL style announcement of ASP [1] ... $ASP[i]$ AS numbers to $ASP[i + 1]$ and conditions;
– 'import:' attributes values of the $ASP[i + 1]$**"aut-num"** object to match the defined in the RPSL style acceptance of ASP [1] ... $ASP[i]$ AS numbers and conditions.

When acceptance or announcement is defined using RPSL "as-set:" or "route-set:" classes, the corresponding **"as-set"** and **"route-set"** objects are queried iteratively from the ROPC DB. Then, 'members:' attributes of the objects are analyzed consequently for presence of the ASP [1] ... $ASP[i]$ AS numbers in them.

Reaching the last element *ASP[P]* of the *ASP[]* set means that the route prefix path has reached the border of the controlled AS. In this case, the value of the *ASP[i +* *1]* should be the controlled AS number. The corresponding **"aut-num"**, **"as-set"**, and **"route-set"** (if required) objects are pre-defined objects that represent the connectivity and routing policies of the controlled AS and stored locally in the ROPC DB.

The analysis concludes once everything is matched and found perfectly after examination of all attributes values of queried objects. The route prefix is marked as 'normal' (set the flag *ANO_R = 0*) if no contradictions have been found during analysis, or 'anomalous' (set the flag *ANO_R = 1*) otherwise.

5. Query the "Anomalous Prefixes DB" for each route prefix *NRLI[k]* to check the history of the route prefix and define the appropriate action that should be taken:

– attach the bit-encoded "blocking" *ACTION* parameter and store the "anomalous" route prefix (marked with the flag *ANO_R = 1*) along with the *NEXT_HOP* parameter if it is not found during the query search.
– attach the bit-encoded "unblocking" *ACTION* parameter and remove the previously stored route prefix if it is marked as "normal" (the flag *ANO_R = 0*).
– provide more checking and attach the additional *ACTION* parameter if needed.

6. Create the "Action Vector" for each route prefix with the *ACTION* parameter attached to it.

'Action Vectors' contain the route prefix subjected to the specific action, the bit-encoded ACTION parameter, priority flag, and the NEXT_HOP parameter with the IP address of the next hop router (external BGP neighbor IP address).

All 'Action Vectors' are placed into the "Action Vectors Queue" for further processing by "Action Modules". The polling time for this queue is 15 s.

If queried objects are not found in the ROPC DB, then they are retrieved from external IRR DBs and stored (cached) in the ROPC DB for further convenience and reduction of time delays. The main purpose of the ROPC DB is to optimize performance of the analysis algorithm by caching and storing all the necessary information and to reduce the load on the external IRR DBs. Cached objects from the IRR DBs are stored in the ROPC DB for 24 h (can be adjusted) and removed afterwards to keep up with the possible changes submitted to the IRR DBs.

"Action Modules" process the "Action Vectors Queue" and activate for each 'Action Vector' according to the bit-encoded *ACTION* parameter in it. The bit-encoded *ACTION* parameter assumes that each action is triggered by setting the corresponding bit of the parameter into 1. Currently, there are several actions encoded and supported:

– Bit 0 – block the route prefix and reconfigure the BGP router
– Bit 1 – generate alert messages (e-mails to network administrators)
– Bit 2 – unblock the route prefix and reconfigure the BGP router

The list of actions can be further improved in its variety by assigning the available bits to new actions similarly.

Blocking and unblocking actions are performed at the specific border BGP router that receives the route prefix subjected to the actions. It is done by modifying the correspondent ACL or filter list (including or excluding the route prefix) applied to the configured external BGP neighbor at the BGP border router. The *NEXT_HOP* parameter is used to retrieve all the additional data from the "BGP-4 Connectivity DB", such as the IP address of the BGP border router that should be reconfigured, router configuration style, set of reconfiguration commands, the ACL or filter-list content, etc. Updated ACLs or filter-lists are stored in the "BGP-4 Connectivity DB" after their modifications.

Successful completion of the specific action sets the corresponding bit in *ACTION* parameter to 0. Once all bits are set to 0, the last "Action Module" that processed the 'Action Vector' removes it from the queue.

The "Anomalous Prefixes Checker Module" performs the periodic check-up of the anomalous route prefixes listed in the "Anomalous Prefixes DB" (time period is set to 2 h). This module helps identify the changes of the route prefixes marked as anomalous and blocked by filters at the border BGP routers of the controlled AS due to the arranged actions for the anomalous route prefixes (and, thus, such route prefixes are not installed in the routing tables and advertised further). All route prefixes listed in the DB are grouped by the IP addresses of the border BGP routers inside the controlled AS by matching the IP addresses of the BGP neighbors (*NEXT_HOP* parameters) configured on the corresponding BGP routers. The necessary information is obtained by querying the "BGP-4 Connectivity DB". The module polls the border BGP routers and extracts the data of the route prefixes received from the BGP neighbors but blocked by filters. Then, the appropriate 'Update Vectors' containing the required data (such as the set of route prefixes *NLRI[ ]*, the set of the AS_PATH path elements *ASP[ ]*, etc.) are created and placed in the "Update Vectors Queue" for their normal processing and analysis.

The solution is developed using Python programming language. All BGP-4 interaction is handled by the incorporated YABGP agent [21]. A NoSQL document-oriented database, MongoDB [22] is used to host and operate all the solution's DBs. Its advanced features allow the effective storage and retrieval of cached IRR DB objects and controlled AS BGP routers topology and configuration data.

3 Testing and Evaluation

Testing and performance evaluation are conducted using the simulated lab network configuration, which resembles closely the actual connectivity between the real ASes as a part of the Internet (Fig. 3).

The simulation is performed with the help of the virtual lab environment EVE-NG [23]. Here, the AS20485 is set as the AS monitored and controlled by the developed solution. This AS has its own uplinks (AS1299 and AS3356) and downlinks (AS21127 with its customer AS31364), as well as the part of the external connectivity relations represented by the AS3267 and AS29581. The IPv4 full-view routing table is reproduced using the ExaBGP toolkit [24] and fed to the UP_AS_1299 and UP_AS_3356 BGP routers to simulate the full-view routing table obtained by AS20485 through its uplink

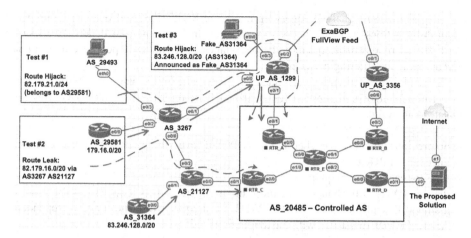

Fig. 3. The simulated network configuration and tested cases.

ASes. The full-view reproduction is done using the pre-recorded data at one of the AS20485 customer sites with further modifications of the AS_PATH attributes to comply with the actual route prefixes propagation through AS1299 and AS3356.

All the BGP routers of the controlled AS are Cisco-based routers and use Cisco-style configuration interface. Thus, the "Action Modules" for blocking and unblocking received route prefixes apply the automatically produced named inbound filtering prefix list for each external BGP neighbor. All the external BGP neighbors are configured to perform inbound soft reconfiguration.

There are four cases tested with the help of the simulated lab network:

- Case #0 – processing a normally propagated route prefix without any anomalies (advertisements with short path and long path);
- Case #1 - route hijacking by advertisement with different AS number;
- Case #2 - route leaking in transit due to transit AS misconfiguration;
- Case #3 - route hijacking by acting as a fake AS with the targeted AS number.

The developed solution demonstrates good performance and ability to detect anomalous routing information for all tested cases (#1–#3). The anomalous prefix is effectively blocked once it is received by the AS20485 and detected by the solution. The time delay (TD) in seconds between the moment the route prefix is received and the moment it is blocked (if anomalous) can be estimated using the following formula:

$$TD = T_{UVQ} + T_{UAM} + T_{AVQ} + T_A,$$

where T_{UVQ} is the 'Update Vector' waiting time in the queue (sec), T_{UAM} is the analysis time for the 'Update Vector' being processed by the "Update Analysis Module" (sec), T_{AVQ} is the 'Action Vector' waiting time in the queue (sec), T_A is the time for the action to take the effect (sec).

All calculations, analysis, querying the cache ROPC DB, and other programmed algorithmic actions are conducted very fast (less than 1 s), so no time delays are considered. The T_A time is measured directly, and it takes 2 s for the BGP peer reconfiguration and updating the filtering prefix-list.

The following assumptions are made for the extreme (or the worst possible) situation (according to the overall solution design and analysis algorithm) when the BGP routing update is analyzed:

– no previous objects are cached in the ROPCDB (everything has to be retrieved from the external IRR DBs);
– time for each external query is 1 s;
– the AS_PATH is the longest for the routing update and its length is 14 (here, it is assumed that all ASes in the AS_PATH are unique, while the longest prepended AS_PATH was registered with the length of 41) [25];
– all vectors in queues are picked up at the last moment (full waiting time).

This way, the estimated TD value should be calculated as:

$$TD = 15\,sec + 14 \times (1 + 5)\,sec + 15\,sec + 2\,sec = 116\,sec.$$

Therefore, there is the 116 s of time delay to identify the anomalous route prefix and exclude it from the BGP exchange for the worst possible situation. However, there is the typical 4–26 s of time delay for the regular cases when no extreme conditions are observed, the ROPC DB is not empty, and the queue is processed quickly.

The default polling time of the "Anomalous Prefixes Checker Module" can be reduced to the desired value. Typically, two hours are enough for many BGP route leaking and hijacking cases to be found and corrected, or those anomalous route prefixes should remain blocked for the investigation that is more detailed.

4   Conclusion

This paper presents the solution that performs the in-depth analysis of the propagated BGP routing information and detects the anomalous route prefixes using the public information about routing policies and connectivity of all ASes on the Internet. The solution is aimed at preventing the frequent cases of the BGP route leaking and hijacking and designed to stop propagating the anomalous routes over the transit area of the controlled AS.

The proposed solution can be easily integrated into existing infrastructure of any telecom operator because it has no requirements for upgrading the BGP routers firmware or routing software. It performs independently as an external tool and interacts with the BGP routers using the regular means to filter the anomalous route prefixes from the active routing tables and advertisement.

Further development of the proposed solution includes full support of the IPv6 BGP routing information, more detailed analysis of the RPSL data, and adaptability to sophisticated network configurations of telecom operators (such as BGP confederations, split

ASes, etc.). Also it is necessary to adopt the external data on detected routing anomalies from external sources that perform independent monitoring and analysis of BGP connectivity and BGP routing activity on the Internet.

Acknowledgements. The authors thank D. S. Schetinin for extensive technical support of this work.

References

1. Rekhter, Y., Li, T., Hares, S.: RFC 4271 - A Border Gateway Protocol 4 (BGP-4) (2006). https://tools.ietf.org/html/rfc4271
2. Patel, K., Meyer, D.: RFC 4274 - BGP-4 Protocol Analysis (2006). https://tools.ietf.org/html/rfc4274
3. Butler, K., Farley, T.R., McDaniel, P., Rexford, J.: A survey of BGP security issues and solutions. Proc. IEEE **98**, 100–122 (2010)
4. Smith, J., Birkeland, K., Schuchard, M.: An Internet-Scale Feasibility Study of BGP Poisoning as a Security Primitive arXiv:1811.03716v5 (2018)
5. Sriram, K., Montgomery, D., US NIST, McPherson, D., Osterweil, E., Verisign Inc., Dickson, B.: RFC7908 - Problem Definition and Classification of BGP Route Leaks (2016). https://tools.ietf.org/html/rfc7908
6. Siddiqui, A.: BGP Security in 2021 (2022). https://www.manrs.org/2022/02/bgp-security-in-2021/
7. Q1 2022 DDoS attacks and BGP incidents (2022). https://blog.qrator.net/en/q1-2022-ddos-attacks-and-bgp-incidents_155/
8. Hiran, R., Carlsson, N., Shahmehri, N.: Does scale, size, and locality matter? Evaluation of collaborative BGP security mechanisms. In: Proceedings of the IFIP Networking, 2016, pp. 261–269 (2016). https://doi.org/10.1109/IFIPNetworking.2016.7497237
9. Kruegel, C., Mutz, D., Robertson, W., Valeur, F.: Topology-based detection of anomalous BGP messages. In: Vigna, G., Kruegel, C., Jonsson, E. (eds.) RAID 2003. LNCS, vol. 2820, pp. 17–35. Springer, Heidelberg (2003). https://doi.org/10.1007/978-3-540-45248-5_2
10. Feamster, N., Balakrishnan, H.: Detecting BGP configuration faults with static analysis. In: Proceedings of Symposium on Networked Systems Design and Implementation, May 2005, pp. 43–56 (2005). https://doi.org/10.5555/1251203.1251207
11. Zhang, M.: On the State of the Inter-domain and Intra-domain Routing Security (2015). https://www.cs.uoregon.edu/Reports/AREA-201512-Zhang.pdf
12. Alkadi, O., Moustafa, N., Turnbull, B., Choo, K.: An ontological graph identification method for improving localization of IP prefix hijacking in network systems. IEEE Trans. Inf. Forensics Secur. **15**, 1164–1174 (2020). https://doi.org/10.1109/TIFS.2019.2936975
13. Lepinski, M., Sriram, K.: RFC8205 - BGPsec Protocol Specification (2017). https://tools.ietf.org/html/rfc8205
14. Li, Q., Liu, J., Hu, Y., Xu, M., Wu, J.: BGP with BGPsec: attacks and countermeasures. IEEE Netw. **2018**, 1–7 (2018). https://doi.org/10.1109/MNET.2018.1800171
15. Hu, Y., Perrig, A., Sirbu, M.: SPV: secure path vector routing for securing BGP. In: ACM 2004 SIGCOMM, Portland, OR, vol. 34, no. 4, pp. 179–192 (2004). https://doi.org/10.1145/1030194.1015488
16. Lepinski, M., Kent, S.: RFC6480 - An Infrastructure to Support Secure Internet Routing (2012). https://tools.ietf.org/html/rfc6480
17. List of Routing Registries in IRR. http://www.irr.net/docs/list.html. Accessed 14 Oct 2022

18. Mansurov, A.V., Schetinin, D.S.: Automatic detection and distribution prevention of incorrect BGP-4 routing information. Mod. Sci.: Actual Probl. Theor. Pract. Ser. Nat. Tech. Sci. **9,** 78–84 (2019). (in Russian)
19. RADb Search. https://www.radb.net/. Accessed 14 Oct 2022
20. Alaettinoglu, C., et al.: RFC2622 - Routing Policy Specification Language (RPSL) (1999). https://tools.ietf.org/html/rfc2622
21. YABGP Project Newspage. https://yabgp.readthedocs.io/en/latest/. Accessed 14 Oct 2022
22. MongoDB. https://www.mongodb.com/. Accessed 14 Oct 2022
23. EVE-NG. https://www.eve-ng.net/. Accessed 14 Oct 2022
24. Mangin, T.: Exa-networks. In: GitHub (2022). https://github.com/Exa-Networks/exabgp/. Accessed 14 Oct 2022
25. BGP Routing Table Analysis Reports. https://bgp.potaroo.net/as6447/. Accessed 14 Oct 2022

Simulation of DDoS Attacks on LTE and LoRaWAN Protocols in the NS-3 Network Simulator

Dmitry A. Baranov$^{(\boxtimes)}$ ⓘ, Aleksandr O. Terekhin ⓘ, Dmitry S. Bragin ⓘ, and Artur A. Mitsel ⓘ

Systems and Radioelectronics, Tomsk State University of Control, 46 Lenina Avenue, Tomsk 634050, Russia
bda@csp.tusur.ru

Abstract. The issue of creating a simulation model for conducting a DDoS attack on an information system is discussed in this paper. This system works on LTE and LoRaWAN protocols. According to the authors, this model makes it possible to evaluate the parameters of reliability and failure for data transmission channels under the influence of DDoS attacks. The evaluation was carried out when creating software and hardware systems for the Internet of Things systems. The scenario proposed in the article makes it possible to create a new simulation model of a data transmission network. The process of initialization and configuration for the network elements of the LTE and LoRaWAN protocols is carried out. This solution allows us to level the disadvantages inherent in each of the protocols available in the simulator. There is also a description of the behavior of network attacks.

Keywords: DDos · LORAWAN · LTE · NS-3 · Wireless security · Flood attacks

1 Introduction

The constant growth of network technologies requires regular improvement for the available research and design tools for telecommunications networks. The following main methods can be distinguished:

- study of parameters for a really working system;
- measurement of parameters on experimental stands;
- analytical modeling;
- simulation modeling.

One of the convenient methods for studying computer networks is simulation modeling. The essential advantages of this method are the economic benefits, as well as the closer to the real behavior of the models in comparison with analytical modeling.

Among the available ready-made software solutions for network simulators and emulators, the discrete event network simulator NS-3 was chosen. The advantages of

V. Jordan et al. (Eds.): HPCST 2022, CCIS 1733, pp. 291–301, 2022.
https://doi.org/10.1007/978-3-031-23744-7_22

this solution are open source code, ready-made implementations for network protocols, and emphasis on implementation for low levels of the network model. However, most of these solutions do not pay enough attention to information security issues.

The main idea of this article is to carry out a DDoS attack using a unified network simulation model of various network protocols. To achieve this goal, the following tasks were identified:

- determine the intensity threshold under the influence of a DDOS attack, where data transmission is still possible in the communication channel;
- reveal the dependence of the system load during data transfer on the intensity of the attack.

Section 2 of this article provides an overview of previous studies on information security issues for the LTE and LoRaWAN protocols. In Sect. 3, a description of the tasks for modeling is given, a unified protocol model using a class add-on, a description of the input (static) parameters for the experiment, and a mathematical description of the parameter estimation after a network attack. Section 4 involves the analysis of the obtained simulation results.

2   Overview of Approaches to Assess Protocol Safety

A review of the available solutions for modeling and assessing the safety of protocols were carried out. It was found that the main niche is occupied by research related to real devices or an analytical approach. In [1], the authors explore the potential security vulnerabilities of networks and devices for the LoRaWAN protocol, emphasizing that insufficient security is currently a serious problem. After analyzing the LoRaWAN protocol network stack, the authors propose the implementation of typical attacks using ready-made commercial equipment.

In [2], the authors draw attention to the dominant development of the LoRaWAN protocol for communication in low-power wide area networks, highlighting three vulnerabilities in the specification that can be used to launch denial of service (DoS) attacks on end devices in the network. The studies were carried out using models of colored Petri nets.

In another paper [3], the authors investigate a specific bit-flip attack on the LoRaWAN security vulnerability. In particular, possible risks are analyzed and a countermeasure is proposed to prevent them.

Article [4] investigates the security aspects of the procedure used in LoRaWAN to allow a device to establish a connection to a network server. Vulnerabilities have been identified in this algorithm. The experiments were carried out using real equipment.

The work [5] is devoted to the study of LoRaWAN protocol key management algorithms in the field of the Internet of Things. To improve the security of the system architecture, experiments were carried out with a network model implemented in a network simulator without taking into account the physical and data link layers. A similar problem was solved by the authors of [6] while studying potential vulnerabilities in the Internet of Things systems. They offer their own threat model for testing.

In 4G LTE networks, research is also being conducted on the subject of information security. The authors of [7] implemented various types of existing DDoS attacks on volumes, protocols, and applications in experimental 4G LTE networks. However, the studies were conducted using real mobile devices.

The authors of [8] analyzed the specifications of the LTE network access protocol, discovered and classified vulnerabilities. The studies were also carried out using real LTE mobile devices in real networks. Most of the received data is related to the power consumption of network nodes. The authors [9] offer their vision of the structure of the information security subsystem. This should increase the control of energy accounting.

In [10], the authors assessed the correctness of the implementation for the LTE security functions, which should protect personal data from compromise. This article used an analytical approach.

The work [11] is devoted to a discussion of the implementation results based on the actual deployment of fraudulent LTE base stations. A previously unknown technology for potentially tracking the location of mobile devices as they move is also discussed. All experiments were also carried out using real mobile devices. The authors of a related article [12, 13] describe experiments and procedures based on commercially available equipment and unmodified open-source software.

The attention of the authors to the low levels of the LTE protocol was attracted in [14], where studies are carried out on the degree of vulnerability of LTE to interference, spoofing and listening to radio frequencies, and various physical layer threats that can affect critical communication networks of subsequent generations. The authors also provide an overview of LTE mitigation and spoofing techniques. All experiments performed in this work relate to real equipment.

The authors of [15] argue that the existing test platforms for LTE networks are either limited in functionality and/or extensibility or too difficult to modify and configure. Therefore, they presented an open-source platform for experiments with LTE.

3 Model of DoS Attack in LORAWAN and LTE Networks

3.1 Description of the Simulated Network

The network consists of the following main elements shown in Fig. 1:

- an object located at a given height and constantly transmitting the same messages over channels that are implemented under the LORAWAN and LTE protocols;
- a device on the ground configured to receive only the LORAWAN protocol;
- a device located on the ground and configured only to receive via the LTE protocol (the exchange is assumed through the base station);
- an attacker device (Elrw, Elte) that implements a denial of service attack for each of the channels individually and together.

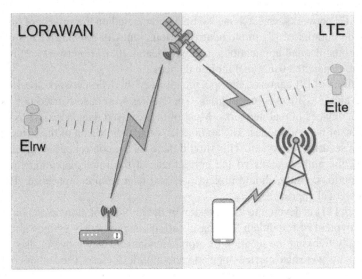

Fig. 1. Graphical representation of network objects.

The NS-3 simulator is software written in the C++ programming language. The process of modeling networks (creating scenarios for the behavior of elements in the system) is reduced to writing a program, with the call of additional functions.

3.2 LORAWAN_Ue Class

The LORAWAN_Ue class was created to implement the initialization, configuration, and behavior of the nodes themselves, operating via the LORAWAN protocol. The presence of such a class is directly related to the initialization of the node itself and the network interface, which is used to set up a data transmission channel following the protocol.

As arguments to the constructor of this class, a pointer to the LORAWAN protocol channel object, the global index of the node in the network, and the name of the node are passed. The first argument is used to determine the network participants between which the data transfer channel is configured. The second argument is intended for the convenient identification of the subscriber and the issuance of the necessary addresses in the future. The latter is an auxiliary parameter for convenient identification of the node when tracing functions work.

After the object is created, the MAC address is assigned to the required network interface. To simplify the address assignment process, depending on the version of the LORAWAN protocol link layer address space extension, an index increment by one is used each time an object of this class is created.

A change in the state of a node on a transmission line is tracked by associating callback state functions, declared earlier as static output tracing functions to the terminal, with the network interface. In the same way, functions are connected that notify of the successful transmission of the packet (represented as the size of the packet in bytes) and confirmation that all data has been received by the addressee.

An important aspect is the position of the object in space. At this stage, a movement model is created with an indication of the initial position in three-dimensional space without the use of movement. This step is related to the operation of the network interface at the physical layer of the protocol.

The completion of the network node configuration and at the same time exiting the class is a confirmation of the changes by associating the network interface with a common (for the script) node.

3.3 LTE_Channel Class

Due to the peculiarity of work for the ready-made LTE module in NS-3, the process of creating network elements differs from the LORAWAN protocol described above. In a separate class LTE_Channel, not only the initialization of the node and network interface is taken out, but a fully configured channel with all network participants. It includes devices for constantly sending and receiving messages, a base station to which subscribers are connected, position models, and data tracing.

Also, as a single argument, the class constructor receives the number of devices that fulfill the goals of the attacker (0 - the absence of such), namely the implementation of a network attack directly on the LTE channel. This step is related to the peculiarity of the LTE module and the NS-3 simulator as a whole since the configuration and operation logic for all nodes must be carried out before starting the simulation process.

Channel creation focuses on initializing the lte_helper object and the nodes involved in the protocol. It is also strictly necessary to separate the nodes into the roles of base stations and subscribers. In this regard, the array that stores subscriber objects is distributed in such a way that the attacker's devices are the first in it, and the original transmit and receive devices are the last two. The distribution in the array is shown in Fig. 2. To simplify the model, one base station was used.

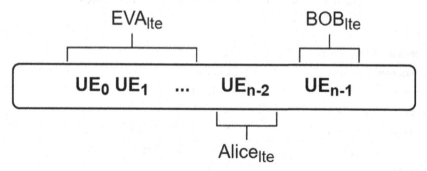

Fig. 2. Distribution of subscriber devices in the array.

Due to the presence of the lte_helper object, the channel configuration process is reduced to creating movement models for subscribers and base stations, indicating the initial position in space, calling the functions for assigning a network interface to a node depending on its role, and indicating the connection of each subscriber to the base station he needs.

The size of the transmitted information is set by the type of traffic used in the transmission (in this work, we used the transmission of SMS messages). And data tracing is carried out by calling the EnableTraces () method of the lte_helper object.

All information about the operation of the physical and data link layer for the LTE protocol is received in text files stored next to the script file.

3.4 Mathematical Model

The elements of the network (the model under consideration) are a group of receivers (serving devices) that are under attack, as well as a group of transmitters, presented as a set of network interfaces for one module.

The receiving device can process only one packet at a certain time, therefore, be in a free or busy state. If the device is busy, then, provided that there is free space in the buffer, the packet is queued.

The model from [16] was used to distribute network elements. The models [17, 18] were taken as the basis for the interaction of network layers for the LORAWAN and LTE protocols, where the data presented in Tables 1 and 2 were used as input static parameters.

The delay attribute for each of the available protocol classes allows you to model the process of processing requests by the receiver. We can set the processing time, which correlates with the performance of the real device.

Table 1. Input static simulation parameters for LORAWAN.

Parameter	Value
Antenna power ue_Alice	27 dBm
Antenna power ue_Bob	27 dBm
Frequency	868 MHz
Number of channels	4
Initial location ue_Alice	X = 0 m, Y = 0 m, Z = 1000 m
Initial location ue_Bob	X = 0 m, Y = 0 m, Z = 1,5 m
Package size	100 B

Table 2. Input static simulation parameters for LTE.

Parameter	Value
Antenna power enb	46 dBm
Antenna power ue_Alice	23 dBm
Antenna power ue_Bob	23 dBm
Frequency	800 MHz
Initial location ue_Alice	X = 0 m, Y = 0 m, Z = 1000 m
Initial location ue_Bob	X = 0 m, Y = 0 m, Z = 1,5 m
Initial location enb	X = 1 km, Y = 1 km, Z = 10,5 m
Package size	100 B

Service of incoming packets occurs according to the normal distribution law. The normal distribution parameters are M_O - mathematical expectation, S_O - standard deviation and R - node performance (operations/seconds). Thus, the service rate at the receiving device is calculated:

$$\mu(t) = \left(\frac{1}{S_o\sqrt{2\pi}} e^{-\frac{(t-M_O)^2}{2S_o^2}} \right) / R \qquad (1)$$

λ – the arrival rate of the input packet stream, so the channel is busy at the attacking receiver:

$$\mu P = \frac{\lambda}{\mu} \qquad (2)$$

Ratio of lost packets:

$$BLER = \frac{N_n(t)}{N(t)}, \qquad (3)$$

where $N(t)$ is the number of packets received by the system during time t, $N_n(t)$ is the number of lost requests during time t.

4 Experimental Setting and Simulation Results

The basis of the experiment was the introduction of attacker devices to implement a DoS attack by sending a large number of false packets over each of the transmission channels. To accomplish the task, a function was implemented that transfers a packet to an address in the network. Its call is carried out using the built-in function Schedule.

In NS-3, the Schedule tool allows you to periodically call any functions or methods throughout the simulation process. The arguments indicate the time after which it is necessary to call the function, a reference to the function, and the arguments of the called function.

4.1   Simulation Results

As a result of the simulation, a dependence graph was obtained for channel occupancy on the intensity of the attack (the number of false packets sent per second) for 10 s. The modifiable modeling parameter is the intensity of the attack (the input stream of packets). The graph is shown in Fig. 3.

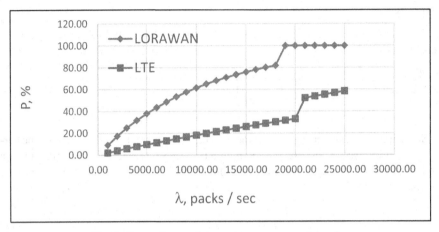

Fig. 3. Dependence of channel occupancy on intensity.

As can be seen from Fig. 3, with an increase in the intensity of the attack, the percentage of channel occupancy increases. When attacking each of the channels separately, LTE shows better fault tolerance compared to LORAWAN. There is also a sharp jump in both cases in the intensity range from 15,000 to 20,000 packs/sec. Above 20,000 packs/sec, the LORAWAN-based channel becomes completely inoperable. Occupancy percentage is 100, but LTE is only 58.31% loaded.

In the presented simulation results, the advantage of LTE is due to the presence of the AMC (Adaptive Modulation and Coding) module. AMC adjusts the MCS parameter, which determines the modulation type and coding rate, depending on the current BLER, thereby providing more stability to the data link.

When disconnecting the AMC module and setting MCS to a constant value (28), the results were obtained, which are shown in Fig. 4.

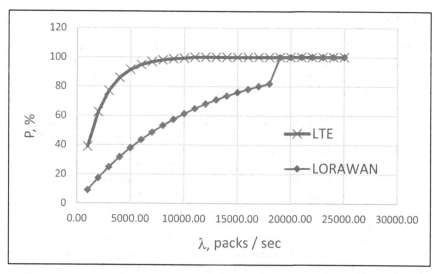

Fig. 4. Dependence of channel occupancy on intensity without AMC LTE operation

As can be seen from Fig. 4, with a simultaneous attack on each of the channels separately, LTE shows less fault tolerance compared to LORAWAN. The channel occupancy spike persists when using LORAWAN alone. Above 10 000 packs/sec, the LTE-based channel becomes completely inoperable. The intensity dependence of BLER in experiments is directly proportional.

5   Conclusion

A unified network simulation model was developed to carry out a DDoS attack on various network protocols. The proposed model can be applied to study the behavior of various information systems under the influence of network attacks.

The use of such models can be useful in implementing a whole class of various network attacks. This makes it possible to flexibly change the conditions and parameters of the studied information network. It will also allow you to evaluate and identify the necessary parameters to prevent such network attacks.

The main advantage of the proposed model is the versatility in the choice of wireless network protocols. In the future, it is planned to introduce other network attacks of various classifications in IoT systems. It is also planned to offer solutions that allow them to be prevented.

Acknowledgements. The article was prepared as part of implantation of the Leading Research Center (LRC) «Trusted Sensor Systems», financial support provided by Ministry of Digital Development, Communications and Mass Media of the Russian Federation, and Russian Venture Company (RVC JSC) (Agreement №009/20 dated 10 April 2020).

References

1. Es, E., Vranken, H., Hommersom, A.: Denial-of-service attacks on LoRaWAN. In: The 13th International Conference, 27 August 2018, Hamburg, Germany, pp. 1–6 (2018)
2. Emekcan, A., Gowri, S., Piers, L., Danny, H.: Exploring the security vulnerabilities of LoRa. In: 3rd IEEE International Conference on Cybernetics (CYBCONF), 21–23 June 2017, Exeter, UK, pp. 1–6 (2017)
3. JungWoon, L., DongYeop, H., JiHong, P., Ki-Hyung, K.: Risk analysis and countermeasure for bit-flipping attack in LoRaWAN. In: International Conference on Information Networking (ICOIN), 11–13 January 2017, Da Nang, Vietnam, pp. 311–321 (2017)
4. Naoui, S., Elhoucine, M., Azouz, L.: Enhancing the security of the IoT LoraWAN architecture. In: International Conference on Performance Evaluation and Modeling in Wired and Wireless Networks (PEMWN), 22–25 November 2016, Paris, France, pp. 1–30 (2016)
5. Stefano, T., Simone, Z., Lorenzo, V.: Security analysis of LoRaWAN join procedure for Internet of Things networks. In: IEEE Wireless Communications and Networking Conference Workshops (WCNCW), 19–22 March 2017, San Francisco, CA, USA, pp. 204–211 (2017)
6. Shelupanov, A., Konev, A., Kosachenko, T., Dudkin, D.: Threat model for IoT systems on the example of OpenUNB protocol. Int. J. Emerg. Trends Eng. Res. **9**(7), 283–290 (2019)
7. Feng, J., Hong, B.-K., Cheng, S.-M.: DDoS attacks in experimental LTE networks. In: Barolli, L., Amato, F., Moscato, F., Enokido, T., Takizawa, M. (eds.) Web, Artificial Intelligence and Network Applications. AISC, vol. 1150, pp. 545–553. Springer, Cham (2020). https://doi.org/10.1007/978-3-030-44038-1_50
8. Shaik, A., Borgaonkar, R., Asokan, N., Niemi, V., Seifert, J.: Practical attacks against privacy and availability in 4G/LTE mobile communication systems. In: Network and Distributed System Security Symposium (NDSS), 21–24 February 2016, San Diego, CA, USA (2016)
9. Nikiforov, D., Konev, A., Antonov, M., Shelupanov, A.: Structure of information security subsystem in the systems of commercial energy resources accounting. J. Phys: Conf. Ser. **1145**(1), 9 (2019)
10. Mjølsnes, S.F., Olimid, R.F.: Easy 4G/LTE IMSI catchers for non-programmers. In: Rak, J., Bay, J., Kotenko, I., Popyack, L., Skormin, V., Szczypiorski, K. (eds.) MMM-ACNS 2017. LNCS, vol. 10446, pp. 235–246. Springer, Cham (2017). https://doi.org/10.1007/978-3-319-65127-9_19
11. Jover, RP.: LTE security, protocol exploits and location tracking experimentation with low-cost software radio. In: ShmooCon, 15–17 January2016, Washington, USA (2016)
12. Lichtman, M., Jover, R.P., Labib, M., Rao, R., Marojevic, V., Reed, J.H.: LTE/LTE-A jamming, spoofing, and sniffing: threat assessment and mitigation. IEEE Commun. Mag. **54**(4), 54–61 (2016)
13. Labib, M., Marojevic, V., Reed, J.H., Zaghloul, A.I.: How to enhance the immunity of LTE systems against RF spoofing. In: International Conference on Computing, Networking and Communications (ICNC), 15–18 February 2016, Kauai, HI, USA (2016)
14. Rupprecht, D., Jansen, K., Pöpper, C.: Putting LTE security functions to the test: a framework to evaluate implementation correctness. In: The 10th USENIX Conference on Offensive Technologies, 8–9 August 2016, Austin, TX, pp. 40–51 (2016)
15. Gomez-Miguelez, I., Garcia-Saavedra, A., Sutton, P.D., Serrano, P., Cano, C., Leith, D.J.: An open-source platform for LTE evolution and experimentation. In: Proceedings of the Tenth ACM International Workshop on Wireless Network Testbeds, Experimental Evaluation, and Characterization, 3–7 October 2016, New York, USA, pp. 25–32 (2016)
16. Baranov, D., Terekhin, A., Bragin, D., Konev, A.: Implementation and evaluation of nodal distribution and movement in a 5G mobile network. Future Internet **13**(12), 321–336 (2021)

17. Khan, F., Portmann, M.: Experimental evaluation of LoRaWAN in NS-3. In: The 28th International Telecommunication Networks and Applications Conference (ITNAC), 21–23 November 2018, Sydney, NSW, Australia (2018)
18. Sabbah, A., Jarwan, A., Al-Shiab, I., Ibnkahla, M., Wang, M.: Emulation of large-scale LTE networks in NS-3 and CORE. In: Military Communications for 21st Century (MILCOM), 29–31 October 2018, Los Angeles, CA, USA (2018)

Transaction-Oriented Approach to the Design of Information Systems Based on Formal Grammar

Anton Yu. Unger[✉] [ID], Yulia S. Asadova, and Aleksandr A. Gololobov

MIREA – Russian Technological University, Vernadsky Avenue 78, 119454 Moscow, Russia
unger@mirea.ru

Abstract. The article discusses some important aspects of designing cloud information systems based on formal grammars. The proposed approach is based on
the formalization of the information exchange unit and the development of the
fundamental principles of client-server interaction. The proposed universal unit
of information exchange is compatible with both relational and non-relational
databases. The data transfer protocol between all components of the information
system is described. A transaction is chosen as an informational message, as some
atomic operation consisting of a context and a descriptor. Two protocols for transferring a transaction between a client and a server are outlined, guaranteeing the
integrity and security of the transmitted data. The approaches differ in the implementation of the access control system and the way the transaction is described.
The advantages and disadvantages of each of the approaches are considered. As
a description of the transaction, a model language is proposed, which is based
on a context-free formal grammar. The language supports a full set of operations
for interacting with the database, as well as control structures, and provides data
type control. A translator of a model transaction description language has been
implemented that generates a target code for programming languages popular with
hosting providers.

Keywords: Formal grammar · Backus-Naur form · Transaction descriptor ·
Relational database · Structured query language · Object-relational mapper

1 Introduction

Currently, the so-called cloud-based resources are becoming more and more popular [1].
Placing information in the cloud allows the resource owner not to worry about data safety,
data availability for users, fault tolerance and security. Fault tolerance is becoming an
essential requirement for modern information systems. It can be provided by duplicating
the resource on several physically and geographically dispersed servers.

A detailed discussion of how fault tolerance is provided in information systems is
not the purpose of this article, but it is important to note here that the central place for
storing resources is the database, which must be able to scale to provide fault tolerance.
All modern relational and non-relational databases have built-in scalability features [2].

© The Author(s), under exclusive license to Springer Nature Switzerland AG 2022
V. Jordan et al. (Eds.): HPCST 2022, CCIS 1733, pp. 302–317, 2022.
https://doi.org/10.1007/978-3-031-23744-7_23

There are various approaches to manage the process of deploying an information system in the cloud. The traditional approach is to create a specific isolated area for an information system in the cloud – a domain in which developers can work as if they were working in a local environment. The traditional approach is to use virtualization as the basis for resource sharing.

The existing virtualization methods are so advanced that they allow the developer to consider a domain as a dedicated cluster with a given number of processor cores, a given amount of RAM and a given storage capacity. The computing power of the cluster can be dynamically scaled depending on the needs of the information system, while all the scaling concerns, as a rule, become the task of the cloud provider, not the system developer. This approach is in line with the *Infrastructure as a Service* (IaaS) ideology [3].

The IaaS approach provides almost complete isolation of individual domains hosted on the same hosting, so it works well in cloud computing practice. In fact, the approach does not imply the creation of a cloud-based information system, but the transfer of the existing local information system to the cloud.

Most modern information systems are built on a modular basis. The main modules include database server, application server, and client application. We also consider a caching server to be a mandatory module, since caching can significantly increase system performance by eliminating repeated requests to the database.

The database server is the core of the entire system, since this is where information is stored and processed. Thus, to ensure the performance of the information system as a whole first of all it is necessary to reduce the load on the database server as much as possible. Caching is the most common solution to this problem.

Unfortunately, there is no ready-made solution for building a high-performance computing cluster. The maintenance of a large information system requires a large number of highly qualified specialists familiar with this particular system.

On the other hand, all information systems (with the possible exception of the largest ones such as *Google* or *Amazon*) are built according to the same classical principles. The information flows that circulate within the system are unique. Most of the program code deals precisely with the management of these information flows. This management code is commonly referred to as the application business logic.

This article introduces a certain *formalism* for describing the business logic of the application. This formalism refers to the heart of the information system, namely the database.

2 Platform as a Service

An alternative approach to develop a cloud information system is the *Platform as a Service* (PaaS) model. This model provides the developer with ready-made components, such as, for example, a database management system (DBMS) or a mailing system. Usually these components are already configured by the provider in the most optimal way [4]. This approach does not require the developer to radically revise the traditional principles of information systems design, but removes such issues as installing and configuring the operating system, installing and configuring the database server, scaling the server in accordance with the increased load on the system, etc. These questions become the task of the cloud provider.

On the other hand, it will not work to take a ready-made locally developed system and transfer it to the cloud "as is", since traditionally either open source projects (for example, PostgreSQL DBMS) or popular commercial projects are used for system development, while cloud providers often offer their own solutions. This may or may not be compatible with popular open source solutions.

The two described approaches to developing a cloud information system IaaS and PaaS require the developer to write code for the business logic of the application. In the case of IaaS, this code can be written in any programming language, since the runtime is the guest operating system. In the case of PaaS, the choice of programming language is determined by the cloud provider. Either way, it takes a skilled developer or a group of skilled developers to write and maintain the code. Otherwise, the components of the information system, fine-tuned by the provider and ready for scaling, will not be used effectively.

3   Software as a Service

The last model adopted in cloud computing practice is *Software as a Service* (SaaS). This model makes the least use of the virtualization. Instead, the developer is provided with a ready-made software product with a clearly described interface. This approach minimizes the amount of custom code on the server side, since communication with the server is possible only through sending requests and receiving responses. The developer just needs to study the interface of the server software and develop a script for the client application to send requests to the server.

This approach is the best in terms of scalability, as evidenced by email services such as *Gmail* and *Microsoft Outlook*, which send billions of emails per day. At the same time, this approach is the least flexible, since it only provides the developer with an interface without the possibility of writing code.

If you go from the IaaS model to the PaaS model and from the PaaS model to the SaaS model, you can see how the share of server software is decreasing, while the share of client software is growing. At the same time, obviously, it is impossible to concentrate all the business logic of the information system entirely on the client's side without serious security issues.

Let us set the following problem. How, by concentrating as much code on the client side as possible, make the information system secure, ensure data integrity and consistency, and retain maximum flexibility for working out any business scenario?

4   Basic Concepts

Let us take a traditional client-server architecture as a basis. This architecture works on a request-response basis. To ensure the integrity and consistency of stored data on the server side, requests must be sent to the database management system. From a developer's point of view, the choice of a specific DBMS should be completely irrelevant. Data integrity is ensured at the *transaction* level [5].

Transactions are supported by both relational and non-relational DBMSs. Without loss of generality, we will further use a relational DBMS for examples. From the point of view of the DBMS, a transaction is a set of changes in the database, which is either executed entirely or not at all. For a relational DBMS, the natural language for describing transactions is *structured query language* (SQL).

It is out of the question to send requests written in SQL directly from the client to the database. An additional level of abstraction is needed.

This level is the *metalanguage* – a formal language for describing transactions, which is directly projected into the SQL query language, but has additional capabilities to ensure access control and verification of requested or changed data.

Before embarking on the development of such a language, it is necessary to develop basic concepts for building an information system. The key components of any information system are:

- data storage;
- data access control;
- data cache.

Traditionally, the data storage unit in a relational DBMS is a record in a table. The table is thus a container for records of the same type. From the point of view of an object-oriented approach, a table corresponds to a certain class, and a record corresponds to an instance of this class. Such translation is typical of the so-called *object-relational mappers* (ORM) and is widely used in the practice of developing information systems [6].

Also, in ORM terminology, the concept of an entity is used, which corresponds to a class associated with a table in a relational DBMS. By definition, an entity is a named set of attributes. The attributes correspond to the columns of the database table. Several attributes are used to link an entity with other entities, which together form the database schema. These attributes are commonly referred to as associations.

Let us take an entity as a data storage unit. Each entity has a name that is unique within the sys-tem, which is uniquely projected onto the name of the database table.

Each instance of this entity must be identified. From the point of view of a relational DBMS, this requirement corresponds to a primary key, which is explicitly or implicitly present in any table. The choice of the type of identifier depends on the DBMS used. For example, in the case of MySQL DBMS, the most efficient is the integer primary key, which is generated by the system automatically when a new entity is inserted. This eliminates possible collisions when the same identifier is assigned to different instances. Without going into technical details yet, we note that a request to create a new instance of this entity must return an identifier.

Let us decide on the configuration of our system. It is quite simple (Fig. 1).

Here, the information system may include, in general, M load balancing nodes, K request processing nodes, and more than one data storage and metadata cache nodes. The information system (the part of it that is located in the cloud) includes:

- entity storage node;
- node for processing and caching user queries;
- node for caching metadata;
- load balancing node.

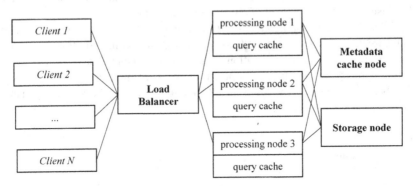

Fig. 1. Main components of an information system.

In Fig. 1 the query processing node and the query results caching node are combined. This is because the query processor needs full control over the cache in order to store the results in the cache and remove out-of-date data from the cache.

However, there is a separate node for caching the metadata. Metadata in our understanding is small pieces of information that are used in almost every request. These include access control lists, list of transaction identifiers, and user sessions. The metadata also includes information about cached entities: the time the entry was created in the cache, the lifetime, the collection identifier, etc.

It is reasonable to cache the results of user queries to disk, as they can generally be quite large. In addition, there may be multiple copies of the same results for load balancing purposes.

On the other hand, metadata about cached responses needs to be stored in memory on a separate node. A separate node will avoid the problems associated with the desynchronization of multiple cached copies. To provide fault tolerance, several such nodes can be combined into a cluster [7]. Such cluster can be implemented on the basis of *Memcached* or *Redis Cluster*.

Caching is an indispensable component of any modern information system, as it can reduce the number of requests to the main data store and greatly increase system performance. However, creating an efficient caching scheme is not a trivial task. You need to answer the following questions:

1. What can be cached?
2. Who is allowed to receive the cached result?

In order to answer the first question, you need to decide what is the result of the query. All client requests to the system are directed to the data warehouse. This statement is a consequence of the fact that the state of the system is completely determined by the state of the database. The only exception is the configuration of the database itself. However, this configuration is not cacheable.

Formally, the result of a database query is the so-called *result set*, which is a set of rows from one or more related tables ordered by a certain criterion. Each table has its own entity. Fetching several related entities at a time is traditionally performed using JOIN queries. Such queries can be executed very efficiently in modern DBMSs, but their use leads to difficulties with horizontal scaling, since related entities must be stored on the same storage node. In addition, the result of a JOIN query is a nested structure that is difficult to identify. It is much easier to work with the result of referring to only one entity. In this case, the result is a collection of instances of some entity, which in the general case can be empty.

So, let us postulate the first provision of our formal description of the principles of information system design. The result of a user request can be a collection of zero or more instances of the requested entity.

This collection can be easily identified. Indeed, in general, a select request contains the following sentences:

- *EntityId* – indicating the entity identifier;
- *CriteriaSet* – indicating a set of criteria for selecting entities;
- *OrderBy* – indicating the sort order of entities in the collection;
- *Offset* – indicating the beginning of the requested collection in the resulting selection;
- *Limit* – indicating the number of entities in the collection.

Thus, any hashing function, such as *md5*, can identify a collection by a concatenated string like this one:

$$CollectionHash = md5(EntityId + CriteriaSet + OrderBy + Offset + Limit).$$

Here, the " $+$ " symbol means concatenation. This hash can be stored in the metadata caching node as a key for an entry that stores all the necessary information about the collection. This information includes not only the creation time of the entry, but also the list of groups that are allowed access to the cached collection.

5 Access Control

Controlling access to data is an essential component of an information system. Access control is implemented by setting a set of restrictions for user groups. Following the principle that all in-formation should be stored in the database, the restrictions themselves are also instances of a certain entity (namely, *Restriction*), access to which is carried out according to the protocol standard for all entities.

The initial rules are set at the time of system installation and can be changed by the administrator. Group-level constraints are the most flexible access control mechanism. Let us show by an example how to describe an access control policy using only entities.

Let us introduce two main entities: *User* and *Group*. These two entities are related by a many-to-many relationship (Fig. 2a), because one user can be a member of several groups, and one group can include several users. A many-to-many relationship is traditionally implemented at the relational database level using an auxiliary table that stores foreign keys to the parent user and group tables. Traditionally entities start with an uppercase letter and database tables with a lowercase letter.

Such an auxiliary table does not follow the principle: each table must be projected onto a specific entity. The use of a many-to-many relationship is perfectly acceptable in ORM practice; however, for our information system it is not acceptable. It is required to enter an auxiliary entity that would link the *User* and *Group* entities. Let us call this entity *Role*, reflecting the fact that group membership can have additional attributes.

So, the *Role* entity is associated with the *User* and *Group* entities by a many-to-one relationship (Fig. 2b).

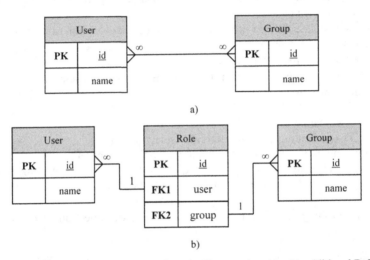

Fig. 2. Base access control list tables: a) direct implementation; b) with additional Role entity.

It is appropriate to mention here that all relationships between entities can generally be described by one or four types of relationships [8]:

- "one-to-one";
- "one-to-many";
- "many-to-one";
- "many-to-many".

Here, only one-to-many and many-to-one relationships are *asymmetric*. This means that there is no uncertainty about which entity will store the foreign key. An entity associated with another entity in a many-to-one relationship is always the foreign key keeper; the table onto which this entity is projected has a FOREIGN KEY column pointing to the parent table. An entity that has a one-to-many relationship with another entity is always projected onto the parent table.

The other two types of one-to-one and many-to-many relationships are *symmetric* in the sense that each can store a foreign key. The many-to-many relationship can always be replaced with the two asymmetric relationships one-to-many and many-to-one, as we just did for the *User* and *Group* entities.

A symmetric one-to-one relationship in the sense of a database schema is no different from an asymmetric many-to-one and one-to-many relationship. Indeed, let two entities, *A* and *B*, have a one-to-one relationship. Then the database schema for them may look like this:

```
CREATE TABLE `a` (
  `id` INT NOT NULL,
  PRIMARY KEY (`id`)
);
CREATE TABLE `b` (
  `id` INT NOT NULL,
  `a_id` INT [NOT NULL],
  PRIMARY KEY (`id`),
  FOREIGN KEY (`a_id`) REFERENCES `a` (`id`)
);
```

As you can see, the choice of the table storing the foreign key is arbitrary and depends only on the context of the problem being solved. Therefore, the decision should be made at the business logic level, not at the database schema level.

With this in mind, we exclude from consideration the one-to-one relationship when designing an information system. From a server software perspective, this type of relationship is indistinguishable from a many-to-one or one-to-many relationship.

Let us go back to our entities *User*, *Group*, *Role*. These entities are sufficient to provide access control. Indeed, a row in the access control list (ACL) is an instance of the *Restriction* entity, and the list itself is projected onto a table with the following structure:

```
CREATE TABLE `restriction` (
  `id` INT NOT NULL,
  `entity` VARCHAR(255) NOT NULL,
  `action` VARCHAR(15) NOT NULL,
  `priority` INT NOT NULL,
  `selector` TEXT,
  `groups` TEXT,
  PRIMARY KEY (`id`),
  INDEX (`entity`, `action`)
);
```

Here the fields have the following meaning. The *entity* field corresponds to the entity unique name, *action* is the action to which the restriction is applied, *priority* is the priority of applying the restriction, *selector* is a set of additional filters for the entity to apply this restriction, *groups* is the list of groups that are allowed access. A specific instance of a restriction is accessed by the primary key *id*.

The list of available actions is limited to the basic operations that are allowed on data in the database, namely: *Create, Read, Update, Delete* (CRUD). A default value is defined for each action. For example, for most information systems, unless otherwise stated, it is reasonable to choose the following as the default actions:

- *read* – allowed for everyone;
- *create* – denied for everyone;
- *update* – denied for everyone;
- *delete* – denied for everyone.

After installing the information system, the *Restriction* table is empty. The *User* table must contain at least one user with administrative privileges, so the user table must contain one record. The *Group* table must also contain one record corresponding to the group of this user. An entry in the *Role* table should reflect the relationship of this user with this group.

In addition, the system must have three virtual groups that are not in the database: the *all* group, which includes all users, the *registered* group, which includes all authorized users, and the *owner* group, which contains the owner of this entity. The owner group is a group for a single user; its purpose is to control access to a resource created by a given user, or a resource for which this user is assigned an owner.

Entities that implement the concept of owner must contain the attribute of the same name – an association with the *User* entity. When a resource is created, its creator is assigned as the owner. Initially, only the system administrator has the right to reassign the owner, but since both the *User* and the *Restriction* are ordinary entities, this right can be transferred to any user using a selector.

An attribute selector of the entity *Restriction* specifies a subset of instances of a given entity that meet a specific set of criteria. For example, let the entity *Car* have the following distinctive fields: color and brand. The selector of this entity can be a string of the form:

```
selector = "color=gray: model=Toyota".
```

This string does not have an entity name because it has a separate field in the restriction table. The selector allows us to further filter entities. In this example, the restriction only applies to gray Toyota cars.

Entity level access restrictions require the ACL to be cached [9]. Indeed, a single request to the server can involve many entities, so storing the ACL in the database would result in redundant re-quests to the restriction table.

It is time to talk about what the request should be.

6 Request as a Transaction

Above, we postulated the provision that the result of executing a read request can return a collection of instances of the same entity. Avoiding JOIN queries will obviously lead to an increase in the total number of database calls. For example, instead of one query, there is:

```
1. SELECT * FROM `category` LEFT JOIN `product` ON cate-
gory.id = product.category_id WHERE category.id IN (<cat-
egories>),
```

that returns all products from all categories mentioned in the < categories > list, you need to make two queries: one that returns a collection of categories, and the other that returns a collection of all products from all requested categories.

```
1. SELECT * FROM `category` WHERE id IN (<categories>);
2. SELECT * FROM `product` WHERE category_id IN (<catego-
ries>).
```

Here, two queries instead of one do not degrade system performance. Indeed, remember that the storage node must ensure data integrity. The logic for working with data is implemented on the client side. Thus, in this case, the client requests a list of categories (possibly all at once) and stores it. Products are loaded dynamically without the need to refer to the category table every time.

Data fetch requests do not change the information stored on the server, therefore, this approach is perfectly acceptable. However, for data modification requests (*create*, *update*, *delete*), it is necessary to ensure consistency that the entities change consistently. At the DBMS level, this means that all changes occur within a *transaction*.

By definition, a transaction is used to change data atomically. In SQL terms, a transaction is simply a sequence of queries to a database. A fetch request can also be thought of as a single request transaction. If it is natural to use an entity as a unit of data storage, then it is natural to use a transaction as a unit of data exchange.

So, let us postulate the following position: the information transmitted from the client to the server is one transaction. It is permissible to transfer several transactions in one request to reduce the number of calls to the server (the so-called RTT – Round Trip Time).

The standard language for describing transactions is SQL. It is a high-level declarative programming language that is the primary means for communication between a custom query processing node and an entity storage node. At the same time, it is obvious that the SQL language is not suitable for the interaction of the client software located on the user's terminal with the query processing node located in the cloud.

The standard protocol for communication between the client and the cloud server is the secure HTTPS protocol. The unit of exchange in this protocol is the HTTP message. Thus, it becomes necessary to wrap the transaction in an HTTP message.

We consider two possible approaches to transfer a transaction from a client to a server [10].

7   Transaction as Unit of Exchange

The first approach involves the transfer in the request body of all the information that is required for the execution of the transaction. In general, the required data can be divided into:

- *transaction schema* – a sequence of calls for creating, updating or deleting entities;
- *transaction context* – data required for execution.

An approach *Transaction as a Unit of Data Exchange* involves the transfer of both a schema and a context. One of the possible exchange formats is shown in Fig. 3. Here, the body of the HTTP message is shown on the left, and the corresponding SQL statements, which will be executed, are shown on the right.

As is well known, data modification on the server must be performed by POST, PUT, DELETE requests. Each action has its own request type:

```
[{
    "entity": "Product",
    "action": "create",
    "fields": {
        "cid": "c1",
        "color": "gray",
        ...
    }
}, {
    "entity": "Product",
    "action": "update",
    "fields": {
        "color": "red",
        ...
    },
    "filter": {
        "id" : {
            "$in": [1,2,3]
        }
    }
}, {
    "entity": "Category",
    "action": "delete",
    "filter": {
        "id": 4
    }
}]
```

```
START TRANSACTION;

INSERT INTO `product` (id,
color, …) VALUES (0, 'gray',
…);

SELECT * FROM `product` WHERE
id IN (1,2,3);

UPDATE `product` SET
color='red' WHERE id IN
(1,2,3);

SELECT * FROM `category` WHERE
id=4;

DELETE FROM `category` WHERE
id=4;

COMMIT;
```

Fig. 3. Request body (left) and corresponding transaction (right).

- *create* – POST;
- *update* – PUT;
- *delete* – DELETE.

However, one transaction can contain actions of all listed types. Therefore, as a standard type of request for transferring a transaction to the server in the body of an HTTP message, we will choose the POST type, as the most popular among web application developers.

Figure 3 shows that, in addition to modification statements, the transaction contains fetch instructions that are not in the request body. This is due to security reasons. Indeed, in order to make a decision about the possibility of updating or deleting instances of an entity, it is necessary to check each instance against all selectors from the restriction table. The check is performed by simply intersecting the list of user groups with the list of groups from all rows of the restriction table with a suitable selector. The action is allowed only if the intersection is not empty.

It is important to note that verification of all actions contained in a transaction can be performed before the first actual data modification. This corresponds to the *deferred data modification* strategy and allows to quickly rollback the transaction in case of an access violation.

There are three possible types of actions. Let us consider each of them in more detail.

Creating a new entity requires passing in the entity name and fields associated with their values. Fields with default values cannot be passed. To maintain the integrity and consistency of data, the identifier for each instance of the created entity must be generated by the database. At the same time, in one transaction, references to an entity that was previously created within this trans-action are allowed. To ensure referential integrity, it is suggested to use the following method.

Each new instance of an entity is supplied with an additional client id (*cid*), which the client can generate at will. The uniqueness of this identifier is important only within the framework of this request. On the server side, once an entity is instantiated and the primary key *id* assigned to it, the relationship between id and *cid* is maintained. If the transaction is successful, a result is returned to the client containing both the temporary *cid* and the unique *id*.

It is allowed to update entities by a unique identifier *id* and bulk update by a set of criteria. These two types of update must be differentiated. From the point of view of the SQL language, both the first update (by identifier) and the second (by a set of criteria) match the WHERE clause. However, from a caching point of view, updating by identifier should result in the deletion of all cached collections in which the given entity instance occurs. Updating by criteria removes all cached collections with a tag that matches the given criteria.

Deleting entities is also allowed by identifier and by a set of criteria, which corresponds to the WHERE clause in the SQL DELETE statement.

The considered approach, in which the transaction scheme and its context are transmitted to the server, is very flexible. The client can send a request describing almost any action with the data stored in the database. Naturally, if the user does not have enough rights to perform some of these actions, the transaction will be rolled back. To ensure the correct operation of the access control system, it is necessary that all entities for which updating or deleting are supposed to be extracted from the database. This can result in excess system load and increased traffic between the query processing node and the data storage node.

In addition, modifications of individual entities in the general case can lead to modifications of other entities according to the business logic of the information system. For example, creating an item in a shopping cart (entity *Order*) should lead to a stock of this item in the warehouse, i.e. to modify the *Product* entity. Obviously, creating an instance of the *Order* entity is allowed for any registered user, however, modification of the *Product* entity is allowed only for system moderators.

This contradiction can be resolved, for example, by using a set of triggers registered for each entity, but we will consider a different approach.

8   Transaction as an Entity

Instead of passing the transaction schema in the user's request, you can store it on the server side. This approach is possible on the assumption that the interface of the information system is quite limited. It includes a set of standard methods by which a client can communicate with a server.

All methods can be classified into one of two types: methods for retrieving data, and methods for modifying data. Both the first and the second type can be divided into *schema* and *context*.

In this approach, the schema is stored on the server side. The client only passes the method identifier and context during the request.

To implement this approach, you need a means of describing a transaction in some *metalanguage*. This metalanguage must support functions for performing database operations, i.e. creation, updating, and deletion of data; as well as flow control statements and a basic set of boolean and arithmetic expressions.

To describe the metalanguage, we will use the apparatus of *formal grammars*. Obviously, a language that includes branching operators (generally nested) cannot be described by a regular grammar [11]. Consider a fragment of a context-free grammar described by the following Extended Backus-Naur form (EBNF).

```
T ::= begin S {;S} end
S ::= I = E | if E then S [else S] | for I in I do S
S ::= update(I, I [,E]) | delete(I [,E])
E ::= create(I, I) | select(I [,E])
E ::= Z[= | > | < | >= | <= | !=]Z | Z
Z ::= T{[+| - | or]T}
T ::= F{[* | / | and]F}
F ::= I | N | not F | (E)
I ::= C | IC | IR
N ::= R | NR
C ::= a | b |...| z | A | B |...| Z | .
R ::= 0 | 1 |...| 9 | .
```

The following designations are adopted here. Non-terminals start with an uppercase letter, terminals start with a lowercase letter, keywords in bold. The language generated by this grammar is denoted by L.

This grammar is intended to describe the transaction schema. As a sequence of database calls, a transaction supports queries to *create* a new entity, *update* an existing entity, *delete* an existing entity, and fetch (*select*) a collection of entities into the current context. The context refers to the information required to complete the transaction. Some of this information is transmitted from the client; some is fetched from the database.

In what follows, a transaction scheme described in the L language will be called a transaction descriptor. As the grammar suggests, the language allows for branching expressions, loops, and function calls. Custom functions are not allowed, and built-in functions are limited to: *create, update, delete,* and *select.* The update and delete functions are actually procedures. The create function returns the identifier of the newly created entity instance. The select function returns a collection of objects, which in general can include zero or more instances of the requested entity.

The L syntax supports the following *data types*:

- decimal integers;
- floating point numbers;
- strings.

These data types are a subset of the types used in the DBMS. Let us make a number of re-marks. First, this subset does not include a boolean data type, since booleans can be replaced with an integer. Second, there is no data type to represent date and time, since date and time can be stored in integer format as a timestamp.

These replacements are made in order to maximally isolate the server part of the information system from the details of data interpretation. The storage node must store data and ensure its integrity. The data processing node must provide controlled access to data and take care of storing the results in the cache. The client part of the information system can interpret the received data according to the business logic of the application. An integer received as a result of executing a query to the database can be interpreted in different ways: for example, as an account number or as a timestamp.

The set of transaction descriptors represents a set of scenarios for working with the information system. In fact, it is the interface of the server side of the information system for working with data. A descriptor is accessed by its unique identifier. All descriptors are stored in the database in a compiled form, so the target code for the L language translator is the *native language* of the server software.

Since the native languages for most cloud information systems are procedural or object-oriented languages, it is natural to translate a transaction descriptor into a *procedure* that takes a transaction context as a parameter.

The context is serialized from the client to the server. A popular format for exchanging data on the web is JSON (as used on the Fig. 3), which is suitable for serializing complex structures. JSON for-mat supports primitive and aggregate data types. Let's remember what type of data is needed to execute a transaction: a set of fields associated with their values (I) for create and update procedures, a set of criteria for filtering entities (E) in select, update and delete functions. Both the first and the second data type can be represented by an object, i.e. an unordered set of key-value pairs, where the key is a variable that can be used in a transaction descriptor, and the value can be any data type, including other objects and arrays. The *L* language supports accessing object properties using dot notation and accessing array elements using square brackets.

9 Conclusions

Each of the considered approaches to transferring transactions from the client to the server has advantages and disadvantages. The main disadvantage of the first approach Transaction as an entity is a complex system for providing secure access to data. Indeed, the client submits a sequence of actions to be performed on the storage node. The data processing node, based on the information received from the user, should allow or prohibit the execution of an action in accordance with the access level for this user. In the worst case, the last action in the sequence may be prohibited. In this case, the transaction must be rolled back.

In this sense, the second approach *Transaction as an Entity* is less complex, since to ensure access control, it is sufficient to associate with each descriptor a list of user groups that are allowed to perform this transaction.

The disadvantage of the second approach is the complexity of writing a translator for the transaction description language. Translation should take into account not only the peculiarities of the native programming language used in the system, but also the peculiarities of the underlying DBMS. On the other hand, the set of programming languages supported by cloud providers, as well as database management systems, is very limited, so that a translator written for a given DBMS-language pair can be injected into the system as a ready-made module.

References

1. Hatton, L.: Empirical test observations in client-server systems. Computer **40**(5), 24–29 (2017)
2. Lenhardt, J., Chen, K., Schiffmann, W.: Energy-efficient web server load balancing. IEEE Syst. J. **11**(2), 878–888 (2015)
3. Linthicum, D.S.: A guide to cloud-enabling your software. IEEE Cloud Comput. **3**(2), 20–23 (2016)
4. Silva, L.M., Alonso, J., Torres, J.: Using virtualization to improve software rejuvenation. IEEE Trans. Comput. **58**(11), 1525–1538 (2019)
5. Mitko, R.: Comparison between characteristics of NoSQL databases and traditional databases. Comput. Sci. Inf. Technol. **5**, 149–153 (2017)
6. Han, W., Whang, K., Moon, Y.: A formal framework for prefetching based on the type-level access pattern in object-relational DBMSs. IEEE Trans. Knowl. Data Eng. **17**(10), 1436–1448 (2015)
7. Bolognesi, T.: Toward constraint-object-oriented development. IEEE Trans. Softw. Eng. **26**(7), 594–616 (2020)
8. Chillion, A.H., Ruiz, D.S., Molina, J.G., Morales, S.F.: A model-driven approach to generate schemas for object-document mappers. IEEE Access **7**, 59126–59142 (2019)
9. Ermakov, N.V., Molodyakov, S.A.: A caching model for a quick file access system. J. Phys: Conf. Ser. **1864**(1), 012095 (2020)

10. Unger, A.Y.: A formal pattern of information system design. J. Phys: Conf. Ser. **2094**(3), 032045 (2021)
11. Korablin, Y.P.: Equivalence of the schemes of programs based on the algebraic approach to setting the semantics of programming languages. Russ. Technol. J. **10**(1), 18–27 (2022)

Author Index

Printed in the United States
by Baker & Taylor Publisher Services